CORPORATE SUSTAINAB
IN THE 21ST CENTURY

Corporate sustainability needs a rethink. We have entered the human-influenced Anthropocene age, and we are witnessing accelerating changes in earth system processes. Businesses' current initiatives, such as product innovation and pollution reduction, are not enough to combat the intensifying social-ecological challenges that face us.

Corporate Sustainability in the 21st Century is an innovative new textbook which provides a fresh conceptual framework for understanding and engaging with sustainability, now and in the future − "Business In Nature." This book critically discusses key concepts and topics related to corporate sustainability, with a focus on corporate sustainability strategies and corporate value chains. Setting itself apart from existing books, it introduces ideas from global ecology and the natural sciences to provide readers with a new language for discussing business and sustainability.

This book maintains an international perspective throughout, with a wealth of examples, case studies and discussion questions. It will be a valuable text for students of corporate sustainability; business, nature and society; and environmental studies, and will also be useful for managers seeking a new perspective on how being "green" can fit with business goals.

Rafael Sardá, Senior Scientist of the Spanish National Research Council (CSIC), and Academic Collaborator of the Operations, Innovation and Data Sciences Department at ESADE Business School, Ramon Llull University, Spain.

Stefano Pogutz, Senior Researcher and Tenured Faculty, Department of Management and Technology, and Sustainability Lab, SDA Bocconi School of Management, Bocconi University, Italy, and Chair of the CEMS (the Global Alliance in Management Education) Faculty Group, "Business and the Environment".

CORPORATE SUSTAINABILITY IN THE 21ST CENTURY

Increasing the Resilience of Social-Ecological Systems

Rafael Sardá and Stefano Pogutz

Routledge
Taylor & Francis Group

LONDON AND NEW YORK

First published 2019
by Routledge
2 Park Square, Milton Park, Abingdon, Oxon OX14 4RN

and by Routledge
52 Vanderbilt Avenue, New York, NY 10017

Routledge is an imprint of the Taylor & Francis Group, an informa business

British Library Cataloguing-in-Publication Data
A catalogue record for this book is available from the British Library

Library of Congress Cataloging-in-Publication Data
Names: Sardá, Rafael, author. | Pogutz, Stefano, author.
Title: Corporate sustainability in the 21st century : increasing the
    resilience of social-ecological systems / Rafael Sardá and Stefano Pogutz.
Description: Abingdon, Oxon ; New York, NY : Routledge, 2019. | Includes index.
Identifiers: LCCN 2018033058 | ISBN 9781138744592 (hbk) |
    ISBN 9781138744653 (pbk)
Subjects: LCSH: Management—Environmental aspects. | Sustainable development. |
    Social responsibility of business.
Classification: LCC HD30.255 .S269 2019 | DDC 658.4/083—dc23
LC record available at https://lccn.loc.gov/2018033058

ISBN: 978-1-138-74459-2 (hbk)
ISBN: 978-1-138-74465-3 (pbk)
ISBN: 978-1-315-18090-8 (ebk)

Typeset in Interstate
by Apex CoVantage, LLC

This book is dedicated to our own kids and wives (Conxita and Elayne) and to all young people, who deserve a better world. We need to change the course of our planetary actions and this book aims to contribute to this purpose. Both of us have two kids, and we wish that Júlia, Laura, Louis and Simon, as well as all the other kids in this world, can live their lives in peaceful harmony with our planet.

CONTENTS

FIGURES

FIGURE BOXES

TABLES

BOXES

FOREWORD

In the interest of full disclosure, I have known the authors (at least one of them) for over 30 years. When the idea for this text germinated, I was there discussing with the authors why a book like this is needed and what gap it would fill once carried through to its fruition. Having watched the idea bloom to publication (not to mention my own participation in some very enjoyable and intense debates) I am more than pleased to see the result and consider it a great honor that the authors have asked me to contribute a foreword to this book.

The authors bring to this book a remarkable combination of expertise – decades of research in the fields of ecology and economics – finely balanced with their years of teaching in prominent business schools across Europe. The book more than epitomizes the authors' attempt at inculcating among the student body that fine balance as expertly captured in the title of this book: *Corporate Sustainability in the 21st Century: Increasing the Resilience of Social-Ecological Systems*.

The debate between the need to protect the environment, meeting national economic development priorities and ensuring that people across the spectrum have the needed societal safeguards is as old as the environmental movement and the world community's desire to work toward sustainability. Beginning with the 1992 Rio Summit on Environment and Development, the world community succeeded in demonstrating that there is no inherent conflict between these three pillars of Sustainable Development and that the business community has a vital role to play.

While this thinking is firmly embedded among policy makers representing the countries in the United Nations and similar fora, it remained a top down approach – CEOs agreeing to principles in international conferences and asking their employees to implement. While it had a salutary impact in some companies, progress across the board is rather limited. To correct this, a number of highly noteworthy initiatives such as the Global Compact, the World Business Council on Sustainable Development, Global Reporting Initiative and many others have begun working on mainstreaming the concepts, increasing dialogues among all relevant parties and finding and propagating workable solutions.

The authors acknowledge that much was achieved but believed and have succeeded in conveying through this book that the world needs a movement, where the champions of the movement are students graduating from business schools taking the reins of leadership from the day they join their chosen path. At the same time, this book is not just for students; there are many corporate managers who, once they get the instructions from their bosses, are tasked with ensuring both sustainability and profitability. They will find in this book case studies where a number of new ideas are tried and tested. In fact, one of the highlights of the book is those case studies.

Even if you are not a student, decision maker or a corporate executive, this book is worth a read for it contains several valuable insights gleaned from a number of related disciplines. The authors succeeded not just in talking about the importance of sustainability but in how one might actually get there by following the approaches outlined here. I highly recommend the book as important course material in business schools across the world that seek to increase social and ecological resilience in their quest for attaining corporate sustainability.

Kilaparti Ramakrishna, PhD
Head of Strategic Planning
Green Climate Fund

PREFACE

Young people deserve a better world. Young people, such as the one appearing on the cover of this book, are the world's most dynamic human resource and they are fully aware that they are inheriting a planet where human living conditions are becoming more and more fragile in front of their eyes and, ecologically speaking, getting worse. During the last few centuries, humans have been acting as a powerful force in nature. Human activity, predominantly the global economic system, has become the prime driver of change reshaping the earth and bringing our planet into what has been recently described as a new geological era, the Anthropocene ("a new phase in the history of both humankind and of the earth, when natural forces and human forces became intertwined, so that the fate of one determines the fate of the other" - Zalasiewicz et al., 2010: 2231).

The Anthropocene defines earth's most recent geologic period as being human-influenced, based on overwhelming global evidence that atmospheric, geologic, hydrologic, biospheric, and other earth system processes are now altered by humans acting inside a social-ecological paradigm of rapid transformations and accelerated trajectories (by a social-ecological paradigm, this book adopts a worldview that recognize the interdependences between human societies and ecological processes that are necessary for the survival of both). The Anthropocene epoch is responsible for great human progress; in many regions of the planet, human life has dramatically improved, life expectation has increased, we enjoy better health and people are growing smarter. Progress was able to take societies out of poverty and bring increased human prosperity and welfare; we - as scientists and scholars - acknowledge it! However, it would have been far more preferable if in addition we were able to control population growth through better education, better redistribution of the welfare we created and if our achievements were not to the detriment of the natural environment. However, we must try to be optimistic since we still have some time available to do all this work (control the population, redistribute welfare, create a resilient functional environment), but we need to do it soon. Reaching a balance between economic development and social-ecological protection has become a priority for humanity and decoupling prosperity from social-ecological impacts a mandatory requirement. Young people are asking us to do it.

Businesses are fundamental stakeholders in societies. The fundamental role of firms was that of providing goods and services to satisfy people's needs and desires while respecting regulations and laws. In recent decades, profit-maximization and a short-term vision were the two guiding principles that defined the boundaries for business action. However, in the last few years, the expectations placed on business have changed. Today, Boards of Directors, Chief Executive Officers, managers, and investors are expected to address a broad range of challenges, from economic to social, to ecological, and, in addition, in the social-ecological paradigm in which we live, all these challenges are widely interlinked and interconnected. Driving these expectations, in September 2015 the United Nations set the Sustainable Development Goals (SDGs), a universal call to action to end poverty, protect our planet and ensure that all people enjoy peace and prosperity. The SDGs identify the challenges to humanity for the coming decades and they ask for further support from the business community and the private sector in providing substantive help to achieve its aspirational targets. Although business may have direct implications in some of them, the reality is that business affects all of them. Back in the 1980s, humanity agreed that Sustainable Development should become a strategic option to provide guidance for human development into the future. Forty year later, the challenges have become even bigger and sustainability, the way in which we can put in practice the strategy, more necessary than ever. Corporate sustainability is becoming mainstream and there is no way back.

Corporate sustainability is a fact. For many years, businesses have been trying to introduce corrective actions in their production mechanisms to reduce negative impacts on the natural environment. Efforts have recently accelerated as firms have started to use some sustainability strategies. Anticipating regulations, diminishing risks by preventing pollution, developing sustainable products or services, or transforming markets by creating new business models and through collaborative schemes, can give companies competitive advantages through cost reduction, innovation, consumer engagement, and differentiation. Nevertheless, the present pace of change is not enough to tackle the intensifying global social-ecological requests. To address the consequences of earth's fragility, corporate sustainability in the 21st century should be profoundly different: a transformational change is needed and only the seeds of this new approach are starting to be seen today. This book advocates for a change in the way companies, both large corporations and small firms, deal with environmental issues and explores the relationships between business and ecosystems in a broad manner. In the book, we take the perspective of the natural environment as the conditions under which humans and other organisms live inside a nested and complex network of social-ecological interactions. These conditions, which have been in balance for thousands of years, are both in serious threat and threaten us.

The environment is threatened. Is business responsible? Should business deal with this? Large corporations and small firms work in fierce competitive systems and they are under pressure from the logics of shareholder value maximization and short-termism. Those probably are not the perfect conditions in which to deal with these environmental threats and, maybe, other agents (policy makers, NGO representatives, people) should be asked to address these problems. In this book, we take a clear and bold perspective. It is time for business to think long-term. It is time to develop a much clearer and more conscious vision. It is time to understand what these threats really mean for companies. The approach that sees these threats only as risks and opportunities for companies needs to be integrated or replaced. For too long, business has not considered the natural environment as a core and constitutive element for its activities, but it is. Nature provides businesses with numerous benefits through goods and services derived from its ecosystem functionality, and it is indispensable to keep this functionality working to ensure healthy long-term business activities. Business should go ahead, but how should companies respond to these challenges? In order to answer this question, this book introduces and discusses a new conceptual framework, "Business In Nature" (BInN), which aims to guide business toward a deep and extended rethinking of its value chains, with the ultimate goal of increasing the resilience of social-ecological systems.

This book departs from scientific evidences of the environmental crisis and warnings from the United Nations. It analyzes the different strategic options and organizational changes that companies operating in a global system need to implement starting from today. It provides new principles, perspectives and frameworks to understand and manage sustainability strategies and value chains. The new sustainability view of the 21st century forces firms to deal with environmental pressures and to enforce innovative changes into strategic decisions. No firm can escape this trend.

Feeling a sense of "urgency" and managing the "impossible"

A sense of "urgency"

Earth is at stake. We are crossing dangerous limits, and we must acknowledge the perils deriving from our actions. In a very recent paper, more than 15,000 scientists worldwide signed a warning to humanity – following a similar initiative that was also launched 25 years ago – stating: "We are fast approaching some limits of what the biosphere can tolerate without substantial and irreversible harm" (Ripple et al., 2017: 1026). Ecological unbalances are causing many people to live in crisis conditions but many others, especially in western countries, do not care. It seems that the message of urgency does not reach people, having or not having responsibilities, but can we survive without a healthy environment?

We reshaped our planet in a few centuries: the atmosphere is warming; freshwater lakes and rivers are shrinking; deltas of rivers are eroding; tropical forests, converted into crops, are disappearing; nitrogen flows are increasing; oceans are acidifying and filled with plastics; other novel chemical entities are appearing; biodiversity is getting homogeneous, the average size of wild species is diminishing, and species are becoming extinct. Many natural processes have lost their evolutionary balances (e.g. the carbon cycle is altered, the nitrogen cycle is altered, we have created a hole in the ozone layer). We are losing the stable Holocene-like conditions in which we used to live. Instead, we have entered the Anthropocene, a novel era full of uncertainties.

In 2008, there were almost 1 billion fewer people on earth than today (the most updated scenarios are pointing to a population of 10 billion in 2050). In the coming decade, 1-2 billion more people will join our club of "normal" consumers. As a consequence, the pressure on earth systems is going to increase drastically. As an example, recently, with the Paris Agreement all the countries on our planet agreed to maintain a temperature within the limit of 1.5°-2°C degrees. However, national strategies to curb carbon emissions do not seem to be ambitious enough, and the likelihood of overpassing what is considered a "safe boundary" for our humanity seems high.

In the famous 2006 movie *An Inconvenient Truth*, while sitting in a plane Al Gore asks himself why politicians use expressions such as "We will deal with this tomorrow" in reference to the climate change debate. The real fact is that even though it is imperative to deal with this problem today itself, as a global society, we still do not feel this sense of urgency. In addition, when scientists, NGO representatives and – unfortunately – few leaders attempt to convey this urgency, there is a recurrent and diffused argument about the fact that this is "too much catastrophism." The consequence of this position produces negative effects, diminishing the impact of these calls to action on people. Companies need to define their role with regard to such urgencies and give clear signals about the necessity to take action as soon as possible.

We need to feel the urgency. We need to change the way we produce, consume and live, and we need to do it quickly because crossing thresholds that affect natural processes that have been stable for thousands of years will bring us enormous challenges and a plethora of uncertainties. It is true that human invention has no limits, that some of the processes with which we are now fighting can be reversible (the hole in the ozone layer serves as a very good example; it is the result of using very harmful products, and we were able to find alternatives that are not as harmful to this layer as before and to stop further degradation). Perhaps we can remove carbon from the atmosphere, perhaps we can restore the flow of the rivers, maybe we can restore and enhance the oceanic productivity, but in any case it is smarter to apply a "precautionary principle" in our way of life and not leave the future in the hands of irreversible processes and "maybes."

Managing the "impossible"

In the book *Catalyzing Change*, the World Business Council for Sustainable Development (probably the most influential organization in the field of business and sustainability) spent some time describing its own history. This book gives particular relevance to a boat meeting on a warm spring night in Norway in 1990 "where more than 140 business leaders, CEOs and diplomats discuss how business might join the global conversation around spearheading economic progress while safeguarding the environment" (Timberlake, 2006: 5). The book you are about to read also has an interesting genesis – maybe not as influential, but we are still proud of it. It was on a cold autumn night in 1996 in the lobby of a hotel in the city of Cologne that professors of eight well-respected European business schools (Bocconi University, Italy; Cologne University, Germany; Corvinus University, Hungary; ESADE-Barcelona, Spain; HEC Paris, France; Louvaine-la-neuve University, Belgium; Saint Gallen University, Switzerland; and Warsaw School of Economics, Poland), under the auspices of CEMS (formerly "Community of European Management Schools," now known as "Global Alliance in Management Education"), got together to share their teaching experiences on environmental management issues and started on a journey that has been uninterrupted to the current day. Since then, we have taught thousands of students in International Management and Business, most of them now holding executive positions.

Just by chance, the people who initially formed the group, and those who later joined the experience, came from a diverse range of disciplines, entailing an integration of concepts that enriched our professional development and our way of thinking and acting. The successive group coordination meetings and the courses carried out have led us to shape the thinking that builds the chapters of this book. One of our colleagues, Sándor Kerekes, founder of the group, had dinner in Budapest in 1994 with one of the most influential scholars in environmental economics, David Pearce. Back in the 1980s, some economists started to talk about the need to introduce market incentives to help solve the environmental pollution problems. One of them, and one of the most important scholars in field, was David Pearce, author of the famous book, *Blueprint for a Green Economy* (Pearce et al., 1989), where he contributed to the definition of the pillars of the emerging discipline of environmental economics. Sándor and David, as kids playing in a football field, were engaged in a bright discussion about business, economics, and the natural environment, which ended in a piece of paper reproduced in Box 0.1. As Sandor said: "What I have learned from David Pearce: There is always some good reason for optimism" – and this is the optimism we need today.

Box 0.1   David Pearce (contribution from Sándor Kerekes)

On February 15 1994, David Pearce visited Hungary. As an expert, he was asked to comment on our paper on environmental problems related to the Gabčíkovo – Nagymaros Dams. At the Faculty of Business Administration of the University of Economics (previously known as Karl Marx University of Economics Budapest) we had been teaching environmental economics for four years already, and we were using the book written in 1990 by David Pearce and Kerry Turner, *Economics of Natural Resources and the Environment*. This book, in addition to Pearce's earlier book, published in 1989 (David William Pearce, Anil Markandya, Edward Barbier, *Blueprint for a Green Economy*), has had a significant impact on me because it was the first book that persuaded its readers to take action. Earlier, whatever I read related to the environment dealt with the "limits to growth," with ozone depletion, with acidification, etc.; therefore, with the catastrophic situation of the natural environment.

   Please do not forget that at that time we were living, at the beginning of what we have called the "transition" phase, we had a huge economic recession, we had no resources for environmental protection, and we had a strange political situation as well. So, we felt the natural catastrophe, but we could not imagine a light at the end of the tunnel. There was relatively little work in social sciences on this topic. Maybe I was misled by my previous profession in chemistry as well. The blueprint was the first comprehensive and relatively optimistic book in my hands. The blueprint encouraged action just by saying that to avoid future disasters, and even to change the operation of the economy, was possible and not so difficult. With immense pleasure and respect, but with a rather pessimistic view of the world, I waited for a meeting with David Pearce. And then it happened. David gave a lecture to my PhD students, and before we went to dinner, I took a bottle of wine to my office and tried to convince David, that the situation was hopeless, at least from the perspective of an eastern bloc environmental scientist. The glass of wine was imbibed slowly while David drew the essence of environmental economics on the flipchart below. Today, I am surprised to find that in 1994 he summarized the core points in three graphs (see Figure 0.1). First is the circular economy, the second is decoupling (or Kuznets' curve) and the third is the flow economy, instead of a stock economy. He took me to the optimistic side in

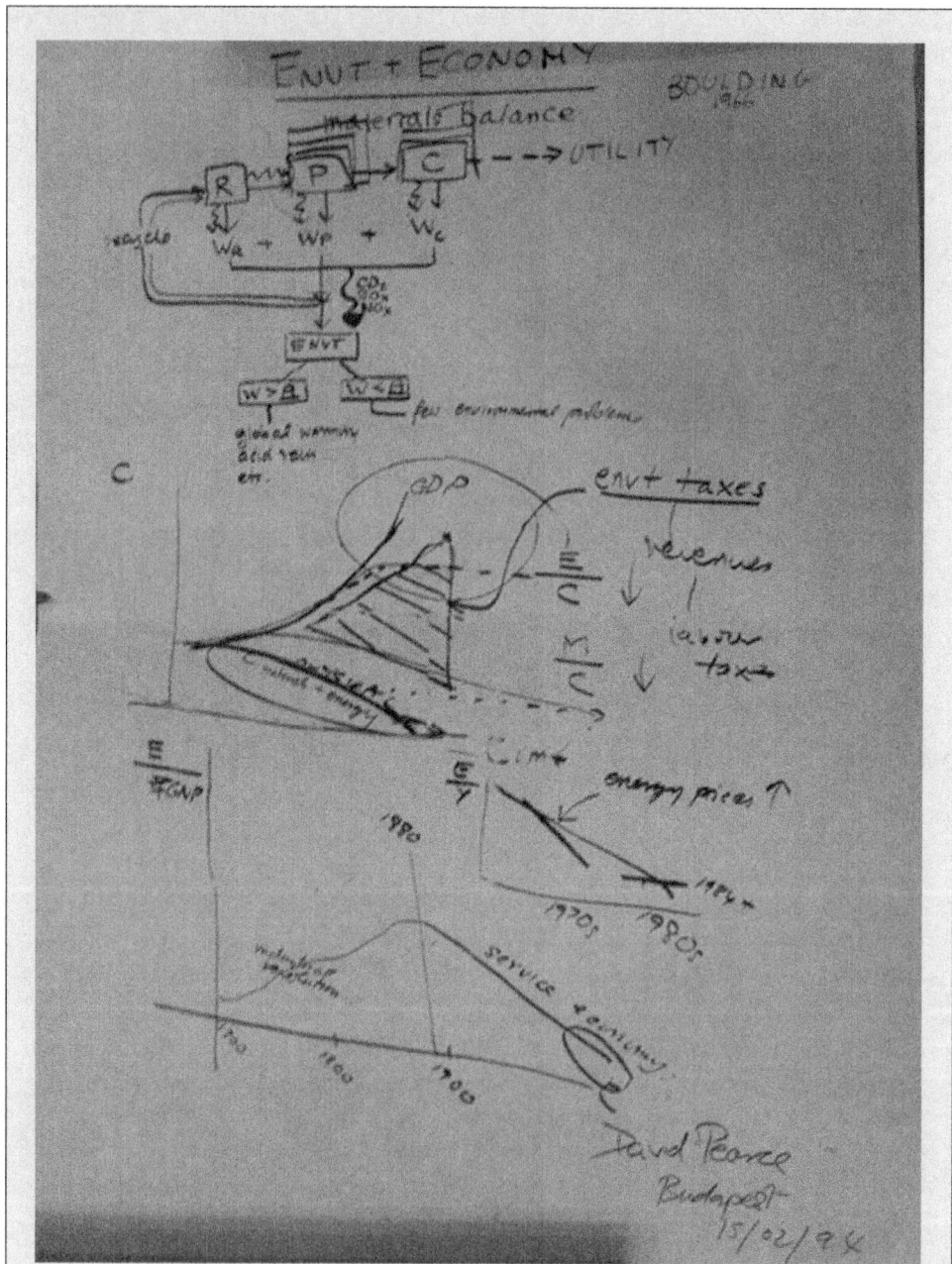

Figure B0.1 Flipchart from David Pearce.
Source: Kind courtesy of Sándor Kerekes.

less than 15 minutes. Since then, our profession is still muddling around these concepts. David, unfortunately, has since passed away, but my colleagues and I have kept him and his vision alive in the CEMS Faculty Group. Teaching university students is a fantastic job, but even a greater responsibility. One must believe, as David Pearce did, that we can shape our future and, of course, even in the Anthropocene there is hope for positive change.

Twenty years later, in March 2005, I invited him again. He passed away in May, but he wrote to me: "All good wishes and I hope to see you again someday."

Source: Sándor Kerekes.

Another of our colleagues, Romain Laufer, dedicated his research to the development of a multidisciplinary approach to management based on the notion of legitimacy and developed a theory about the management of major risks, how to deal with the breakup of our framework of reference and with the uncertainties that lay behind it. The impressive rise of environmental stressors is conditioning us, and we live in a society that needs to deal with such major risks. The move in our planet from Holocene-like conditions into Anthropocene-like conditions has resulted in a state-shift change that could potentially transform events we thought impossible into possible, and we need to be ready to tackle them. As Romain indicated, we need to learn not just to manage the possible or the improbable, but also to manage the impossible. "It is necessary to have a system of norms to act but it is impossible to have any relevant ones. It is then necessary to repair the failures of the symbolic system by the procedures of the type 'authorizing laws,' 'ethics committee,' 'security standards'" (Laufer, 1993); we need to learn how to manage these risks, how to transform and adapt to them, but it would be much smarter if we do not reach these thresholds by mitigating drastically our impacts right now.

Goals and structure of this book

The goal of this book is to provide a knowledge-based platform to illustrate how corporate sustainability should be addressed during the coming decades. This book intends to offer a new perspective to support managers dealing with environmental and social challenges, connecting theories, concepts, and frameworks with practice. Corporate sustainability is mostly addressed today building on concepts such as the triple bottom line (profit, people, and planet) and measuring the economic, social, and environmental performance of firms. One problem with this approach is that these three separate accounts do not capture the real interdependence between business and the natural environment. At the turn of the century, we are facing a systemic crisis on earth that calls for a new strategic and operational approach to facilitate the management of companies in our small but precious planet. More specific objectives of this book are to:

- understand how sustainability in the 21st century should be addressed and managed by companies and how it should be integrated into strategy and organizational structures,
- introduce the "Business In Nature" view, a new concept that positions business organizations inside social-ecological systems, and to acknowledge the active role of firms in contributing to the maintenance – or the enhancement – of social-ecological resilience,

- provide two conceptual frameworks on sustainability strategies and value chains, in order to guide managers in dealing with sustainability as a novel driver of companies' strategy, innovation, and profit,
- introduce new constructs and terms from global ecology and natural sciences to the community of management practice, and to contribute in framing a new language in business – a new jargon that facilitate a comprehensive representation of the relationship between business and the natural environment,
- help the reader develop skills and competencies that build on sustainability knowledge, and, to explain how "being green and socially responsible" fits with business goals.

This book builds on a broad knowledge base created thanks to more than 25 years of experience in teaching (from Bachelor to Executive Education courses), developing of a multitude of research projects, hundreds of meetings with managers and companies, and participating in many conferences, workshops and events. The book is ambitious, and it is aimed at spreading the "Business In Nature" view at multiple levels and in different communities. This includes students, who are getting a first glance at this topic; managers and executives, who require a novel system of thinking coherent with future business challenges; and the public at large interested in corporate sustainability.

The book is structured into three themes (or sections).

- The first theme is aimed at introducing the "Business In Nature" (BInN) concept. In the first chapter, we introduce the calls of attention from the scientific community and the calls for action from the United Nations about the need to change our course. We explain some jargon that can be used in the frontier between the domains of business and ecology. In the second chapter, we explore the evolution of corporate sustainability; we illustrate why companies address sustainability and discuss the business case for sustainability. Finally, the third chapter is dedicated to describing the BInN concept and its associated principles.
- The second theme aims to provide an extended illustration of how companies can address sustainability under the BInN framework. We discuss four basic approaches that can guide managers in rethinking the value chain of companies: making production systems sustainable in Chapter 4; managing sustainable supply chains in Chapter 5; designing sustainable products and services in Chapter 6, and finally, in Chapter 7, innovating business models for sustainability.
- The third and final theme of the book provides the reader with a solid framework that can support future and present managers in implementing corporate sustainability under the "Business In Nature" (BInN) concept.

The book is organized to facilitate the learning experience thanks to several pedagogical features.

- Every chapter starts with clear learning objectives that help to focus the reader's attention.
- Every chapter includes a short summary at the beginning that introduces the main themes and a conclusive summary at the end that helps refresh the reader's memory.

- In-text boxes illustrate examples of interesting and real-world best-practice companies and organizations. This broad series of case studies shows the application of the concepts discussed in the chapters to the business context and demonstrates the relevance of these topics to today's managers and business leaders.

- At the end of each chapter is an annex with a proposed list of open questions. These questions can be useful both for teachers to raise class discussions and for students to facilitate the learning experience, checking the level of comprehension and self-awareness of the key concepts analyzed.

- Each chapter includes some final assignments to help teachers with ideas on individual papers, team projects and exercises that can be proposed to students. These assignments cover a broad range of topics illustrated in the text and help consolidate the key concepts illustrated.

- Links to additional web-based resources (websites, videos, etc.) are also provided at the end of each chapter in order to stimulate further study on specific concepts and arguments and extend the understanding of corporate sustainability.

- At the end of each chapter, the authors have provided a list of case studies that can be used to deepen the topics discussed in the book.

- Finally, three extended case studies – Pinkton Hotels, Ikea and Enel – are provided with specific questions that can be used in class for discussion with students and for assignments.

Our ambition with this book is to make "BInN" the guidance for managing companies in the new century. Open your hearts to long-term-oriented business and give to the young people the world they deserve.

References

Laufer, R. (1993). *L'entreprise face aux risques majeurs: A propos de l'incertitude des normes sociales*. Paris, France: L'Harmattan.

Pearce, D. W., Markandya, A. & Barbier, E. (1989). *Blueprint for a green economy*. London, UK: Routledge.

Pearce, D. W. & Turner, R. K. (1990). *Economics of natural resources and the environment*. Baltimore, MA: John Hopkins University Press.

Ripple, W. J., Wolf, C., Newsome, T. M., Galetti, M., Alamgir, M., Crist, E., Mahmoud, M. I., Laurance, W. F. and 15,364 scientist signatories from 184 countries. (2017). World scientists' warning to humanity: A second notice. *BioScience*, 67(12): 1026-1028.

Timberlake, Ll. (2006). *Catalizing change: A short history of the WBCSD*. Switzerland: WBCSD c/o Earthprint Limited.

Zalasiewicz, J., Williams, M., Steffen, W. & Crutzen, P. (2010). The new world of the Anthropocene. *Environmental Science and Technology*, 44: 2228-2231.

ACKNOWLEDGEMENTS

In a way, we can say that we started the journey of this book many years ago with the CEMS network (Global Alliance in Management Education). Colleagues, friends and students from different universities and business schools have had a profound impact on this project, somehow contributing to shaping our ideas and organizing the main themes. We would like to thank the people that have been sharing their knowledge and experience with us for almost 25 years, teaching "business and the environment" issues. We express our sincere gratitude to Sándor Kerekes, Werner Delfmann, Adam Budnikowski, Romain Laufer, Daniel Tyteca, Marie Paul Kestemont and Christy Degen (our initial group), as well as many others with whom we spent hours and days together: Rolf Wuestenhagen, Gyula Zilahy, Ägnes Zsóka, Jost Hamschmidt, Minna Halme, Carlos Romero, Stefanie Hillie, Yuri Blagov, Adriana Budeanu, Renato Orsato and many others more recently.

We are particularly grateful to our good friends Kilaparti Ramakrishna and Anjaly. All the time spent with us and the help to frame the ideas of this book have been key to the completion of this project. Friends forever.

We also would like to offer our special appreciation to Monika Winn and Antonio Tencati for sharing long days of research, and for their fundamental contribution regarding the definition of some of the concepts and conceptual frameworks that are embedded in this book. Thanks, dear friends!

Our work was also deeply influenced by other brilliant scholars and great friends. At Bocconi University, Milano, we thank Francesco Perrini, Matteo Di Castelnuovo, Clodia Vurro, Angelo Russo, Nicola Misani, Alberto Grando, Filippo Giordano and Manlio De Silvio; Marco Frey at Scuola Superiore Sant'Anna; Mike Russo at University of Oregon; and John Morelli at Rochester Institute of Technology. At ESADE Business School, Barcelona, we would like to thank Miguel Angel Heras, Alberto Gimeno, Juan Ramis, Josep-Francesc Valls, Josep Rucabado, Daniel Arenas, Davide Cannarozzi, Oriol Pascual, David Murillo and Alice Bisiaux. In addition, we would like to thank many other people and friends from the Centre d'Estudis Avançats de Blanes (CEAB-CSIC).

Special thanks also to Manlio De Silvio, who kindly contributed to the editing of the figures and to Giorgia Rizzi, former student at Bocconi University and CEMS, who kindly contributed to this project helping with the editing of three cases: Pinkton, Ikea and Enel.

We also want to acknowledge the permission given by Will Steffen and Sarah Cornell from the Stockholm Resilience Center in Sweden, as well as some companies and organizations that facilitated the use of some of their materials such as Patagonia, Puma, Barilla and the MIT Sloan Management Review Journal.

ABOUT THE AUTHORS

Rafael Sardá

Rafael Sardá is Senior Scientist of the National Council of Research of Spain (CSIC) and Academic Collaborator of ESADE Business School in Barcelona. He holds a PhD in Biology (Extraordinary Award of the University in 1984) in the University of Barcelona and an MBA for ESADE. He has been working with the Business and the Environment Faculty Group of the CEMS (The Global Alliance in Management Education) since its formation. He is developing research and applications in the frontier between social and natural systems, how they work and interact, how to cope with present and emerging, local and global environmental problems and the role, if any, that science and regulations might play in it. He is author of around 200 research papers, responsible leader or co-leader in more than 20 national and international research projects, and director of 12 Doctoral Thesis and a large number of Master Thesis. He represents the CSIC in international expert group consultations on coastal and marine research affairs.

Stefano Pogutz

Stefano Pogutz is tenured researcher in the Department of Management and Technology, at Bocconi University, Milan, Italy and professor at SDA Bocconi School of Management. He holds a PhD in International Management from the Scuola Superiore Sant'Anna, Pisa. He is the Chair of the Business and the Environment Faculty Group of the CEMS (The Global Alliance in Management Education). He has launched and directed for five years the Bocconi University Specialized Master in Green Management, Energy and CSR. Dr. Pogutz's interests and expertise are in corporate sustainability, sustainable supply chain, innovation management and corporate social responsibility. His research has been published in several books and academic journals. He has a broad teaching experience including undergraduate, graduate and executive programs in several CEMS network business schools.

Theme I
Business In Nature

1   Welcome to the Anthropocene

Learning objectives

- Illustrate the theoretical discourse about the new world of the Anthropocene.
- Introduce new knowledge about earth system and sustainability science to guide our future development.
- Explain the social-ecological paradigm and provide a basic overview of its framework.
- Introduce new concepts from global ecology and natural sciences to the community of management practice.
- Provide basic knowledge about the ecological footprint and the provision of ecosystem goods and services.
- Describe what is new in the way the United Nations is addressing sustainable development at the turn of the century.
- Introduce the world's largest corporate sustainability initiative, the United Nations Global Compact initiative.
- Explain what the Sustainable Development Goals (SGDs) are.
- Illustrate the concept of resilience.

Chapter in brief

Awareness of global environmental problems among people is largely increasing today as mankind is reshaping the planet. Environmental threats such as climate change, population growth, habitat loss, and the more traditional concerns such as air and water pollution, forest loss and land degradation are bringing communities and nations together to agree on and adopt a common framework for action. Businesses now operate inside a new social-ecological paradigm and they need to learn how to manage the interdependences and interconnections in this new operating environment. This chapter advances the theoretical foundation of the book as it advocates for a paradigmatic shift associated with the "Business In Nature" concept. We will explain the reasons why we need to make a fundamental change as we explore the repeated calls from the scientific community and from the United Nations. These explorations lead us to rethink the way humanity inhabits the planet before irreparable damage changes our way of life and more broadly destroys our civilization as we know it.

We begin with an introduction to the new era of the Anthropocene and give an overview of the critical social-ecological changes occurring in this era. We illustrate new theoretical approaches taken from global ecology and natural science, such as planetary boundaries, great acceleration, global connectivity and the social-ecological system approach. Then, we review the findings from global assessments (Millennium Ecosystem Assessment) and calls for action (Sustainable Development Goals) reached toward this end. We provide the basis for understanding why companies need to go beyond the triple bottom line approach and adopt transformative changes to address the social-ecological crisis we find ourselves in.

Introduction

In one sense this book is about the process of globalization of environmental issues and the mainstreaming of environmental values within all sectors and policies, and how this process has been accelerated in recent decades. Main drivers include increased connectivity through trade, finance, travel, migration, communication, innovation and technological change. While many countries and companies have benefited from globalization through rapid economic growth, all of them have been affected by a host of adverse impacts, including those brought about by climate change, water stress, pollution, and ecosystems degradation, as well as rising inequality within and between countries, recurrent economic and financial crises, political instability and threats of global epidemics.

From the social sciences we hear of our entrance into the Fourth Industrial Revolution, a technological revolution that will fundamentally change everything we do:

> The First Industrial Revolution used water and steam power to mechanize production. The Second used electric power to create mass production. The Third used electronics and information technology to automate production. Now a Fourth Industrial Revolution is building on the Third, the digital revolution that has been occurring since the middle of the last century. It is characterized by a fusion of technologies that is blurring the lines between the physical, digital, and biological sphere.
>
> (Klaus Schwab, foreignaffairs.com, December 2015)

On the other side, from the natural sciences we hear of our entrance into the Anthropocene, "an era in which our species is having undeniable impacts on the environment at the scale of the Planet as a whole" (Steffen et al., 2007). We are starting to understand that man is able to modify the suite of interacting physical, chemical, and biological global-scale cycles and energy fluxes that provide the life-support system, and at the same time we are ready for another industrial revolution. Maybe it is time to rethink in a better way what we want to do. Still, we have not decided if this new Anthropocene era should start at the beginning of the first or the third Industrial Revolution, but the truth is that social and natural scientist are telling us that we are moving our planet from "Holocene-like conditions" to "Anthropocene-like conditions," and this could have fundamental implications for our way of life.

The transformations associated with the outcomes of these changes are produced at a high speed and have planetary implications in their scope and impact. From social aspects, we can dream of a future based on the rise of income and life quality but we can also forecast a future with increased inequalities, disrupted labor markets, and non-desirous migration patterns. From the natural perspective, very few indicators are giving us hope, and most of them point out large patterns of degradation (Rockström et al., 2009; Steffen et al., 2015). Running companies today – and in the coming future – is and will be a thought journey. Business is facing – and will face – major shifts in supply and demand, major changes in consumer expectations about new products, services and responsibility issues, and major organizational changes. As we are going to see throughout this book, from small-medium enterprises to large multinational conglomerates, firms will need to re-examine the way they operate, the way they produce and make people consume their products and services.

As previously stated, mankind is reshaping our planet, not merely in social dimensions (e.g., the economy and our social-cultural achievements), but also in its ecological dimensions – we are changing biospheric processes that allow many species to live. A major characteristic of the Anthropocene epoch is that it introduces new responsibilities and offers new opportunities. The Fourth Industrial Revolution needs to be carried out by taking into consideration the value of our relationship with nature. There is no doubt that we are facing a crossroads in the history of humanity and we must overcome it in an intelligent and responsible way. We know now that past Industrial Revolutions were unsustainable. We must move on to another one; a better one. Acknowledging the Anthropocene gives mankind the responsibility of enhancing welfare and prosperity while not doing harm to the planet and maintaining its functional integrity.

Welcome to the Anthropocene

The new world of the Anthropocene

Earth systems have undergone steady shifts in the past over different (usually thousands of years) timescales. However, for the last 11,000 years we have been living in the present interglacial epoch, in the geological era of the Holocene, where warm and accommodating global conditions prevailed, enabling the emergence and flourishment of human civilizations. Amazingly enough during the last 100 years, we have dramatically changed these conditions; increased human disturbances of critical earth system processes (interactions of land, ocean, atmosphere and life that together provide the space upon which species depend) have raised the risk of a state shift in the planet. Man, especially after the post-world war period of the last century and accelerating in the last decades, acts as a drastic global forcing agent, transforming the ways in which we (as humans) have been relating with our earth-system processes throughout millenniums. These changes can be observed in the soils and, as a consequence, a group of scientists suggested the idea of naming a new geological era for the planet: The Anthropocene.

The Anthropocene is a new geological epoch (for more details see Annex A.1.3 to this chapter). In the Anthropocene, humans are not just occupying and changing more and more of the planet, they are the most powerful contributor to the observed changes in how earth systems work. It was Paul Crutzen who formally addressed the idea that we are living in a different kind of geological era, one shaped basically by people: "The Anthropocene could be said to have started in the late eighteenth century, when analyses of air trapped in polar ice showed the beginning of growing global concentrations of carbon dioxide and methane" (Crutzen, 2002: 23). Since then, much research has been carried out to establish facts to describe this period, and geologists are trying to date its initial beginning and agree that the best way to distinguish geological eras is through the analysis of fossil records. The Anthropocene could easily be recognized by taking today's cities as the remains of our civilizations, our fossils for the future (Zalasiewicz et al., 2010, 2012; *The Economist*, 2011). Sediment layers accumulated in recent years have plenty of concrete and brick from roads and cities, and other novel entities such as radionuclides and microplastics that accumulate in soils. The Anthropocene defines this geological period of time as being human-influenced, and scientists are presenting overwhelming global evidence that many earth system processes have been altered in this new epoch, so the term has quickly begun being used. The Anthropocene is synonymous with how humans behave today in the planet. Behavior that is no longer sustainable asks for changes to be introduced in order to alleviate the pressure on earth system processes to avoid very negative and uncertain consequences. Basically, man forgot that we are a part of nature and that we need to live within it.

Back in the 1980s, scientists gathered together in a network of projects to develop collaborative research on global change and sustainability under one basic objective, to describe and understand the interactive physical, chemical and biological processes that regulate the total earth system and the manner in which they are influenced by human activities. This resulted in the creation and establishment of different research platforms including the International Geosphere-Biosphere Programme-IGBP; DIVERSITAS; the International Human Dimensions Programme-IHDP; and the World Climate Research Programme-WCRP. Today all these platforms have merged into one called Future Earth that reinforces this collaborative approach (for more details see Annex A.1.3 to this chapter). Under this umbrella, the new concept of the Anthropocene opens a large number of different perspectives coming from different disciplines trying to be integrated into a unique way to see the world. The Anthropocene reframes the ever-evolving relationship between humans and their non-human environment. In the Anthropocene we will be talking about climate change, but also about history and migration, about technology and labor, all of this in an integrated way.

The Anthropocene represents the emergence of a new way to see the earth. We recognize that humans are an integral part of the earth system and that we are reshaping its future. The Anthropocene can be considered as a coin with two sides; one side is about progress – humanity is richer and more prosperous than before; the other side is about costs; it is about creating social-ecological disequilibrium. We need to understand that the Anthropocene is affecting all functional systems on earth, "a new phase in the history of both, humankind and of the earth, when natural forces and human forces became intertwined, so that the fate of one determines the fate of the other" (Zalasiewicz et al., 2010: 2231). It is clear that the future must decouple prosperity and welfare from such external negative costs; progress should not continue to rely on large social inequalities and ecological degradation. We need to change this situation, and businesses cannot escape from this. A major challenge for the future that is going to affect the ways in which we produce, but also the way in which our clients consume, is the way in which we manage issues in firms. In other words, we need to manage companies acknowledging that we are in the Anthropocene. In the following section, we present four different messages that are going to shape our relationship with the earth in our new epoch.

The clock is ticking: some general messages

Of all the global environmental problems that are challenging humans these days, climate change is probably the most commented one; however, climate change, even though it influences much, is part of and a major contributor to what is being called global change by the natural sciences. Global change refers to the impact of human activity on the key processes that govern the functioning of the biosphere, and encompasses many different changes in the state of the planet (atmospheric composition, habitat loss, water cycle, global nutrient cycling, loss of soil, UV increase, desertification, etc.). In the paragraphs that follow, we introduce four different fundamental concepts that describe the global change phenomenon and that are impacting everything we do.

Great acceleration

The great acceleration is a concept that describes the rapid non-linear dynamics that affect the global socio-economic activity, structure, and function of the earth system over time. The concept gained visibility when the former IGBP program and the Stockholm Resilience Centre (see the link in Annex A.1.3 for detailed information) published a dashboard of 24 indicators which depict the dramatic acceleration in human enterprise and the impacts on the earth system over the last two centuries, giving to these trends the name "the great acceleration" (see Figure 1.1). Besides these iconic 24 charts, other syntheses of dynamic trends of social and ecological variables have found their way in scholarly literature. The Worldwatch Institute's Vital Signs provides one such global perspective; for quite some time Vital Signs compiled and provided a collection of indicators in a comprehensive, user-friendly manner showing key trends. With a better focus on atmospheric issues, the National Aeronautics and Space Administration-NASA provides up to date data streams about global warming and climate change.

The 12 socio-economic trends and the 12 earth system trends altogether have given us a clear signal; both series of trends mirror one another and they provide us with an unmistakable message that significant earth system processes are now inter-related with our patterns of human consumption and production. In addition, the analysis of Steffen et al. (2015a) suggests that consumption in the wealthy western countries is more important in terms of impacts on the earth system than population growth (although the latter is not insignificant). The study of these 24 trends also shows that the beginning of this great acceleration in many of these charts is around the 1950s, after World War II, and many scientists agree that this should be the starting date of the Anthropocene.

Socio-economic trends

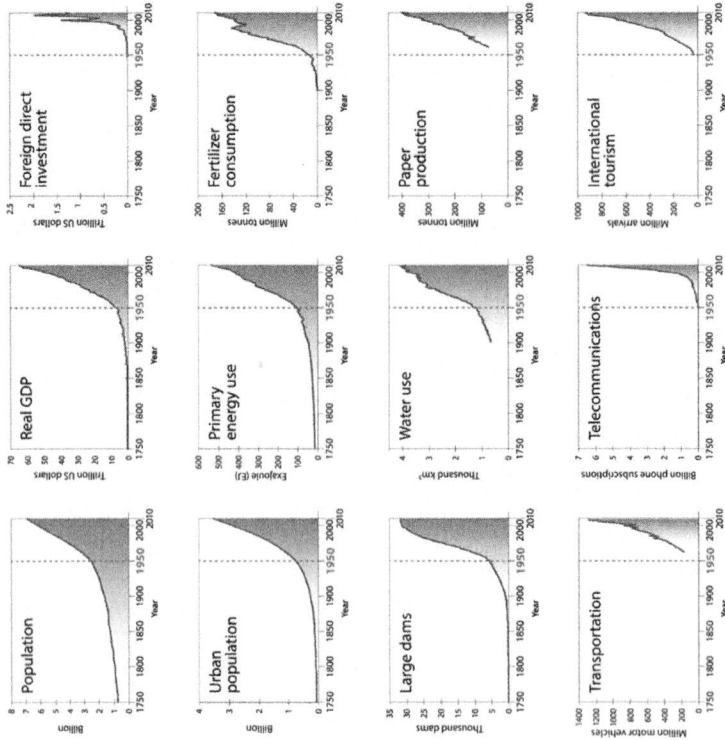

Earth system trends

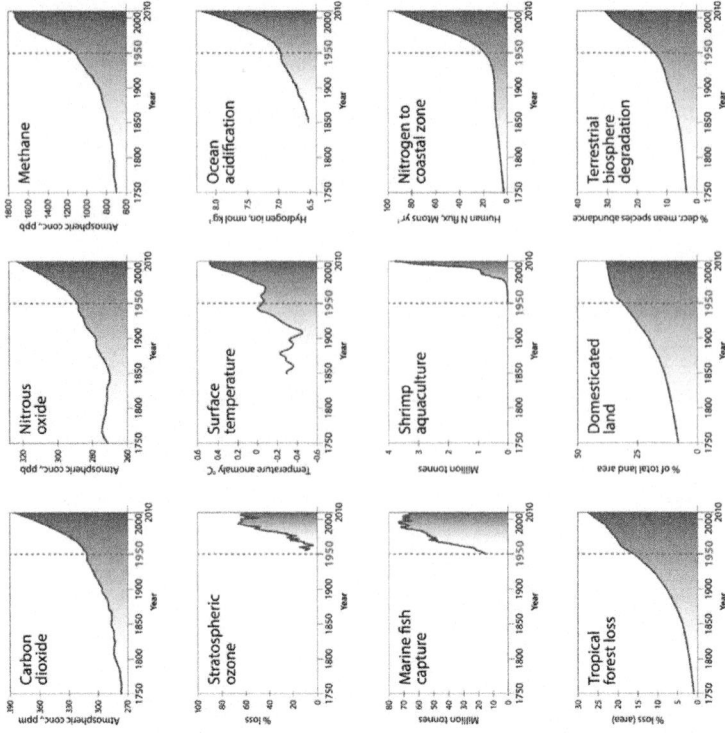

Figure 1.1 The great acceleration graph charts.
Source: adapted from the graphic published in Steffen et al, *The Anthropocene Review*, 2015a.

Most of these dynamics are not sustainable and if we do not reverse these trends rapidly there is the possibility that they will cause societies to collapse (Steffen et al., 2015a). We need to find sustainable solutions that take into consideration earth system dynamics and social inequalities (recently, some of these 24 charts have been split into three different groups of countries: OECD, BRICS, and the rest of the world – all showing different dynamics). In more recent years, some efforts resulted in progress; for example, we see significant improvement in the levels of stratospheric ozone; we notice signs of slowing construction of large dams. That said, for many other indicators we are still moving largely in an unsustainable path (see the planetary boundaries section that follows). However, the open question is (as exemplified by Steffen et al., 2015a): can the great acceleration in its present form can continue indefinitely? There is no clear answer yet but the question remains:

> Will the next 50 years bring the Great Decoupling or the Great Collapse? The latest 10 years of the great acceleration graphs show signs of both but cannot distinguish between these scenarios, or other possibilities. But 100 years on from the advent of the great acceleration, in 2050, we'll almost certainly know the answer.
>
> (Steffen et al., 2015a: 94)

No matter which way the scientific community answers this question, the great acceleration is calling on us to create new, more sustainable ways of living and organizing our societies.

Planetary boundaries

The science of sustainability was deeply influenced by an article about the concept of "planetary boundaries" produced by scientists of different disciplines led by the Stockholm Resilience Centre (Rockström et al., 2009). The planetary boundaries framework has received considerable attention during recent years, not only in science but also in policy and business circles. The framework applies rigorous methods to evaluate global change and earth system science under a resilient thinking approach. The planetary boundaries concept identified nine global priorities related to human-induced changes to the environment (see Figure 1.2). For those priorities, it provides some precautionary biophysical boundaries within which humanity can thrive and boundaries that we should not cross. This rigorous synthesis of detailed sustainability science is now able to send a powerful message to alert people of what we should not surpass to maintain good conditions for human societies on earth. The planetary boundary framework departs from the Holocene-like earth conditions as its desired state and then,

- identifies nine key processes that regulate the stability of these conditions,
- chooses variables as indicators to measure such nine processes,
- defines the range of uncertainty within which a threshold effect could occur, and
- propose a precautionary boundary to define a "safe operating space."

The message behind the planetary boundaries is that we have limits in our existence within the earth system and if we desire a stable, functional and resilient earth system (the global

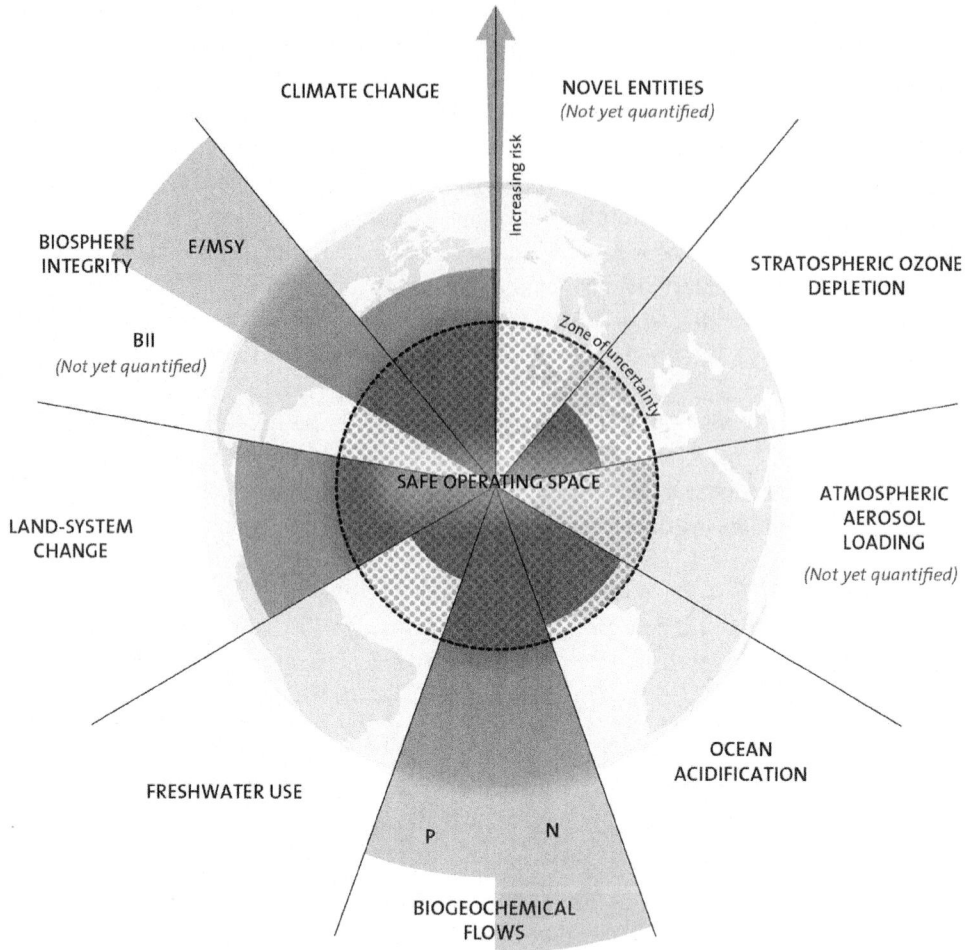

Figure 1.2 Planetary boundaries.
Source: Azote Images/Stockholm Resilience Centre, adapted from the graphic published in Steffen et al., *Science*, 16 January 2015b.

regulation seen on earth as a consequence of interactive physical, chemical and biological processes), we need to address these issues with a great deal of urgency. These limits put pressure on every societal agent (including business communities – both large and small) on earth and everyone should act accordingly.

Planetary boundaries were reviewed recently (Steffen et al., 2015b), updating the information initially given in Rockström et al. (2009). Unfortunately, from this revised version we could see that four of these nine planetary boundaries have now crossed the threshold as a result of human activity: climate change, loss of biosphere integrity, land-system change and altered biogeochemical cycles (phosphorus and nitrogen). With the knowledge that we have now it is safe to say that two of them – climate change and biosphere integrity – can act as a "core" boundary having a large influence on the others, and potentially they could bring us to a new shift of the earth system. For the nine key processes, the "planetary boundaries" concept provides a safe boundary zone, a zone of increasing risk with potential uncertainties and a zone of high risk of unpredictable consequences, a latest zone for which we do not have a framework of reference. We can quickly review the control variables associated with these nine global priorities, as follows.

- **Climate change** [atmospheric CO_2 concentration in parts per million]: safe zone condition (350ppm) and zone of uncertainty (350–450ppm). A global agreement (Paris agreement, UNFCCC, 2015) was adopted with the objective to avoid crossing the 450ppm threshold. At current rates of greenhouse gas emissions and today's concentration of 407.6ppm (January 4, 2018), the world community will reach 450ppm in less than 15 years. Climate change, one of today's most pressing issues, represents the greatest and widest-ranging market failure ever seen (Stern, 2006). Greenhouse gasses are accumulating in the atmosphere resulting in a global warming effect with potential consequences on all other variables.

- **Biosphere integrity** [extinction per million species-years, E/MSY]: safe zone condition (10) and zone of uncertainty (10–100). We now know that we exceeded these numbers and our planet is now in the midst of its sixth mass extinction of plants and animals (Barnosky et al., 2011; Ceballos et al., 2015). Species are becoming extinct at a significantly faster rate than for millions of years before and the deep influence of this extinction on biosphere functionality is still uncertain. The vast majority of planet's ecosystems reflect the presence of people and the entire planet is becoming more and more homogenized with high rates of habitat loss.

- **Stratospheric ozone depletion** [stratospheric O_3 concentration]: safe zone condition (276 dobson units-DU) and zone of uncertainty (276–261 DU). Probably the only key process for which we expect better numbers in the near future. Addressed globally in the Protocol of Montreal (Vienna Convention for the Protection of the Ozone Layer, September 1987) and tightened regularly ever since.

- **Ocean acidification** [carbonate ion concentration, average global surface ocean saturation state with respect to aragonite]: safe zone condition (above 80%) and zone if uncertainty (+80%–+70%); presently under safe zone but moving down. Ocean acidification refers to a reduction in the pH of the ocean over an extended period of time, caused primarily by uptake of carbon dioxide (CO_2) from the atmosphere, highly related to climate change.

- **Biogeochemical flows** [Phosphorous-P, flow from freshwater systems into the ocean, and Nitrogen-N, industrial and intentional biological fixation of N]: safe zone condition (P.- 11Tg yr^{-1}; N.- 62Tg yr^{-1}) and zone of uncertainty (P.- 11–100Tg yr^{-1}; N.- 62–82Tg yr^{-1}). Nitrogen flows are clearly exceeding these boundaries. The increase in phosphorus and nitrogen addition rates are too high to be managed in the biogeochemical natural processes have multiple consequences and has been especially detrimental to water quality.

- **Land-system change** [area of forested land as % of original forest cover]: safe zone condition (75%) and zone of uncertainty (75%–54%). The present value is 62%, located in the zone of uncertainty. Land systems are the result of human interactions with the natural environment. Land system changes are central to food security issues and the built-environment expansion, however transformations in social-ecological systems are part of a trade-off that needs to be sustainably managed.

- **Freshwater use** [maximum amount of consumptive blue water use]: safe zone condition (4000km^3 yr^{-1}) and zone of uncertainty (4000-6000km^3 yr^{-1}). Safe use of water can be critical for the future as many conflicts can arise as water is a fundamental and valuable resource for society. Although still in the safe boundary, we have many local conflicts due to freshwater use.
- **Atmospheric aerosol loading** [aerosol optical depth]: there is not a clear global measure. Aerosols are minute particles suspended in the atmosphere. When these particles are sufficiently large, we notice their presence as they scatter and absorb sunlight. Aerosols interact both directly and indirectly with earth's radiation budget and climate but a complete global response is not clear yet.
- **Novel entities:** not quantified yet. The formally called chemical pollution boundary is now renamed as novel entities, substances, and materials created entirely by humans (synthetic organic pollutants, radioactive materials, genetically modified organisms, nanomaterials and micro-plastics, among others). They pollute the environment but the ultimate consequences are largely unknown.

The planetary boundaries framework suggests a need for adaptive governance in an era that brings a multitude of uncertainties to all societies inhabiting the planet. Defining rigorous environmental boundaries for planetary functionality can help us in the mission of protecting habitable conditions for living. The clear message that we can get from the planetary boundaries framework and its trends is that continued inaction and lack of policy implementation will reduce the options for fair conditions for humans on earth. We believe that business, as one of the most powerful agents of change, should re-evaluate its future strategic plans and managerial decisions according to this new knowledge.

Population growth

Around 7.46 billion people are living on the planet today (March 2018); just 12 years ago we were 1 billion fewer. Although most forecasted scenarios are calling for a stabilization of the world population around 2050, during the last 12 years, yearly human growth rate has been more or less constant between 0.87% and 1.33% (1.12% on average) (United States Census Bureau). It has been estimated that the world population reached one billion in 1804, then it took 123 years to get two billion, only 33 years to get to three billion in 1960, and today it is still growing exponentially. Given the current global population another American organization, the Population Reference Bureau (PRB), estimated that people alive today may represent about 7% of the total number of humans who have ever lived.

All of these data are just statistics. This is true but you need to take this into account. Some decades ago, we were captured by a sentence of Paul Ehrlich and Anne Ehrlich in the well-known book *Valuing the earth* (Daly and Townsend, 1993):

> One of the toughest things for a population biologist to reconcile is the contrast between his or her recognition that civilization is in imminent serious jeopardy and

the modest level of concern that population issues generate among the public and even among elected officials.

(Ehrlich and Ehrlich, 1993: 55)

We do not want to enter into the controversy about population issues, about the previous and criticized work of these authors *The population bomb* (Ehrlich and Ehrlich, 1968), other Malthusian theories, or the newly revisited work in which they suggest that the basic message of the 1968 book is even more important today than it was 40 years ago (Ehrlich and Ehrlich, 2009). This book is not written to fundamentally understand population dynamic theories, but it is really important to understand that we live in a world of 7.5 billion. A planet that could soon be supporting as many as 10 billion human beings has to work differently from the one that held 1 billion people, 200 years ago.

In the prologue of this book, we clearly stated the need to understand this "sense of urgency." We need to come up with better worldwide mechanisms to deal with what the United Nations is bringing to the large table, the new Sustainable Development Goals (SDGs) and its plans to deal with human population issues (see coming section), eradicate poverty and hunger, and ensure health, education, sanitation and gender equality.

Global connectivity

Increased numbers of people and increased numbers of people's activities are the main contributors to the observed destabilization in earth system processes. These coupled dynamics have made environmental risks due to growth gain in prominence during recent decades, and this trend seems to continue at an even greater rate. The latest global risk report presented at the Davos Conference (WEF, 2018) indicates that from the five environmental risks selected, three of them lead as the highest risks for humanity (extreme weather events, natural disasters, and failure of climate-change mitigation and adaptation) and the other two are in the top 15 list. In addition, the global risk report is also alerting us that there is a high interconnectedness between these risks and other economic and social processes that makes human populations more vulnerable and exposed. As an example of such interconnections we can see Box 1.1. with details about the pork industry.

Box 1.1   The pork industry

Pork is the most consumed animal protein in the world. Globally, pig farmers produced 108.2 million metric tons (mmt) of pork in 2017 and China alone accounted for almost half of this, an amount that is twice the quantity of meat produced in the 27 EU countries. Other southeastern Asian countries (Vietnam, Philippines, Thailand or South Korea) are also among the top 15 world producers. In these countries pork is now an essential meat.

Since about 40 years ago, when China liberalized agriculture, pork consumption has increased sevenfold, reaching a quantity of 53.4 mmt in 2017. The amount of pork meat production in China is expected to reach 93 mmt for 2020. As a consequence, the Chinese diet, which basically consisted mainly of vegetables just 40 years ago, turned into a diet that is rich in meat protein. However, in order to produce 1 kg of pork you need 6 kg of feed, mostly soy and corn feed (the tropic chain pyramid) and the Chinese demand for this food is highly dependent on world crops for these plants (it is expected that in a near future half of the world's feed crops will be used just to feed Chinese pigs). This growing demand is putting pressure on land use changes as one of the causes of the biospheric tendency to homogenization. In Brazil, for example, more than 25 million hectares of land are used to feed pigs, resulting in habitat loss and increasing species extinction. On the other hand, farmers in China need to increase pesticide usage to satisfy the demand for more production in crops, and pig producers use more antibiotics to ensure that the animals are healthy. The problem of abundant fertilizer usage and pig's manure has a big impact on water pollution, and use of antibiotics creates future problems as these substances are entering into the natural environment (see Planetary boundaries – Novel entities, page 12) without knowing the potential effects for the future. In addition, we should not forget that more animal growth means more emissions of methane into the atmosphere. But it is not just that.

One of the established climate impacts includes the increasing number and persistence of heat waves. Russia, as another important world grain producer as alongside the US and Australia, had massive heat waves during the period from 2010 to 2012. The Russian heat wave of 2010 diminished production by 40% when compared to 2009. Russia banned grain export, triggering restrictions elsewhere. The long-term impact of the grain export ban issued by the Russian government during 2010–11 showed that while the ban did not bring prices down in Russia, it contributed to increased prices of grain internationally. One of the consequences of these heat waves was an increase in the Food and Agricultural Organization (FAO) food price index (a measure of the monthly change in international prices of a basket of food commodities), an issue that put in danger the living conditions of thousands of people elsewhere.

With a booming economy and increasing population, the demand for pig food will continue increasing in China and dietary meat requirements are likely to become even more expensive as the price of soy and corn, used in animal feed, also increases in price. If this trend correlates with an increase in heat waves as a consequence of climate change, its impact on crop productivity could have fatal consequences for destabilization of whole food systems. Opening up more space for even more crops may cause other natural destabilizations to occur as well.

Sources: adapted and elaborated from different sources (*The Economist*, Moore et al., 2017).

In the example of growing pigs, as in many others, observed changes and its consequences are connected by common drivers: a powerful combination of the growing human population and the increased per capita consumption of resources (e.g., land and biodiversity). When we think ahead about the rise of the global economy (globalization, geopolitical landscapes, migration, aging societies, new technologies) everything is connected, covering significant issues that touch on environmental, societal and macroeconomic risks. However, at the same time this opens up opportunities to deal with these risks and to understand how companies can be successful going forward.

The natural capital and the concept of social-ecological systems

Natural capital can be defined as the world's stock of natural assets which include geology, soil, air, water and all living things. The classical view about natural capital is that it is made up of ecosystems and, as a classical definition, an ecosystem consists of a biological community (biotic component) together with its abiotic environment, interacting as a whole (Chapin et al., 2002). It is not the intention of this book to elaborate on this ecological concept, but in a very simple way ecosystems can be described as entities formed by units (its structure and ecosystem components) and what these units do (its functions and integrity). Ecosystems are complex (structurally and functionally) adaptive systems that are controlled both by internal and external factors providing a variety of goods and services upon which people depend - our natural precious resources.

As we saw in the previous section, in the 21st century natural resources cannot be analyzed separately from humans; they are dependent on the social and economic systems with which they interact. We now live in this Anthropocene-like state and a framework for analysis that separates the human and the natural environment is not helpful to deal with our present environmental problems. To drive this new thinking and knowledge that man is present and can alter every ecosystem on earth, we opted for using the term social-ecological systems - linked systems of people and nature, emphasizing the fact that humans must be seen as a part of (*humans in nature*) not apart from (*humans with nature*) nature (Berkes and Folke, 1998) to integrate and show all these interactions and interdependencies. Whether it is Cape Cod, Costa Brava or Lake Como, whether it is the Arctic circle or the Antarctic continent, geographically bounded areas can work as a social-ecological system, a system in which two sub-systems interact: the social and the ecological. Systems that are normally geographically nested in a hierarchical functional order and systems in which businesses operate as part of societies, also in an integrated way. Social-ecological systems (see Figure 1.3) are complex adaptive systems in which the dynamics of both dimensions (sub-systems) are strongly linked.

Understanding social-ecological systems as co-evolving sub-systems should allow managers to respond to environmental feedbacks when we consider the goods and services that we obtain as benefits from the environment. Redman et al. (2004) defined social-ecological systems as "coherent systems of biophysical and social factors that regularly interact in a resilient, sustained manner, working at several scales and using critical resources in a dynamic and continuous adapted way" (Redman et al., 2004: 163). These social-ecological systems are usually defined geographically based on the interest of the party in use, but we need to consider that despite these artificial boundaries, these systems are linked together in a nested way. Another inherent property of these systems is that, whether one considers it in isolation or as a whole, the assets that we have in the natural sub-system are the foundations that allow social sub-systems to grow and generate other types of human capital.

SOCIAL-ECOLOGICAL SYSTEM

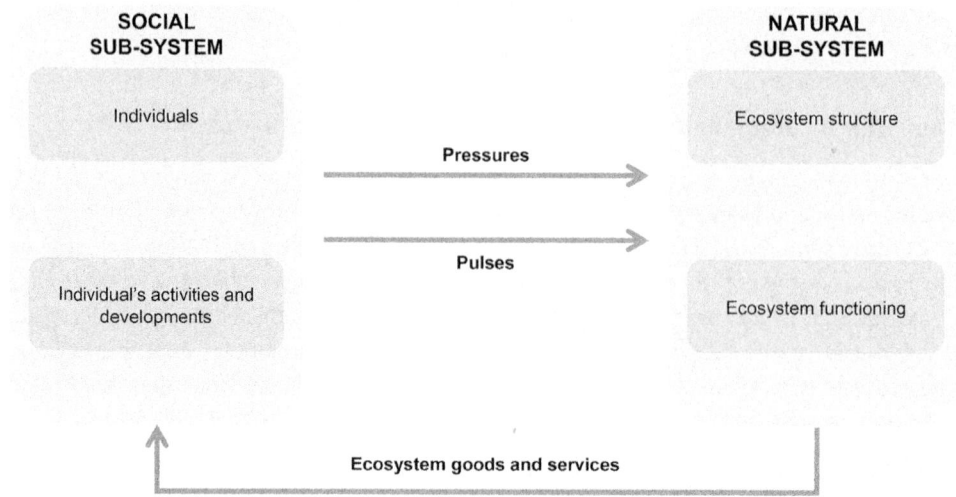

Figure 1.3 Components (structural and functional) of a social-ecological system.
Source: authors' elaboration.

In a social-ecological system, the natural sub-system can be described as some units (components) and their functionality (functions), while the human sub-system can be dis-tinguished between humans and societies (the human capital) and all the activities, actions and developments they enact (human-made capital). Between these two sub-systems we can find a lot of interactions and interdependencies. On one side, social sub-systems are impacting natural sub-systems (generally in a negative way) and this is the main reason why we have a huge environmental crisis these days. On the other side, we are largely dependent on a complete array of ecosystem goods and services coming from the natural sub-system. When we analyze these interconnections as a whole for the entire planet, they will only be "acceptable" if natural resources are used at the rate that they can regenerate themselves to maintain through time the provision of ecosystem goods and services in a resilient man-ner (see section about resilience later, pages 30-31). However, the present dynamics of the planet seen in the previous section results in a world population that steadily increases, and economic activity that grows every year; more people consuming and more consumption by each person, altogether putting excessive pressure on the natural sub-system that becomes degraded over time.

Present degradation of natural capital and the services it provides is widely known (MA, 2005). It is also accepted that a large part of the problem is due to a general failure in the development of global regulations (rules) and norms. In order to understand how it all fits together, a general framework for the analysis of social-ecological systems sustainability was proposed by Elinor Ostrom (2007, 2009). The framework was intended to organize the analy-sis of different attributes of the system: the resource system (the sea); the resource units gen-erated by that system (fish); the users; actors of the system (fishermen); and the governance system (rules). All of these are affected by interactions resulting in outcomes achieved at a particular time and place. The framework should also enable us to organize how these attributes may affect and be affected by the larger socio-economic, political, and ecological settings in which they are embedded, as well as smaller settings (Ostrom, 2007). A revised version of the framework was presented by McGinnis and Ostrom (2014) to put emphasis in the nested structure of these social-ecological systems. It is not the purpose of this book to validate, modify or criticize such a framework; rather, our approach is to make this framework more understandable and usable, and to bring this formulation to the general public and particularly the business community. Without the implication and collaborative action of private companies to deal with sustainable management of natural resources, it will be difficult to tackle the problem and contribute to solutions to address the natural resource depletion.

Human impact and the ecological footprint

In 1996, Mathis Wackernagel and William Rees first published a book that became a "classic" in the ways in which we are observing our planet *Our ecological footprint: Reducing human impact on the earth* (Wackernegel and Rees, reprinted 1998). In their book, the authors developed a new way to measure the impact of humans on earth and their ecological footprint. The book opened with a tale about the relationship between tiny insects and a mushroom with a limited carrying capacity for the populations that relied on it for their food supply. In ecological terms, carrying capacity is the number of organisms of a particular species (animal or plant) that an area (the earth) can support without irreversibly reducing its capacity to support them in the future at the desired level of living within the limits of the ecosystem. The carrying capacity is based on three factors: (a) the amount of resources available in the ecosystem; (b) the size of the population; and (c) the amount of resources each individual is consuming. If size and consumption exceed the amount of resources the system has, populations tend to crash. At the end of the tale, a few adult insects were able to escape the completely consumed mushroom to start the same cycle again in another one. Wackernagel and Rees relate the problem to people and the fable concludes with the moral that if humans consume the entire habitat, there is "no evidence of another earth-like mushroom in our galactic forest" (Wackernegel and Rees, 1998: xi).

Wackernagel and Rees moved this concept into an accounting tool. The tool enables us to calculate the amount of productive land needed for the consumption that humans' need. They defined the ecological footprint as "an accounting tool that enables us to estimate the resource consumption and waste assimilation requirements of a defined human population or economy in terms of a corresponding productive land area" (Wackernagel and Rees, 1998: 9). Although most of the examples used came from the study of cities or national ecological footprints, they use the analogy of a pasture land in their calculations: "[H]ow big a pasture is necessary to support that economy to produce all its feed and absorb all its waste?" (Wackernagel and Reese, 1998: 12). The rationale behind the concept is that if we can calculate the land that is required to enable typical human enterprises for a given number of people and then compare that to the amount of available land, we can determine whether or not we are living within our means. If we're not, we now have a powerful tool to help us alter our lives accordingly.

The ecological footprint of a particular person, society or economy will be calculated as the area of ecologically productive land (and water) that in various classes is required on a continuous basis to provide all energy and resources consumed and to absorb all the wastes discharged. The ecological footprint tracks the use of six categories of productive surface areas:

- plant-based foods and fiber products (cropland),
- animal-based foods and other animal resources (grazing land),
- fish-based food products (fishing grounds),
- provision of physical space for shelter and other infrastructures (built-up land),
- timber and other forest products (forest area), and
- absorption of CO_2 emissions (carbon demand on land).

The ecological footprint is calculated in hectares (see Annex 1 to this chapter to explore in detail the ecological footprint concept in the tools and resources of the Global Footprint Network). The first example of an ecological footprint calculation was the calculation of the ecological footprint of cities. The marine and terrestrial footprint of the city of Vancouver, Canada in 1991 was estimated as a productive output area more than 200 times larger than its political area (Rees and Wackernagel, 1998). At that point, we highly recommend the introduction of the concept of the personal ecological footprint by undertaking the assignments in Annex 1.2 to this chapter.

The ecological footprint has been widely used in national calculations in comparison with biocapacity. Biocapacity is a measure of the existing biologically productive area capable of regenerating natural resources in the form of food, fiber and timber, and of providing carbon dioxide sequestration. It is measured in relation to five categories of use: cropland, grazing land, fishing grounds, forest land and built-up land. Together, these satisfy human demand in the six footprint categories (WWF, 2016). Biocapacity can be used as an ecological benchmark against the ecosystem footprint (Galli et al., 2014; Wackernagel et al., 2014). Then, we can compare the needed assets of productive land able to satisfy the demand of people (consumption and waste) with the productive capacity that there is in that ecosystem. If a population's ecological footprint exceeds the region's biocapacity, that region runs an ecological deficit; if a region's biocapacity exceeds its ecological footprint, it has an ecological reserve.

Since 1998, the World Wide Fund for Nature (WWF; formerly the World Wildlife Fund) has been producing its Living Planet Earth report - one of the world's leading, science-based analyses on the health of our planet and the impact of human activity. In these bi-annual reports, WWF measures human pressures through a global ecological footprint and calculates a Global Living Planet index. This index measures biodiversity by gathering population data of various vertebrate species and calculating an average change in abundance over time. In their latest publication (WWF, 2016), they made an extrapolation about global ecological footprint and global biocapacity in 2020:

> [U]nder a business-as-usual path for the underlying drivers of resource consumption, assuming current population and income trends remain constant, human demand on the Earth's regenerative capacity is projected to continue growing steadily and to exceed such capacity by about 75 per cent by 2020.
>
> (WWF, 2016: 83)

The ecological footprint is a resource accounting tool for natural capital. It is able to measure in a defined space how much ecological capacity is available and how much of it is used in order to explore the dependence on natural capital, the ecologically productive force of Nature. Today, we are 60% above in the use of resources when compared to the available productive space of the planet; we are creating an accounting debt (deforestation, carbon accumulation in the atmosphere, overfishing, etc.), and this annual demand on resources which exceeds what earth can regenerate each year is known as the ecological overshoot. When we put together all the information we have about our ecological footprint and the biocapacity of the planet, it is evident that we are facing a clear invitation to change in order to avoid a global ecological bankruptcy.

Ecosystem goods and ecosystem services

During the last two decades, the term "ecosystem services" has gained acceptance as a framework for addressing the relationships between the human domain and the rest of the natural environment. The basic idea behind the use of the term is the understanding of how ecosystems contribute to humanity's needs through products and services. Since the publication by the United Nations of the International Millennium Ecosystem Assessment (MA, 2005), the interest in ecosystem service assessment has grown exponentially in environmental science and policy. Ecosystems provide a large array of goods, environmental services, and social services. Goods can be thought of as drinkable water, different foods or domesticated animals, and many others; some of these goods can enter into markets and can be exploited, hopefully in a renewable way. Environmental services are items such as the air we breathe, climate regulation, the protection of soils, etc., all basically granted for free. Social services are related to an ecosystem's use for recreational and aesthetic purposes, but also as spiritual and sacred areas.

The notion that ecosystems support human societies through natural resources (goods and services), and that these resources should not be seen as discrete entities, is not new – in fact, a variety of terms was offered in the past to describe this issue. It was first mentioned in the report "Study of Critical Environmental Problems" (SCEP, 1970; reviewed by Daly, 1997) as "environmental services;" such services would decline if the functionality of the ecosystem also declines. The concept was slightly modified by Ehrlich et al. (1977) as "public services of the global ecosystems", then by Westman (1977) as "nature's services" and finally it was named "ecosystem services" (Ehrlich and Ehrlich, 1981; de Groot, 1992; Costanza et al., 1997; Daly, 1997; Gilbert and Janssen, 1998; de Groot et al., 2002; Boyd and Banzhaf, 2007). In 1997 Daily proposed the following definition "conditions and processes through which natural ecosystems, and species that make them up, sustain and fulfill human life" (Daly, 1997: 3) but the concept was later popularized and formalized in the Millennium Ecosystem Assessment (MA, 2005) in a much simpler way "the benefits that people obtain from ecosystems" (MA, 2005: 53). Both definitions established the linkages between ecosystem services and human well-being by recognizing that ecosystems, if sustainably managed and protected, benefit current and future people and societies. This brought the idea that the concept of ecosystem services shows the flow of benefits from nature to people, providing a framework that can be used in the management of public goods.

The concept of ecosystem services is a human-centered concept; services are understood as benefits that people get from nature. Ecosystem services are essential for people and societies - and also for companies, as we will see in the coming chapters - but we should not forget that ecosystem services are nothing other than the result of ecosystem functions, and ecosystem functions are the result of thousands of years of earth evolution. Man used to play with these functions in many ways, as sometimes some of these ecosystem functions are favorable to man, and some related services can also end up being considered "disservices" - something we (in our centered view) do not want. It can be pleasant for people to lie on the beach for leisure practices, but unpleasant if the beach is full of vegetable wastes, even if those are necessary to support beach biodiversity; it can be pleasant to have the beach as close as possible to our home but unpleasant if we need to walk through dunes and salt marshes to reach it. Then, we make decisions and we modify nature based on these trade-offs. We need to understand that this can be risky and simply evaluating nature on the basis of the benefits that society obtains from it can eliminate functional processes that will put the resource itself in danger. These days, beaches are largely eroded because we didn't dedicate enough attention in the past to its protective function (biotic and abiotic components of our environment have many different functions and sometimes we are not paying attention to all of them), and this can have profound consequences in large industries such as tourism. Knowing about ecosystem functions and the way in which they provide us with services is a new "must" in this century.

The United Nations Millennium Ecosystem Assessment (MA, 2005) clearly determined the importance of ecosystem services for human well-being and classified ecosystem services into four categories: provisioning services, regulating services, supporting services, and cultural services. In Table 1.1, we show this classification following the global initiative of the Economics of Ecosystems and Biodiversity (TEEB, 2010).

Table 1.1 Ecosystem services classification with definitions.

Provisioning services	**Definition**. The goods or products obtained from ecosystems.
	• **Food:** ecosystems provide the conditions for growing food. Food comes principally from managed agro-ecosystems (crops and livestock) but marine and freshwater systems, even aquaculture or wild foods from forests, also provide food for human consumption. • **Raw materials:** ecosystems provide a great diversity of materials for construction and fuel, including timber and other wood products, fibers and resins, animal skins, past-living materials, sands and other cultivated materials, and ornamental resources. • **Freshwater:** ecosystems play a vital role in the global hydrological cycle as they regulate the flow and purification of water. • **Medicinal and pharmaceutical resources:** ecosystems and biodiversity provide many plants and animals that can be used as traditional medicines, as well as providing the raw materials for the pharmaceutical industry.

Regulating services	**Definition**. The benefits obtained from an ecosystem's control on natural resources. • **Climate regulation:** ecosystems regulate the global climate by storing and sequestering greenhouse gases (GHGs). Trees, plants and phytoplankton grow and remove GHGs from the atmosphere, which regulates climate. • **Maintenance of air quality:** ecosystems, through trees or other plants, play an important role in regulating air quality by removing pollutants from the atmosphere and emitting chemicals to it. • **Water regulation:** ecosystems have a large influence on the timing and magnitude of water runoff, flooding, and aquifer recharge and water storage. • **Water purification and waste treatment:** ecosystems play an important role in decomposing organic wastes and pollutants, as well as the detoxification of compounds through soil and subsoil processes. • **Erosion control:** ecosystems play a crucial role in retaining and replenishing soil and sand deposits. • **Moderation of extreme events:** ecosystems and living organisms create buffers against natural disasters, thereby preventing possible damage (e.g., wetlands can soak up flood water, trees stabilize slopes, coral reefs and mangroves protect coastlines from storm damage . . .). • **Disease and pest mitigation:** ecosystems can have an influence on pests and pathogens. • **Pollination:** many animals living in ecosystems pollinate plants and trees, which is essential for the development of fruits, vegetables, and seeds.
Cultural services	**Definition**. The non-material benefits obtained from ecosystems. • **Recreation and mental and physical health:** ecosystems can play a role in maintaining mental and physical health. • **Eco-tourism:** recreational pleasure that people obtain from natural or cultivated ecosystems. • **Aesthetic appreciation and inspiration:** ecosystems, language and knowledge have been intimately related throughout human history; ecosystems and natural landscapes have been the source of inspiration for much of our art, culture and, increasingly, science. • **Spiritual and ethical values:** ecosystems can host spiritual, religious, sacred, intrinsic ("existence") or other values important to entire communities.
Supporting services	**Definition**. The natural processes that maintain other ecosystem services. • **Habitats for species:** ecosystems provide spaces that maintain species populations and protect the capacity of communities to recover from disturbances. • **Maintenance of genetic diversity:** ecosystems maintain genetic diversity, allowing different breeds or races to survive and evolve, providing a gene pool for further developing commercial crops and livestock. • **Nutrient and water cycling:** ecosystems facilitate the flow of nutrients and water in different forms.

Source: authors' elaboration based on MA, 2005; TEEB, 2010; and Hanson et al., 2012.

Governance for the planet

Earth system governance is a recently developed paradigm. It is something that is needed but still far from reality, something that will be necessary to deal with the global change issue,

> a system of formal and informal rules, rule-making mechanisms and actor-networks at all levels of human society (from local to global) that are set up to steer societies towards preventing, mitigating, and adapting to global and local environmental change and earth system transformation, within the normative context of sustainable development.
>
> (Biermann et al., 2009: 4)

It needs to be constructed, and the United Nations represents the main actor that should be in charge.

United Nations at the turn of the century

We can begin our consideration of the United Nations and the environmental crisis by going back to the Stockholm Conference (United Nations Conference on Human Environment in 1972), where industrialized and developing nations were brought together for the first time to acknowledge the rights of humanity to a healthy and productive environment, or even 20 years later to the United Nations Conference on Environment and Development (UNCED) of Rio de Janeiro (Rio Conference, June 1992) where a transformation of people's attitudes and behavior was raised as a solution to bringing the necessary changes to remediate a worsening situation. But it may be most useful to begin this section with a few particular developments observed at the close of last century. As the 20th century ended, rapid economic acceleration and the globalization process witnessed a succession of turbulent ecological, climatic or sociopolitical developments: the convergence of some developing, emerging and developed economies increased human rates of environmental degradation; the growing unbalances in the global economy led to increased worldwide societal inequalities; and the absence of global governance mechanisms seemed to have accelerated a social-ecological global crisis. As a response, the United Nations called world leaders to adopt, in its headquarters of New York, the United Nations Millennium Declaration (September 2000), committing their nations to a new global partnership to reduce extreme poverty and setting out a series of time-bound targets – with a deadline of 2015 – that have become known as the eight Millennium Development Goals (MDGs). Since then, the implementation of the MDGs and a careful retrospective analysis of the underlying factors and policy decisions that framed these major events led the United Nations to work with scientific panels and advisors on developing reports on which to base future global decisions. These reports were written to guide the implementation of 21st-century policies for sustainable development. From all of these reports, two of them deserve to be discussed here.

The Millennium Ecosystem Assessment report (MA)

The Millennium Ecosystem Assessment (MA), launched in 2005, assessed the consequences of ecosystem change for human well-being, and the scientific basis for action needed to enhance the conservation and sustainable use of those systems, as well as their contribution to human well-being. It was a complete assessment of the benefits that biodiversity gives to people. The MA involved the work of more than one thousand experts from 2001 to 2004 and produced five technical volumes and six synthesis reports. The MA focused on linkages between ecosystem services and human well-being and the influence of direct and indirect drivers for its change. It constituted a state-of-the-art scientific appraisal of the condition and trends in the world's ecosystems and the ecosystem services they provide, as well as the options to restore, conserve, or enhance the sustainable use of ecosystems.

The MA put together all current knowledge, scientific literature, and data about biodiversity and ecosystem services. The main findings of this report are outlined in Box 1.2. The MA is known to give priority to a new environmental jargon that put the focus on the assessment of ecosystem services and their link to human well-being and development. The assessment identified how changes in ecosystems influence human well-being, as suggested in Figure 1.3, and provided information in a form that decision-makers can weigh alongside other social and economic information. The MA concluded that many ecosystem services are in degradation today, and it raised awareness about non-linearity issues and thresholds that should not be crossed, as well as the observed speed in several problems related to environmental degradation such as biodiversity, nutrients and pollution.

Box 1.2   The Millennium Ecosystem Assessment Report: main findings

The Millennium Ecosystem Assessment Report brought to us four main general findings, as follows.

- Ecosystems are changing these days at the highest rates for any comparable time of human history and, as a consequence, we are noticing a drastic and irreversible loss in Earth's biodiversity.
- Although human well-being and economic development have improved globally during the last decades, these gains have been based on the degradation of many ecosystem services and are based on non-linearity models and the use of non-renewable assets. These trends are putting in danger the possibility of future generations to benefit from ecosystems.
- If we do not change the course of our actions, ecosystem service degradation will continue and will make it impossible to reach our millennium development goals (today rebuilt into Sustainable Development Goals).
- To reverse this situation, significant changes in policies, institutions and practices that are not currently under way should be carried out. Putting ecosystem services on the developmental path for humanity is essential.

Sources: extracted from the Millennium Ecosystem Assessment organization.

The Intergovernmental Panel on Climate Change Assessment reports (IPCC-ARx)

The Intergovernmental Panel on Climate Change (IPCC) is a scientific panel designed similarly to the MA. Following the United Nations Framework Convention on Climate Change (UNFCCC) with the main objective to achieve the "stabilization of greenhouse gas concentrations in the atmosphere at a level that would prevent dangerous anthropogenic interference with the climate system," the convention created a scientific panel in order to provide the decision-makers and the other stakeholders interested in climate change with an objective source of information about climate change. The IPCC is a scientific intergovernmental body set up by the World Meteorological Organization (WMO) and by the United Nations Environment Programme (UNEP). The IPCC publishes a series of reports every five to six years constituting the basic science for the global climate policy. The Panel launched its fifth Assessment Report (IPCC, 2013) in September 2013; it was the most important world-wide document dealing with this issue; see Box 1.3 for the four main general findings in its policy summary.

Box 1.3   The IPPC Fifth Assessment Report: main findings

The policy summary of the Fifth Assessment Report concluded the following.

- Human influence on the climate system is clear, and recent anthropogenic emissions of greenhouse gases are the highest in history. Recent climate changes have had widespread impacts on human and natural systems and it is extremely likely that man is the dominant cause of climate change since the mid-20th century.
- Continued emission of greenhouse gases will cause further warming and long-lasting changes in all components of the climate system, increasing the likelihood of severe, pervasive and irreversible impacts for people and ecosystems. Limiting climate change would require substantial and sustained reductions in greenhouse gas emissions which, together with adaptation, can limit climate change risks.
- Adaptation and mitigation are complementary strategies for reducing and managing the risks of climate change. Substantial emissions reductions over the next few decades can reduce climate risks in the 21st century and beyond, increase prospects for effective adaptation, reduce the costs and challenges of mitigation in the longer term and contribute to climate-resilient pathways for Sustainable Development.
- Many adaptation and mitigation options can help address climate change, but no single option is sufficient by itself. Effective implementation depends on policies and cooperation at all scales and can be enhanced through integrated responses that link adaptation and mitigation with other societal objectives.

Sources: extracted from the IPPC report.

Both the Millennium Ecosystem Assessment (Biodiversity) and the Intergovernmental Panel on Climate Change (Climate Change) responded to the two main conventions developed in the Rio Conference. Through these conventions, the United Nations can warn each one of us about the severity of these problems.

The United Nations Global Compact initiative

The United Nations Global Compact (UNGC) is a worldwide initiative based on a shared set of principle, a kind of code of conduct to encourage businesses to take up more sustainable and responsible policies as well as report their implementation. The UNGC was initially launched in the year 2000 as another initiative to balance the rapid transformation observed at the end of the last century. It is the world's largest corporate sustainability initiative and 9,678 companies are now engaged with the UNGC. The UNGC constitutes a call to companies and corporate leaders to align strategies and operations with universal principles on human rights, labor, environment and anti-corruption, and take actions that advance societal goals (see Box 1.4). When companies incorporate these principles into their strategies, policies and procedures, and establish a culture of integrity, they can move ahead into what today is needed for long-term success.

The UNGC was a decisive action from the United Nations to talk directly with companies rather than just with governments in order to advance into the future and build a more sustainable global economy. The Global Compact supports companies to: (a) promote responsible business by aligning their strategies and operations with the ten principles of this code, and (b) take strategic actions to advance broader societal goals, such as the UN Sustainable Development Goals (see the section that follows), with an emphasis on collaboration and innovation.

Box 1.4   The ten principles of the United Nations Global Compact

Human Rights

Principle 1: Businesses should support and respect the protection of internationally proclaimed human rights;
Principle 2: make sure that they are not complicit in human rights abuses.

Labor

Principle 3: Businesses should uphold the freedom of association and the effective recognition of the right to collective bargaining;
Principle 4: the elimination of all forms of forced and compulsory labor;
Principle 5: the effective abolition of child labor;
Principle 6: the elimination of discrimination in respect to employment and occupation.

Environment

Principle 7: Businesses should support a precautionary approach to environmental challenges;

Principle 8: undertake initiatives to promote greater environmental responsibility;

Principle 9: encourage the development and diffusion of environmentally friendly technologies.

Anti-corruption

Principle 10: Businesses should work against corruption in all its forms, including extortion and bribery.

Sources: extracted from the United Nations Global Compact organization.

"A life of dignity for all" and the Sustainable Development Goals

With input from the High-level Panel of Eminent Persons on the Post-2015 Development Agenda, the Sustainable Development Solutions Network, the Global Compact Office, and the United Nations System Task Team on the Post-2015 United Nations Development Agenda, in July 2013 the Secretary-General, Ban Ki-moon, released his report "A life of dignity for all: accelerating progress towards the Millennium Development Goals (MDGs) and advancing the United Nations development agenda beyond 2015."

The report identifies policies and programs that have driven success in the achievement of these Goals and can contribute to accelerating it. These include emphasizing inclusive growth, decent employment, and social protection; allocating more resources for essential services and ensuring access for all; strengthening political will and improving the international policy environment; and harnessing the power of multi-stakeholder partnerships.

UNSG's report draws on the lessons learnt from implementing MDGs and proposes key elements for the development agenda beyond 2015, which include: (a) universality, to mobilize all developed and developing countries and leave no one behind; (b) sustainable development, to tackle the interlinked challenges facing the world, including a clear focus on ending extreme poverty in all its forms; (c) inclusive economic transformations ensuring decent jobs, backed by sustainable technologies, to shift to sustainable patterns of consumption and production; (d) peace and governance, as key outcomes and enablers of development; (e) a new global partnership, recognizing shared interests, different needs and mutual responsibilities, to ensure commitment to and means of implementing the new vision; and (f) being "fit for purpose" to ensure that the international community is equipped with the right institutions and tools for addressing the challenges of implementing the sustainable development agenda at the national level.

What is particularly relevant to this book are the 15 transformative and mutually reinforcing actions that include: eradicating poverty in all its forms; tackling exclusion and inequality; empowering women and girls; providing quality education and lifelong learning; improving health; addressing climate change; addressing environmental challenges; promoting inclusive and sustainable growth and decent employment; ending hunger and malnutrition; address demographic challenges; enhancing the positive contribution of migrants; meeting the challenges of urbanization; building peace and effective governance based on the rule of law and sound institutions; fostering a renewed global partnership; and strengthening the international development cooperation framework. Clearly in each of these transformative actions, the role of the business community is vital.

The Secretary-General's call to adopt a universal post-2015 development agenda, with sustainable development at its core, and the international system, including the United Nations, to embrace a more coherent and effective response to support this agenda resulted in the adoption of seventeen transformative goals, called the Sustainable Development Goals, at a summit level meeting at the United Nations in 2015. With the adoption of SDGs, it is clear that the United Nations system will continue to reform and make itself "fit for purpose" to respond to the challenges of this new path to sustainable development. But is it not just for governments and the United Nations to act; the business community has a huge role in it as well.

The number of the current global processes reflects a growing recognition of the importance of facilitating an integrated approach towards sustainable development, especially with regards to implementation; such an integrated approach must draw on the inter-linkages among the issues at stake, which will require a range of diverse stakeholders and experts to come together.

The Sustainable Development Goals and the business community

The Sustainable Development Goals (SDGs) succeed the previous MDGs, expanding the challenges of sustainable development. The SDGs are a universal call to action to end poverty, protect the planet, and ensure that all people enjoy peace and prosperity. In a world with huge inequalities, almost one billion people in extreme poverty, and approaching a state-shift in its earth's biosphere (Barnosky et al., 2012), the UN call was clear. What then are the SDGs to accomplish? Who should be applying them? The design elements of SDGs as Mr. Ban Ki-moon outlined with the central theme of "leave no one behind" should be the core message of SDGs. The recognition that everyone has a role to play and that businesses are called to develop a central role in its implementation is also clear. The seventeen SDGs (see Figure 1.4 and Table 1.2) and their associated 169 targets constitute the most important objectives to address for humanity in decades to come, and each company, from small enterprises to large multinational conglomerates, needs to make significant contributions to reach such goals.

Figure 1.4 The Sustainable Development Goals.
Source: adapted from United Nations.

Table 1.2 The Sustainable Development Goals.

People		
	1.	**No Poverty.** End poverty in all its forms everywhere.
	2.	**Zero Hunger.** End hunger, achieve food security and improved nutrition and promote sustainable agriculture.
	3.	**Good Health and Well-being.** Ensure healthy lives and promote well-being for all at all ages.
	4.	**Quality Education.** Ensure inclusive and equitable quality education and promote lifelong learning opportunities for all.
	5.	**Gender Equality.** Achieve gender equality and empower all women and girls.
	6.	**Clean Water and Sanitation.** Ensure availability and sustainable management of water and sanitation for all.
Prosperity	7.	**Affordable and Clean Energy.** Ensure access to affordable, reliable, sustainable and modern energy for all.
	8.	**Decent Work and Economic Growth.** Promote sustained, inclusive, and sustainable economic growth, full and productive employment, and decent work for all.
	9.	**Industry, Innovation and Infrastructure.** Build resilient infrastructure, promote inclusive and sustainable industrialization, and foster innovation.
	10.	**Reduced Inequalities.** Reduce inequality within and among countries.
Planet	11.	**Sustainable Cities and Communities.** Make cities and human settlements inclusive, safe, resilient and sustainable.
	12.	**Responsible Consumption and Production.** Ensure sustainable consumption and production patterns.
	13.	**Climate Action.** Take urgent action to combat climate change and its impacts.
	14.	**Life Below Water.** Conserve and sustainably use the oceans, seas and marine resources for Sustainable Development.
	15.	**Life on Land.** Protect, restore and promote sustainable use of terrestrial ecosystems, sustainably manage forests, combat desertification, halt and reverse land degradation, and halt biodiversity loss.
Peace	16.	**Peace, Justice and Strong Institutions.** Promote peaceful and inclusive societies for Sustainable Development, provide access to justice for all and build effective, accountable and inclusive institutions at all levels.
Partnership	17.	**Partnership for the Goals.** Strengthen the means of implementation and revitalize the global partnership for Sustainable Development.

Source: adapted from United Nations Sustainable Development Goals.

Business is called to play a major role in the SDGs implementation. In a speech addressed to the Private Sector Forum in September 2015, the former UN Secretary-General Ban Ki-moon was clear in his message to the business community:

> The Sustainable Development Goals were forged from the most inclusive policy dialogue we have ever organized. Governments must take the lead in living up to their pledges. At the same time, I am counting on the private sector to drive success. Now is the time to mobilize the global business community as never before. The case is clear. Realizing the Sustainable Development Goals will improve the environment for doing business and building markets. Trillions of dollars in public and private funds are to be redirected towards the SDGs, creating huge opportunities for responsible companies to deliver solutions.
>
> (United Nations, 2015)

Recent reports launched for the business community are introducing a new rule in the game for this century, one about sustainability. As a guide for business action on the SDGs, the SDG Compass aims to introduce companies to this new jargon and to familiarize them with its principles by guiding companies through a five step process (GRI, UN Global Compact and WBCSD, 2015):

- **Understanding the SDGs.** Raise the baseline responsibilities of business in this new century. In doing that managers may develop a new business case identifying new business opportunities and enhancing the value of corporate sustainability.
- **Defining priorities.** After understanding present positive/negative contribution into the SDGs, companies may define priorities. At that level, it is valuable to broaden the analysis of the value chain to identify other potential areas of influence, as well as to engage stakeholders.
- **Setting goals.** Adopting a goal setting approach based on previous identification will bring to the use of selected key performance indicators that will facilitate commitments. Selecting the most appropriate goals to work with and raising the level of ambition can increase innovation and creativity.
- **Integrating.** The sustainability function needs to be integrated into all the other functions of business, and organizations will need to understand that cooperation and collaboration are important values in dealing with all of these aspects.
- **Reporting and communication.** As disclosing information is becoming a mantra in these days, communicating and effective reporting on SDG performance will increase its importance.

In this book we are introducing a new view to deal with corporate sustainability in the firm. This view is aligned with the earlier UN call in the recommendation to align the company's core mission with UN SDGs. Businesses need to be involved in this roadmap:

> [T]his can be through core business operations and value chains, social investments, philanthropic contributions and advocacy efforts. While we recognize that there are some who question the linkage and even the legitimacy of the for-profit private sector and the development agenda, we believe that responsible business is central to growth, productivity, innovation and job creation – all drivers for progress at scale.
>
> (BFP, Harvard Kennedy School and SDGF, 2015)

The purpose of this book is to help in showing the way.

Resilience

Resilience has been generally addressed as the capacity of a system to cope with disturbances without shifting into a qualitatively different state. Although resilience has been investigated by a number of disciplines and from diverse theoretical perspectives (see Winn and Pogutz, 2013, for a review), for the purpose of our book we will be dealing with the concept of resilience adapted to social-ecological systems. Carl Folke, in his seminal paper about this type of resilience, emphasized that when talking about this concept applied to the dynamics of social-ecological systems, the classical concept of the capacity to absorb shocks and still maintain function needs to be taken together with the capacity for its renewal, re-organization and development, another essential element for the sustainability discourse (Gunderson and Holling, 2002; Berkes et al., 2003; Folke, 2006). Folke described the concept of resilience as "the capacity of a system to absorb disturbance and re-organize while undergoing change so as to still retain essentially the same function, structure, identity and feedbacks" following Walker et al. (2004). These authors identify three attributes that can be used to describe social-ecological system dynamics: resilience, adaptability and transformability. On these schemes, adaptability refers to the capacity of the actors within the social-ecological system to influence resilience, especially the collective capacity of humans to manage resilience intentionally, while transformability is the capacity to create a fundamentally new system when the functions, structure, identity and feedbacks of the existing system become untenable. However, there is an important difference when we consider these three attributes together; resilience and adaptability have to do with the dynamics of a particular system, or a closely related set of systems, while transformability refers to fundamentally altering the nature of a system.

Humans and nature are strongly coupled when interacting inside a social-ecological system. As we illustrated, in every natural system we can feel the presence of people and people need nature because of the ecosystem services it provides; however, many times people and nature gets basically disconnected in our societies. Resilience is therefore an attempt to create a new understanding of how humans and nature interact. In the social-ecological system structure of Figure 1.3, the social sub-system can pressure the natural sub-system to a point of no-return, transforming itself into another different sub-system, or to a point in which it can cope with the disturbance, regenerate or re-organize, maintaining through time its original structure and function. Sometimes, even if we manage the system with the goal of regeneration in mind, gradual pressures can coalesce into another point of no-return, and the natural sub-system also transforms. Occasionally, the nested structure of social-ecological systems pressures this system in a way that impedes its resilience attributes. In the history of man, these transformations have been a common pattern in local environments, sometimes transforming natural sub-systems into less- or non-desirable ones that could affect societal welfare and prosperity. However, today these transformations are becoming massive and they can promote a planetary state shift (Barnosky et al., 2012) that could involve no resilient earth system planetary processes with potentially global catastrophic consequences.

Businesses are important actors that play a major role in many different social-ecological systems and they can have large impacts. As these impacts are evolving into state shifts in our biosphere, the implication of the resilience theory is becoming more and more important. Corporate sustainability in the 21st century should address the resilience of the systems with which companies interplay and not just find possibilities to adapt to the current conditions where companies work, but also to achieve positive transformations for more desirable sustainable development paths.

Summary

In this first chapter of the book, we have explored planetary trends that are occurring inside a global change phenomenon. This phenomenon mirrors the idea that humankind has transformed the world and this world evolved into a new geological era, the Anthropocene. The Anthropocene represents the emergence of a new way to see the earth, a way deeply influenced by human activities, and that requires new capacities and skills to learn how to better manage and protect our planet. The warnings from the scientific community and from United Nations should make us react in order to further develop and incorporate this vision in our production and consumption patterns.

Scientists convened under the Future Earth platform. Future Earth builds on science to accelerate transformations to sustainability. Future Earth, as a global network of scientific partners, works to incorporate the latest research findings into government, business, and community decisions and policies. In a former message from its executive director:

> We urgently need a radical shift in how global society interacts with Earth. Human activities today are disrupting our planet in ways that threaten our air quality, water reliability, food security, and the stability of our climate and ecosystems. Real transformations to overcome these challenges will only happen if we work together – science, technology, business, government, and civil society – as one.
>
> (Amy Luers, Executive Director, Future Earth (www.futureearth.org/about)

Recently the United Nations also made some warnings about facing the huge economic, social and ecological challenges of our Planet. It launched 17 Sustainable Development Goals (SDGs), a group of global aspirational priorities for 2030, and it aligned such goals with the previous Global Compact initiative. The United Nation Global Compact explained why these SDGs matter for businesses and why businesses should help to reach its visionary targets.

Finally, we have defined resilience as "the capacity of a system to absorb disturbance and re-organize while undergoing change so as to still retain essentially the same function, structure, identity and feedbacks" (Walker et al., 2004). We have explained the utility to work with the concept of social-ecological systems and the need to move into future resilient conditions for the earth. In the coming chapters, we will introduce a new view for corporate sustainability that should have resilient conditions as the ultimate goal.

Chapter 1 annexes

Annex 1

A.1.1: Questions for discussion

- Welcome to the Anthropocene. Explain the Anthropocene concept and discuss the "great acceleration."
- Discuss the concept of planetary boundaries. What are the main environmental challenges for the 21st century? How could you imagine the planetary boundaries framework affect and being used by business?
- What is the role of science in defining the future challenges for our society and our economy?
- Explain the concept of ecosystem and ecosystem services. What types of ecosystem services do you know? How ecosystem services relate to human wellbeing? How ecosystem services relate to business?
- How would you proceed to assess and calculate the environmental footprint of your country? What are the pros. and cons. of the utilization of this methodology?
- How would you explain the concept of social-ecological systems?
- According to the Millennium Ecosystem Assessment report (2005), has the quality of ecosystem services improved over the last decades?
- What is the role of policy in addressing the challenges of Sustainable Development?
- What is the Intergovernmental Panel on Climate Change? How does this organization contribute to the science of climate?
- What is the United Nation Global Compact?
- What are the Millennium Development Goals and how do they relate to the Sustainable Development Goals?
- What is ecological resilience? How can resilience, adaptability and transformability contribute to Sustainable Development? What is the difference among these concepts?

A.1.2: Assignments

Assignment 1: The Anthropocene and the planetary boundaries

The assignment starts with the view of Johan Rockstöm's video "Let the environment guide our development" (Rockström, 2010): www.ted.com/talks/johan_rockstrom_let_the_environment_guide_our_development/transcript?utm_campaign=&utm_content=ted-androidapp&utm_source=direct-on.ted.com&awesm=on.ted.com_hOLIA&utm_medium=on.ted.com-android-share.

The video introduces the concept of the Anthropocene and its planetary boundaries providing an overview of the major environmental challenges. In its first part (Rockström, 2010; 0:55-7:08) he delivers on the Anthropocene era where humans are a major force in the biosphere. In its second part (Rockström, 2010; 7:09-12:50) introduces the planetary boundaries framework, and in a third part (Rockström, 2010; 12:50-17:52) he presented different approaches toward sustainability.

The assignment recommends the video to be coupled with some team activity. Students can be assigned in groups and to work with specific companies to check their webpages and sustainability reports. Then, some questions can guide a discussion: (a) How did these companies address the main environmental challenges listed in the planetary boundaries concept?; and (b) In what type of management context can these illustrated concepts be discussed (mission/values, sustainability/CSR strategy, innovation, supply chain management . . .)?

Other similar videos can be found to make the students interpret the concept (www.youtube.com/watch?v=V9ETiSaxyfk).

Assignment 2: The ecological footprint

The exercise starts by playing the Mathis Wackernegel video: www.youtube.com/watch?v=94tYMWz_la4. The video introduces the concept of the ecological footprint as well as the concept of biocapacity and links these concepts with the concept of natural capital. The video can be used as an introductory resource in class for the topic and to guide explanations about how the ecological footprint is calculated. This can be reinforced by a discussion in class.

After displaying the video, you can work individually in the calculation of your personal ecological footprint. We recommend going to the web page of the Global Footprint Network (GFN) and using its calculator to obtain your personal ecological footprint: www.footprintnetwork.org/our-work/ecological-footprint/.

Apart from obtaining your personal ecological footprint, this calculation will allow you to calculate your ecological overshoot. Every year, the GFN raises awareness about global ecological overshoot with the Earth Overshoot Day campaign. This event highlights the day on the calendar when humanity has used all the resources that it takes the planet a full year to regenerate (in 2017, 2 August). You will be able to calculate what the date would be if everyone was acting like you.

Back in the class, share your personal numbers in order to get the class average, followed up by a discussion about the different items in the calculation. The exercise will reinforce a concept that will be used later in this book at a company level.

Assignment 3: Business risks and opportunities arising from ecosystem change

Companies often fail to make the connection between the health of ecosystems and the business bottom line. Many companies are not fully aware of the extent of their dependence and impact on ecosystems and the possible ramifications. The overall assignment is about how some selected companies can develop a business case by taking care of different ecosystem services provided by biodiversity. To focus more your assignment you can work on the relationship of the selected company with a particular environmental problem listed in the planetary boundaries. A list of companies will be given. After selection, you will work following the indications provided by the Corporate Ecosystem Service Review guide (Hanson et al., 2012).

Do an analysis of the three first steps of the Corporate Ecosystem Service Review Guide for your selected company and environmental problem.

- Select the scope of your analysis, define clear boundaries for it. For a business with just one major product, service or market, the ESR scope could be the entire company. However, for a business with multiple products, services or markets, a more manageable scope would be a particular business unit of the company.
- Identify priority ecosystem services. Try to come up with something similar to Boxes 11 or 12 in the Corporate Ecosystem Service Review guide.
- Select the ecosystem service you find the most important for this company in relation to the analyzed problem. Analyze trends in the provision of this ecosystem service by answering the questions in the Guide.

A.1.3: Further readings and additional resources

- The Natural Capital Project is an organization whose mission is to integrate the value nature provides to society into all major decisions. Its ultimate objective is to improve the well-being of all people and nature by motivating greater and more targeted natural capital investments (www.naturalcapitalproject.org/).

InVEST (ecosystem valuation of ecosystem services and tradeoffs) is the flagship tool of the Natural Capital Project with models for mapping and valuing ecosystem services provided by land- and seascapes

- The Anthropocene (www.athropocene.info) is an educational web portal that gives a wide range of information and resources about the concept of the Anthropocene. It aims to inspire, educate and engage people about the interactions between humans and the planet.
- Future Earth (www.futureearth.org/) builds on more than three decades of global environmental change research through the World Climate Research Programme (WCRP), the International Geosphere-Biosphere Programme (IGBP), DIVERSITAS and the International Human Dimensions Programme on Global Environmental Change (IHDP). Today it is a global community that facilitates research, mobilizes networks, sparks innovation, and turns knowledge into action through the work of 20 Global research projects.
- Vital Signs: The Worldwatch Institute's Vital Signs (http://vitalsigns.worldwatch.org/) provides one such global perspective. For quite some time Vital Signs compiled and provided a collection of indicators in a comprehensive, user-friendly manner of key trends. With a better focus on atmospheric issues, the National Aeronautics and Space Administration – NASA (https://climate.nasa.gov/vital-signs/) provides up-to-date data streams about global warming and climate change.
- The Population Reference Bureau (PRB) informs people about population, health and the environment. They also empower others to use that information to advance the well-being of current and future generations (https://interactives.prb.org/annualreport/public/pdf/report.pdf).
- The United Nations Global Compact represents the world's largest corporate sustainability initiative – a call to companies to align strategies and operations with universal principles on human rights, labor, environment and anti-corruption, and take actions that advance societal goals. In its web page, you can find plenty of supporting tools to engage companies in sustainability as well as to help companies in its roadmap to the Sustainable Development Goals (www.unglobalcompact.org/).

Annex 2

A.2.1: References

Barnosky, A. D., Matzke, N., Tomiya, S., Wogan, G. O. U., Swartz, B., Quental, T. B., Marshall, C., McGuire, J. L., Lindsey, E. L., Maguire, K. C., Mersey, B. and Ferrer, E. A. (2011). Has the Earth's sixth mass extinction already arrived? *Nature*, 471(7336), 51-57.

Barnosky, A. D., Hadly, E. A., Bascompte, J., Berlow, E. L., Brown, J. H., Foretelius, M., Getz, W. M., Harse, J., Hastings, A., Marquet, P. A., Martinezz, N. D., Mooers, A., Roopnarine, P. Vermeij, G., Williams, J. W., Gillespie, R., Kitzes, J. Marshall, C. Matzke, N., Mindeli, D. P., Reville, E. and Smith, A. B. (2012). Approaching a state shift in Earth's biosphere. *Nature*, 486(7401), 52-58.

Berkes, F. and Folke, C. (1998). Linking social and ecological systems for resilience and sustainability. In F. Berkes and C. Folke (Eds.), *Linking social and ecological systems: Management practices and social mechanisms for building resilience* (pp. 15-25). Cambridge: Cambridge University Press.

Berkes, F., Colding, J. and Folke, C. (2003). *Navigating Social-Ecological systems: Building resilience for complexity and change*. Cambridge: Cambridge University Press.

BFP, Harvard Kennedy School and SDGF. (2015). Business and the United Nations: Working together towards the sustainable development goals: A framework for action. Retrieved from www.sdgfund. org/sites/default/files/business-and-un/SDGF_BFP_HKSCSRI_Business_and_SDGs-Web_Version.pdf.

Biermann, F., Betsill, M. M., Gupta, J., Kanie, N., Lebel, L., Liverman, D., Schroeder, H. and Siebenhüner, B. (2009). *Earth System Governance: People, places and the planet. Science and implementation plan of the earth system governance project*. Earth System Governance Report 1. Retrieved from www.ihdp. unu.edu/docs/Publications/ESG/IHDP_ReportNo20_ESG_ReportNo1.pdf.

Boyd, J. and Banzhaf, S. (2007). What are ecosystem services? The need of standardized environmental accounting units. *Ecological Economics*, 63(2-3), 616-626.

Ceballos, G., Ehrlich, P. R., Barnosky, A. D., García, A., Pringle, R. M. and. Palmer, T. M. (2015). Accelerated modern human-induced species losses: Entering the sixth mass extinction. *Science Advances*, 1(5), e1400253.

Chapin, F. S., Pamela A. M. and Mooney, H. A. (2002). *Principles of terrestrial ecosystem ecology*. New York, NY: Springer.

Costanza, R., d'Arge, R., de Groot, R., Faber, S., Grass, M., Hannon, B., Limburg, K., Naeem, S., O'Neill, R. V., Paruelo, J., Raskin, R. G., Sutton, P. and van den Belt, M. (1997). The value of the world's ecosystem services and natural capital. *Nature*, 387(6630), 253-260.

Crutzen, P. (2002). Geology of mankind. *Nature*, 415: 23.

Daly, H. E. and Townsend, K. N. (1993). *Valuing the Earth: Economics, ecology, ethics*. Cambridge, MA: The MIT Press.

Daly, G. C. (1997). *Nature's services*. Washington, DC: Island Press.

de Groot, R. (1992). *Functions of nature: evaluation of nature in environmental planning, management and decision making*. Groningen: Wolters-Noordhoff.

de Groot, R., Wilson, M. A. and Boumans, R. M. J. (2002). A typology for the classification, description and valuation of ecosystem functions, goods and services. *Ecological Economics*, 41(3), 393-408.

Ehrlich, P. R., Ehrlich, A. H. and Holdren, J. (1977). *Ecoscience: Population, resources, environment*. San Francisco, CA: W. H. Freeman and Company.

Ehrlich, P. R. and Ehrlich, A. H. (1968). *The population bomb*. New York, NY: Ballantine Books.

Ehrlich, P. R. and Ehrlich, A. H. (1981). *Extinction: the causes and consequences of the disappearance of species*. New York, NY: Random House.

Ehrlich, P. R. and Ehrlich, A. H. (1993). Why isn't everyone as scared as we are? In H. E. Daly and K. N. Townsend (Eds.), *Valuing the earth: Economics, ecology, ethics* (pp. 55-67). Cambridge, MA: The MIT Press.

Ehrlich, P. R. and Ehrlich, A. H. (2009). The population bomb revisited. *The Electronic Journal of Sustainable Development*, 1(3).

Folke, C. (2006). Resilience: The emergence of a perspective for social-ecological systems analyses. *Global Environmental Change*, 16, 253-267.

Galli, A., Wackernagel, M., Iha, K. and Lazarus, E. (2014). Ecological footprint: Implications for biodiversity. *Biological Conservation*, 173, 121-132.

Gilbert A. J. and Janssen, R. (1998). Use of environmental functions to communicate the values of a mangrove ecosystem under different management regimes. *Ecological Economics*, 25, 323-346.

Global Reporting Initiative GRI, United Nation Global Compact and World Business Council Sustainable Development WBCSD. (2015). *SDG Compass. The guide for business action on the SDGs*. Retrieved from www.globalreporting.org/resourcelibrary/GSSB-Item-29-SDG-Compass-Meeting5Nov15.pdf.

Gunderson, L. H. and Holling, C. S. (2002). *Panarchy: Understanding transformations in human and natural systems*. Washington, DC: Island Press.

Hanson, C., Ranganathan, J., Iceland, C. and Finisodre, J. (2012). The Corporate Ecosystem Service Review Guidelines for identifying business risks and opportunities arising from ecosystem change. Version 2.0. Washington, DC. World Resource Institute 39 pp. www.wbcsd.org/Clusters/Natural-Capital-and-Eco systems/Resources/Guidelines-for-identifying-business-risks-and-opportunities-arising-from-ecosys tem-change.

Intergovernmental Panel of Climate Change-IPCC. (2013). IPCC Fifth Assessment Report. www.ipcc.ch/report/ar5/

McGinnis, M. D. and Orstrom, E. (2014). Social-ecological system framework: Initial changes and continuing challenges. *Ecology and Society*, 19(2), 30.

Millennium Ecosystem Assessment (MA). (2005). *Ecosystems and human well-being: Synthesis*. Washington, DC: Island Press.

Moore, F.C., Baldos, U., Hertel, T. and Diaz, D. (2017). New science of climate change impacts on agriculture implies higher social cost of carbo. *Nature Communications*, 8(1), 1607.

Ostrom, E. (2007). A diagnostic approach for going beyond panaceas. *Proceedings of the National Academy of Sciences*, 104(39), 15181-15187.

Ostrom, E. (2009). A general framework for analyzing sustainability of social-ecological systems. *Science* 325(5939), 419-422.

Redman, C. T., Grove, J. M. and L. H. Kuby. (2004). Integrating social science into the long-term ecological research (LTER) network: Social dimensions of ecological change and ecological dimensions of social change. *Ecosystems*, 7(2), 161-171.

Rees, W. and Wackernagel, M. (1995). Urban ecological footprints: Why cities cannot be sustainable – and why they are a key for sustainability. *Environmental Impact Assessment Review*, 16(4-6), 223-248.

Rockström, J., Steffen, W., Noone, K., Persson, A., Chapin, F. S., Lambin, E. F., Lenton, T. M., Scheffer, M., Folke, C., Schellnhuber, H. J., Nykvist, B., de Wit, C., Hughes, T., van der Leeuw, S., Rodhe, H., Sörlin, S., Snyder, P. K., Costanza1, R., Svedin, U., Falkenmark, M., Karlberg, L., Corell, R. W., Fabry, V. J., Hansen, J., Walker, B., Liverman, D., Richardson, K., Crutzen, P. and Foley, J. A. (2009). A safe operating space for humanity. *Nature*, 461, 472-475.

Schwab, K. (2015, December 12). The Fourth Industrial Revolution: what it means, how to respond. Retrieved from Foreign Affairs website www.foreignaffairs.com/articles/2015-12-12/fourth-industrial-revolution.

Steffen, W., Crutzen, P. and McNeill, J. R. (2007). The Anthropocene: Are humans now overwhelming the great forces of nature. *Ambio*, 36(8), 614-621.

Steffen, W., Broadgate, W., Deutsch, L., Gaffney, O. and Ludwig, C. (2015a). The trajectory of the Anthropocene: The great acceleration. *The Anthropocene Review*, 2(1), 81-98.

Steffen, W., Richardson, K., Rockström, J., Cornell, S. E., Fetzer, I., Bennett, E. M., Biggs, R., Carpenter, S. R., de Vries, W., de Wit, C., Folke, C., Gerten, D., Heinke, J., Mace, G., Persson, L. M., Ramanathan, V., Reyers, B. and Sörlin, S. (2015). Planetaryboundaries: Guiding human development on a changing planet. *Science*, 347(6223), 1259855.

Stern, N. (2006). Stern Review Report on the Economics of Climate Change. HM Treasury, London.

The Economics of Ecosystems and Biodiversity, TEEB. (2010, July). The Economics of Ecosystems and Biodiversity for Business – Executive Summary. Retrieved from www.teebweb.org/publication/teeb-for-business-executive-summary/.

The Economics of Ecosystems and Biodiversity, TEEB. (2010). *The economics of ecosystems and biodiversity - Ecological and economic foundations*. Washington, DC: Earthscan.

The Economist. (2011). Welcome to the Anthropocene. *The Economist*. 26 May.

United Nations, UN (2015, August 12). Outcome of the Third International Conference on Financing for Development – Report of the Secretary-General. Retrieved from www.un.org/esa/ffd/wp-content/uploads/2015/08/70GA_SGR_FfD-3_outcome_AUV_12-08-15.pdf.

United Nations Environment Programme UNEP and Ozone Secretariat. (2003). *Handbook for the international treaties for the protection of the ozone layer*. Nairobi: UNON.

United Nations Framework Convention on Climate Change, UNFCCC. (2015). *Adoption of the Paris Agreement* (Report No. FCCC/CP/2015/L.9/Rev.1). Retrieved from http://unfccc.int/resource/docs/2015/cop21/eng/l09r01.pdf.

Wackernagel, M. and Rees, W. E. (1998). *Our ecological footprint: Reducing human impact on the earth*. Philadelphia, PA: New Society Publishers, Ltd.

Wackernagel, M., Cranston, G., Morales, J. C. and Galli, A. (2014). Ecological footprint accounts. In G. Atkinson, S. Dietz, E. Neumayer and M. Agarwala (Eds.). *Handbook of Sustainable Development* (2nd edn), (pp. 371-398). Cheltenham: Edward Elgar Publishing.

Walker, B. H., Holling, C. S., Carpenter, S. R. and Kinzig, A. P. (2004). Resilience, adaptability and transformability in social-ecological systems. *Ecology and Society* 9(2), 5.

Westman, W. E. (1977). How much are nature's services worth? *Science*, 197(4307), 960-964.

Winn, M. and Pogutz, S. (2013). Business, ecosystems, and biodiversity: New horizons for management research. *Organization & Environment*, 26(2), 203-229.

World Wild Fund for Nature-WWF (2016). Living planet report 2016: Risk and resilience in a new era. Retrieved from http://awsassets.panda.org/downloads/lpr_living_planet_report_2016.pdf.

Zalasiewicz, J., Williams, M., Steffen, W. and Crutzen, P. (2010). The new world of the Anthropocene. *Environmental Science and Technology*, 44: 2228–2231.

Zalasiewicz, J., Crutzen, P. and Steffen, W. (2012) The Anthropocene. In F. M. Gradstein, J. G. Ogg, J. G. Schmitz and G. Ogg (Eds.), *A Geologic Time Scale 2012* (pp. 1033–1040). Amsterdam: Elsevier.

2 Corporate sustainability in the 21st century

Learning objectives

- Establish a connection between environmental thinking paradigms and corporate sustainability.
- Illustrate the IPAT fundamental equation and understand how it can be utilized at the corporate level.
- Describe the concept of corporate sustainability and its evolution.
- Understand why companies engage in corporate sustainability, exploring external forces and internal drivers.
- Understand the importance of CEOs and leadership in transforming organization towards sustainability.
- Unravel and discuss the relation between the business case and sustainability, building the theoretical foundations for increasing sustainable value.
- Identify the drivers of the business case for sustainability.

Chapter in brief

This chapter explores the evolution of the concept of corporate sustainability during the last decades and sets the foundations of what is in need today. Sustainability has become part of the managerial jargon and several new concepts have emerged. In the initial part of this chapter, a brief review of these approaches is provided; we set up the conditions to understand why the transformations that occur in social-ecological systems become powerful drivers that shape, and are shaped by, business organizations. Then we analyze "why" business organizations address sustainability, and we investigate the role of external forces such as policy makers and governments, the financial community, NGOs and civil society, customers and competitors. We examine the importance of internal drivers focusing on the role of CEOs and companies' leaders in guiding corporate changes. The last section is dedicated to unravelling the business case for sustainability. This approach has been central to the business and the environment debate for more than two decades. The prevalent approaches are illustrated and discussed, including competitive advantage (differentiation and efficiency/cost), reputation and legitimacy, and risks reduction.

Introduction

In his famous book, the *Wealth of Nations*, Adam Smith, following classical economist authors, discussed the different concepts of value introducing the diamond-water paradox (Smith, 1776): even if life cannot exist without water and it can exist without diamonds, people consider diamonds, by far, much more valuable and precious than water. As a theoretical explanation, the marginal-utility theory of value resolves this paradox; the scarcity of goods is what causes humans to attribute value. Since water is in such large supply in the world, the marginal utility of water is low. In other words, each additional unit of water that becomes available can be applied to less urgent uses as more urgent uses for water are satisfied. Almost 250 years later, in his popular book *Earth in the balance* (1992), Nobel Prize winner Al Gore used this type of paradox to a weighting comparison between gold bars and the entire planet and exemplified in this way a collision between our civilization and the earth.

> On one side, we have gold bars. . . . On the other side of the scales, the entire planet! Hmm? I think this is a false choice for two reasons. Number one, if we don't have a planet. The other reason is that if we do the right thing, then we are going to create a lot of wealth and we are going to create a lot of jobs, because doing the right thing moves us forward.
>
> (*An Inconvenient Truth* transcript from film,
> Al Gore, 2006: 77:37 to 78:34)

For centuries, man has tried to conquer our planet, to eliminate whatever conflict we may have had with the other organisms that share it with us and to use its plentiful natural resources and planetary services. We seemed to have a world with limitless resources, and we used and abused them without collective care. Now, we are discovering that the planet has limits as we have illustrated in Chapter 1, and we need to learn how to exist and prosper with them. As with other activities involving mankind, businesses have been part of the problem and need to become part of the solution. However, the same was suggested more than 25 years ago by Stephan Schmidheiny in his celebrated book *Changing course: A global business perspective on development and the environment* (Schmidheiny, 1992) and still we search for these solutions. We should not delay more action - we need to move from a state of humans using the world as if it could easily be replaced ("humans with nature") to another state in which we need to adjust our common behavior to the functioning of a planetary resource that is unique and not changeable ("humans in nature").

We need to change the way we do things, we need to find the right things to do, we need technological transformations and innovative business models, we need new kind of lifestyles, and we need new ways to bring prosperity for everybody respecting nature and nature's rules.

Corporate sustainability in the 21st century

From global environmental challenges to corporate sustainability: four dominant paradigms

Dominant social paradigm

From the time of Adam Smith, back into the late 18th century, with some exceptions (e.g., the economist David Ricardo and his theory about resource scarcity and economic growth that followed Thomas R. Malthus [Ricardo, 1817 in Sandmo, 2015]), no man thought that the earth could have limits to supporting its civilization. A contemporary, the French naturalist and evolutionary theorist Jean-Baptiste Lamarck, quoted, "Animals that live in the sea are protected from human destruction of their species. Multiplication is so rapid and large that there is no possibility of collapsing any of these species" (Lamarck, 1809 cited Elliot, 1914). It took us 200 years to realize that this perception was wrong and to start to discuss global planetary resource scarcity. Furthermore, over the last 20 years, we have formally recognized the issue that man has been able to change the earth's biophysical environment and is capable of altering some of the most fundamental natural processes and functions that have allowed us to populate the world along with millions of other organisms.

Our species needs an environment to live in - a friendly one, if possible, in which the species does not waste all its energies simply to survive. That environment is our biosphere, the nature that has remained with us since our appearance on earth during the late middle Pleistocene when *Homo sapiens* originated in Africa around 100,000 years ago (Groucutt et al., 2015). However, at that time, the environment was hostile to man. It was only 10,000 years ago that man learned how to modify and use nature, during the Holocene. Around 10,000 year ago, we started to transform the land, to have an impact on ecosystems and to use other species for our own purposes. Since then, and till the mid-20th century, a worldview of "limitless resources" and poor ecosystem understanding and forward planning has driven the dominant social paradigm empowering man to use nature as an infinite resource with no worries about the possible impact of their behavior being expressed (see Figure 2.1).

New environmental paradigm

Around 60 years ago, there was a change in the prevailing worldview about the natural environment and the recognition of the need for a better usage of natural resources was raised. This raised conscious developed in parallel to the introduction of environmental responsibility aspects into the business world under the emerging concept of Corporate Social Responsibility (CSR) (see Box 2.1). It was at this time that the publication of the book *Silent spring* by

American biologist and conservationist Rachel Carson (Carson, 1962) started to modify the way some people viewed our relationship with nature. *Silent spring* presented the case of the entry into our plentiful biosphere of different man-made chemical compounds like pesticides, mostly DDT (dichlorodiphenyltrichloroethane), which not only killed bugs (its ultimate purpose) but traveled up the food chains to threaten both animal populations and people. The book initially presented the myth of a harmonious coexistence of modern man inside nature, but rapidly the same man poisoned nature, and this finally poisoned man. *Silent spring* also meant the end of linear thinking when social and ecological issues were analyzed in a parallel manner without integration. The discovery of the DDT pesticide in 1939 by Paul H. Müller, a Swiss chemist, allowed man to bring under control several insect-borne diseases. In 1948, he received the Nobel prize for Physiology or Medicine for this discovery. *Silent spring* marked the beginning of the modern environmentalism movement, creating different and novel values that had the potential to restructure society and form alternative institutions and lifestyles (see Figure 2.1).

Five years after Rachel Carson published her notorious book, Garret Hardin wrote his popular article "The tragedy of the commons" (Hardin, 1968), discussing population and resources. The paper presented the thesis that a finite world can support only a finite population. It was followed by a claim for collective action rather than leaving mankind to individual choices and back to Adam Smith's laissez-faire doctrine of the invisible hand. The paper argued that if we just allow individuals to pursue private interests, we will not reach a collective interest which is for the welfare of everyone. Hardin used a herdsman metaphor:

> Each man is locked into a system that compels him to increase his herd without limit in a world that is limited. Ruin is the destination toward which all men rush, each pursuing his own best interest in a society that believes in the freedom of the commons.
>
> (Hardin, 1968)

Around 50 years ago, "The tragedy of the commons" highlighted the conflict between private interests and collective needs and asked for better management to solve the paradox between private interest and global resources in a limited world. This metaphor started a series of publications that put forward the growing demand for food and water and the issue of planetary resource limitations as the main challenges for the future of mankind in the decades to come.

A few years later, in 1972 a group of authors under the auspices of the Club of Rome published the book *The limits to growth* (Meadows et al., 1972) which predicted that our civilization would probably collapse sometime in the future as a consequence of scarce planetary resources. The authors modeled rapid industrialization, increased population, demand for food, use of resources, and pollution by constructing scenarios. Under the scenario of a growing population and demands for material wealth, more industrial output, use of resources, and pollution were expressed. On the horizon, as resources become more expensive to obtain and more capital goes towards resource extraction, industrial output per capita will start to fall and finally will collapse. The *Limits to growth* manuscript was highly criticized by other people (Cole, 1973, as an example), but it introduced to the public and discussed in depth several aspects of critical importance: the carrying capacity concept,

1900s	1960s	1080s	2000s
DOMINANT SOCIAL PARADIGM	NEW ENVIRONMENTAL PARADIGM	DOMINANT SUSTAINABILITY PARADIGM	SOCIAL-ECOLOGICAL PARADIGM
Up to the mid-20th century, a worldview of limitless resources, poor ecosystem understanding and forward planning	Change in prevailing worldview and a recognition of the need for sustainable resource use and integrated thinking (Gaia)	Globalization of environmental issues and mainstreaming of environmental values within all sector and policies	A worldview recognizing the mutual inter-associations between human societies and ecological processes that may be necessary for the survival of both

1961
Silent Spring

1968
The Tragedy of the Commons

1972
Limits to Growth

1986
Brundtland Report

1990
Governing the Commons

1992
Earth in Balance

1992
Changing Course

1997
Cannibal with Forks

2005
Millennium Ecosystem Assessment

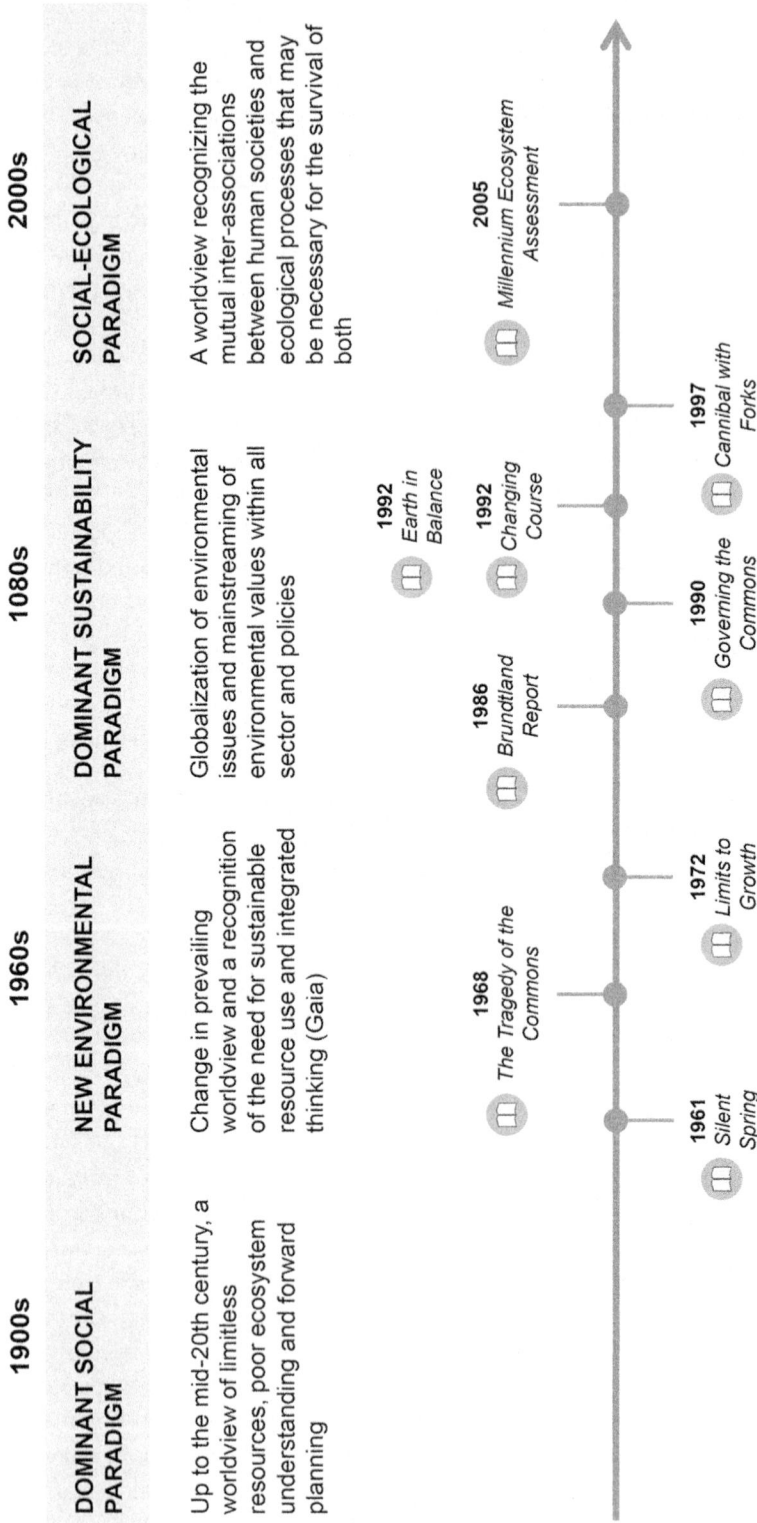

Figure 2.1 Environmental "thinking" timeline: paradigms, events and main publications.
Source: authors' elaboration.

the non-isolation of the economy, society and the environment, and the introduction of the "business-as-usual" scenario. The renewed interest in topics around sustainability revived the discussion between economic growth and the environment when considering future development (Nørgård et al., 2010).

The environmental movement of the 1960s had a very limited impact on the business sector, but it sowed the seeds of one of the most important holistic strategies that man has pushed forward globally. In her book, Rachel Carson also introduced a new environmental vision that would drive the following decades, the idea that man was acquiring the power to alter nature in a globalized way:

> [T]he history of life on earth has been a history of interaction between living things and their surroundings. To a large extent, the physical form and the habits of the earth's vegetation and its animal life have been molded by the environment. Considering the whole span of earthly time, the opposite effect, in which life actually modifies its surroundings, has been relatively slight. Only within the moment of time represented by the present century has one species – man – acquired significant power to alter the nature of his world.
>
> (Carson, 1962: 323)

Box 2.1   The debate on Corporate Social Responsibility

Before Rachel Carson, Garret Hardin or Donella and Dennis Meadows and their colleagues published their manuscripts and theories on the environmental crisis, the introduction of environmental responsibility aspects into the business world were already being considered under the emerging concept of Corporate Social Responsibility (CSR).

In the 1950s, CSR was first introduced as a way of taking public interests into account in the decision-making process of organizations (Drucker, 1954). At that time, Howard R. Bowen, who is considered the father of CSR, in his landmark book *Social Responsibilities of the Businessman*, investigated the responsibility that businessmen were reasonably expected to assume toward society (1953). In his famous quote, Bowen defines social responsibility as "the set of moral and personal responsibilities that the employer must follow, considering the exercise of policies, decisions or course of action in terms of objectives and values desired by society" (1953: 6).

During the 1960s, the CSR concept and practices started to consolidate, thanks to the contribution of scholars such as Keith Davis (1960) and William C. Frederick (1960). Essential elements of this concept were voluntarism and a focus on improving working conditions, customer relations, and philanthropy.

The attention to CSR accelerated in the 1970s with the emergence of opposite positions among scholars and practitioners. On the one hand, the Committee for Economic Development (CED) in 1971 published a landmark document on the Social

Responsibilities of Business Corporations. The CED underlined that business's "basic purpose is to serve constructively the needs of society – to the satisfaction of society" (CED, 1971: 11). In the same vein, Harold Johnson wrote in one of the first books of the decade on CSR: "A socially responsible firm is one whose managerial staff balances a multiplicity of interests. Instead of striving only for larger profits for its stockholders, a responsible enterprise also takes into account employees, suppliers, dealers, local communities, and the nation" (Johnson, 1971: 50). On the other hand, Milton Friedman, winner of the Nobel Prize for Economics in 1976, provided a much narrower view of CSR. His position that builds on the neo-classical approach is first outlined in his early book *Capitalism and freedom* (1962), in which he outlines that "few trends could so thoroughly undermine the very foundations of our free society as the acceptance by corporate officials of a social responsibility other than to make as much money for their stockholders as possible" (1962: 133). In 1970, in a ground-breaking article titled "The social responsibility of business is to increase its profits" published in the *New York Times*, Friedman summarizes his famous position against CSR observing that "there is one and only one social responsibility of business—to use its resources and engage in activities designed to increase its profits so long as it stays within the rules of the game, which is to say, engages in open and free competition without deception or fraud" (Friedman, 1970). Friedman also made statements about business, business-men, and responsibility, questioning: "What does it mean to say that 'business' has responsibilities? Only people have responsibilities. A corporation is an artificial person and, in this sense, may have artificial responsibilities, but 'business' as a whole cannot be said to have responsibilities, even in this vague sense." (Friedman, 1970). Fueled by these contrasting positions, a long-lasting debate on the role of firms in society and their responsibility towards the natural environment populated the discussion in academic journals and among practitioners for decades.

During the 1980s and 1990s, the CSR ideas evolved with multiple streams of research. Two major theoretical approaches were the institutional theory and the stakeholder theory (see review in Barrena-Martinez et al., 2016). The institutional theory (North, 1990; Meyer & Rowan, 1991; Scott, 2007) investigated business adaptation processes facing external and internal pressures, looking at the institutional context. According to this theory, organizations are driven by society and the community of practices to incorporate the set of shared norms, beliefs, values, and principles. This process allows organizations to achieve support and backing for their activities, therefore acquiring legitimacy and license to operate. Stakeholder theory developed starting from the publication of R. Edward Freeman classic book in 1984 (Freeman, 1984; Freeman et al., 2010) and over the years has become a fundamental integrative field regarding CSR. Stakeholder theory and CSR, in fact, have many aspects in common, starting with underlying moral and ethical principles. The consolidation of the stakeholder theory in the fields of management and organization studies has contributed to enlarging the responsibility of the firm from shareholders' to many actors' needs and demands. This includes employees and unions, customers, suppliers, government, investors, media, competitors, NGOs or communities, and the natural environment.

After many years of debate on the responsibility of the firm, we can now say that this discussion was fundamental to preparing a future call to action for business to address the challenges of the Anthropocene epoch, assuming a leading role in both tackling social-ecological problems and providing innovative solutions.

Source: authors' elaboration using multiple sources.

Dominant sustainability paradigm

The globalization of environmental issues and the mainstreaming of environmental values framed the introduction of a groundbreaking concept: the Sustainable Development paradigm (see Figure 2.1). Transposing the Sustainable Development idea within all sectors and policies announced the entrance into a new era in our relationship with the environment. Sustainable Development was defined by the World Commission on Environment and Development (WCED) in 1987 as "development that meets the needs of the present without compromising the ability of future generations to meet their own needs" (Brundtland Report, 1987: 44).

The WCED provides some explanation to this definition. Two key elements are contained: (a) the concept of "needs," particularly, the essential needs of the world's poor, to which overriding priority should be given; and (b) the idea of limitations imposed by the state of technology and social organization on the environment's ability to "meet present and future needs" (Brundtland Report, 1987). Later, these concepts become known as intra- and inter-generational justice. Although the idea of needs and limits were already included in the concept of carrying capacity developed in the discipline of ecology (see Chapter 1), this new paradigm grounded the development debate in a global framework, within which a continuous satisfaction of human needs is the ultimate goal of life conditions. Besides this important concept of needs and limits, the Sustainable Development paradigm clearly rests on the fact that Sustainable Development must be the organizing principle for meeting human development goals while at the same time sustaining the ability of natural systems to provide the natural resources and ecosystem services upon which the economy and society depends:

> [A] society may in many ways compromise its ability to meet the essential needs of its people in the future – by overexploiting resources, for example . . . At a minimum, sustainable development must not endanger the natural systems that support life on Earth: the atmosphere, the waters, the soils, and the living beings.
>
> (Brundtland Report, 1987; Chapter 2; pp. 44–45)

Starting in the 1990s and applying the Sustainable Development global paradigm to the business level, the first attempts to link management actions with societal and environmental challenges were observed (Post, 1991; Shrivastava and Hart, 1992; Roome, 1992). While preparing for the Rio Conference (1992), a group of business leaders got together to establish the World Business Council for Sustainable Development (WBCSD) (see the link in Annex 1.3 to this chapter to find more information on this organization) and wrote its famous book, *Changing course: A global business perspective on development and the environment* (Schmidheiny, 1992). This groundbreaking publication showed to the broad business community for the first time how companies could combine environmental protection with economic performance and introduced the term "eco-efficiency," a term that is still widely used in the world by practitioners and scholars today (see the chapters that follow). The WBCSD is today a highly reputed organization, involving over 200 leading corporations with the aim to accelerate the transition to a sustainable world by helping make its member companies more successful and sustainable by focusing on the maximum positive impact for shareholders, the environment and societies.

Social-ecological paradigm

At the turn of the Century, the United Nations (UN) launched the Millennium Ecosystem Assessment (MA) (see Millennium Ecosystem Assessment, 2005a–e) with the objective to assess the consequences of ecosystem change for human well-being and to develop the scientific basis for action needed to enhance the conservation and sustainable use of those systems (see Chapter 1 for further explanation). The work emphasized the utilization of social-ecological models as a framework for conceptualizing and managing the resilience and sustainability of human-nature systems (Berkes et al., 2003; Stokols et al., 2013), a change in paradigm on top of the Sustainable Development one. The new paradigm focused on a new view of the "world" in which the interdependences between individuals and societies and the natural environment were emphasized; Sustainable Development should recognize such interconnections and take care of its functional processes. Lately and as we saw in Chapter 1, a new resolution of the United Nations in 2015 elaborated the 2030 Agenda for Sustainable Development to transform the world and launched a call to action towards 17 Sustainable Development Goals (SDGs).

Corporate sustainability: an evolving concept

There is no universally agreed definition on what sustainability means. Sustainability relates to the concept of Sustainable Development, the new paradigm that started to become common language with the world's first Earth Summit held in Rio in 1992. In our view, sustainability means translating Sustainable Development into action. Depending on the field of application, we can talk about personal, home, city, and regional sustainability. When this approach is applied to companies, the expression "corporate sustainability" has been popularized and consolidated, becoming a commonly used term. We acknowledge that other theoretical perspectives have been extremely important regarding business and society relations. Corporate Social Responsibility (CSR), which we have briefly analyzed in Box 2.1 (see earlier), and corporate citizenship are examples of these theories. The attention in this book is to sustainability and its implication for business.

For decades, before social and ecological problems became planetary challenges for humanity, most companies were following a profit maximization view. Those were the "rules of the game" when Milton Friedman articulated his famous assumption "the business of business is business" (Friedman, 1962; 1970). Now, times are under change; even more, the planet is under an accelerated pattern of change, and the environmental crisis is pressuring us for action. Societal values and beliefs are evolving and demanding companies to take care about the direct and indirect consequences of their actions. The rules of the game for business are changing, and responsibility and sustainability (two sides of the same coin) are starting to play an important role in these rules. Carrying out socially and ecologically responsible behavior is becoming a question of legitimacy, competitiveness and survival for many companies in different industries and markets.

In an early paper, Dyllick and Hockerts (2002: 131) proposed an initial but very straightforward definition of corporate sustainability that incorporated the stakeholder approach: "meeting the needs of a firm's direct and indirect stakeholders (such as shareholders, employees, clients, pressure groups, communities, etc.), without compromising its ability to meet the needs of future stakeholders as well." This definition focused basically on people and the human environment and did not directly contemplate the natural part of the environment.

Today, mirroring this new social-ecological paradigm, we have recognized that a framework for analysis that separates the human and the remaining natural environment is not adequate to deal with the present problems. Rather, we opted for the term "social-ecological systems" to integrate both things (see Chapter 1 for a further explanation of social-ecological systems). However, in the world of business practices, many companies still do not understand what a social-ecological system is and, many times, what sustainability is. Therefore, we can ask: What are the implications of the interdependence between businesses activities and social-ecological systems for the firm strategy and competitiveness? How can the degradation of social-ecological systems affect their long-term success?

Corporate sustainability has been with us for quite a long time but still there is a big disconnect between sustainable business on a strategic and organizational level and Sustainable Development on a global level. In other words, observing the world of practice, we find large discrepancies between how companies run their businesses and their responses and the current directions and trends of the environmental crisis (e.g., increasing ecological footprint, threatened carrying capacity and depleting natural capital), altogether running into a wide deterioration of the states of social-ecological systems and, finally, our entire planet.

Several surveys and studies have explained that managers sometimes lack common knowledge about the issue of corporate sustainability. A recent research study by MIT Sloan Management Review in collaboration with The Boston Consulting Group (Kiron et al., 2017) illustrates how corporate sustainability is at a crossroad today. Despite some progress in addressing environmental and social challenges, companies taking large responsibilities are still a minority and are unevenly distributed across industrial sectors and regions. At the same time, few companies clearly show that sustainability issues can be drivers for innovation, efficiency, and value creation. The MIT report gives clear indications of the corporate sustainability situation identifying eight key lessons that orient business practices around the world (see Table 2.1).

Table 2.1 Research report "Corporate Sustainability at a crossroads" (Kiron et al., 2017).

Key lesson 1	Set your sustainability vision and ambition: 90% of executives see sustainability as important, but only 60% of companies have a sustainability strategy.
Key lesson 2	Focus on material issues. Companies that focus on material issues report up to 50% added profit from sustainability. Those that do not focus on their material issues struggle to add value from their sustainability activities.
Key lesson 3	Set up the right organization to achieve your ambition. Building sustainability into business units doubles an organization's chance of profiting from its sustainability activities.
Key lesson 4	Explore business model innovation opportunities. Nearly 50% of companies have changed their business models because of sustainability opportunities.
Key lesson 5	Develop a clear business case for sustainability. While 60% of companies have a sustainability strategy, only 25% have developed a clear business case for their sustainability efforts.
Key lesson 6	Get the board of directors on board: 86% of respondents agreed that boards should play a strong role in their company's sustainability efforts, but only 48% say their CEOs are engaged, and fewer (30%) agreed that their sustainability efforts had strong board-level oversight.
Key lesson 7	Develop a compelling sustainability value creation story for investors: 75% of executives in investment companies think sustainability performance should be considered in investment decisions, but only 60% of corporate executives think investors care about sustainability performance.
Key lesson 8	Collaborate with a variety of stakeholders to drive strategic change: 90% of executives believe collaboration is essential to sustainability success, but only 47% say their companies collaborate strategically.

The recent study by MIT Sloan Management Review in collaboration with The Boston Consulting Group provides a snapshot of the state of the art in the current debate on corporate sustainability, but how has corporate sustainability evolved over the years? What types of managerial practices have been adopted by companies? Dyllick and Muff (2015), in a recent paper, reviewed the main approaches and developed a typology of corporate sustainability behavior (in the paper the authors use the term business sustainability as a synonymous of corporate sustainability) with a focus on effective contributions to Sustainable Development. This typology ranged from business sustainability 1.0 (refined shareholder value management), to business sustainability 2.0 (managing for the triple bottom line) and to business sustainability 3.0 (true sustainability). The truly sustainable business approach introduced by Dyllick and Muff (2015) will help us to move up into what we are proposing as a novel framework for addressing corporate sustainability in this century, the "Business In Nature" concept (BInN businesses), which we will introduce in Chapter 3.

Business sustainability 1.0: refined shareholder value management

The first attempt to introduce sustainability into the business paradigm in a formalized way resulted from the acknowledgment that there were new business challenges that came from risks and opportunities outside the traditional markets forces. The basic idea was that companies could increase shareholder value just by better managing the relationship with some stakeholders (e.g., NGOs, media, policy makers) but without changing the basic business purposes and strategies. Novel social-ecological emergencies (e.g., air pollution, waste generation, resource scarcity, child and forced labor) challenged companies by introducing new risks and creating opportunities to secure, or to increase, their share of overall value added. Following that, sustainability became a viable business approach for companies as it expanded the possibilities to growth and therefore enhanced their competitive position. Moreover, companies that pursuit this idea of sustainability converted themselves into an attractive investment because they were offering superior returns in the form of shareholder value creation.

The initial idea of business sustainability confronted the traditional business view in companies that had seen the protection of the natural environment only as a cost factor. This discussion drives us back into the nineties, when some pioneer companies were starting to show that it was possible to be profitable while reducing pollution and behaving responsibly, and when some scholars in strategy and industrial economics started to raise questions like "Does it pay to be green?" In April 1991, Harvard Business School strategy professor Michael Porter published in Scientific American a one-page commentary on the debate between environmental protection and economic competitiveness (Porter, 1991). This short article challenged the prevalent approach that considered the investment in environmental protection unproductive, arguing that this approach was wrong due to a static view of competition. Building on his model of "dynamic competitiveness," Porter contended that well-designed and strict environmental regulations can enhance competition. In another cutting-edge paper titled "Toward a new conception of the environment-competitiveness relationship " (Porter and van der Linde, 1995) they explained, based on several case studies, the underlying logic that linked the environment, resource productivity, innovation, and competitiveness. The reduction in ecological impacts translates into an increase in resource productivity, which in turn can create a competitive advantage. Based on these argumentations, the so-called Porter hypothesis (Porter, 1991; Porter and van der Linde, 1995) asserted that firms can benefit from environmental regulations. It argues that well-designed environmental regulations stimulate innovation which, by enhancing productivity, increases firms' private benefits. (Ambec and Barla, 2002; Wagner, 2003; Berchicci and King, 2007), this innovative perspective stimulated a broad debate among scholars and practitioners. Protecting nature, adding new regulations, and increasing environmental standards can be used to encourage innovation and resource productivity, translating into better products, reducing firm's consumption, and improving economic performance.

In this vein, several studies followed exploring the relations between low cost, differentiation and niche strategies with environmental performances (Hart, 1995; Shrivastava, 1995; Reinhardt, 1998). In a nutshell, environmental strategies "could benefit [firms] by reducing costs through ecological efficiencies, capturing emerging 'green' markets, gaining first mover advantage in their industries, ensuring long-term profitability, establishing better community relations, and improving their image" (Shrivastava, 1995: 937).

Under these initial ideas, business sustainability was defined by the PwC and SAM Group (2006) as follows.

> Corporate sustainability is an approach to business that creates shareholder value by embracing opportunities and managing risks deriving from economic, environmental and social developments.
>
> (PwC and SAM Group, 2006).

Despite the effects of corporate activities on social-ecological systems, this approach to sustainability prioritized the economic factors and mostly followed environmental regulations. As Dyllick and Muff (2015: 9) commented, under this logic, "business success still is evaluated from a purely economic view and remains focused on serving the business itself and its economic goals. The values served may be somewhat refined, but still oriented toward the shareholder value."

Business sustainability 2.0: managing for the triple bottom line

In the mid-90s, a new framework to manage sustainability and corporate performance was introduced. This framework, popularized as the "Triple Bottom Line" (TBL) went beyond the traditional measures of profits, return on investments, and shareholder value to include environmental and social dimensions. The framework is due to John Elkington's book *Cannibal with forks* and following literature (Elkington, 1994; 1997), and it was proposed as a new language, a new jargon for business, in order to improve the response to the calls of action on sustainability. Following this framework, new business statements as the "win-win-win business strategies" or the 3P formulation "people, planet, and profits" entered into the language of corporations. This is the origin of the new business sustainability 2.0, which broadens the business perspective, introducing a TBL approach in which value creation goes beyond shareholder value and includes societal and environmental values.

The TBL relies on the idea that only when companies measure their social and environmental impact they can move forward into responsible organizations. Following the TBL, companies should develop three different bottom lines:

- a traditional one, which refers to the measure of corporate profits, the "bottom line" of the profit and loss account,
- a social one, which refers to how responsible an organization has been regarding society and its constituencies throughout its operations - this is also known as the "people account"
- one that responds to how responsible they have been regarding the natural environment (e.g., natural resources, ecosystem services, biodiversity), the "planet account".

The TBL was thought to reflect the essence of sustainability allowing companies to not only measure all impacts of their organizations including profitability and shareholder value but also their impact on the social and ecological capital. The introduction of the TBL in corporations went in parallel to the development of environmental management systems (EMS) and the diffusion of the eco-efficiency concept (see Chapter 4 for more information). TBL facilitated transparency and reporting that could be associated with the definition of objectives, the management of activities and programs, as well the measurement of achievements. As was pointed out by Dyllick and Muff (2015: 9), a definition of this type of business sustainability could be the one introduced by the Network for Business Sustainability back in 2012:

Business sustainability is often defined as managing the triple bottom line—a process by which firms manage their financial, social and environmental risks, obligations and opportunities. These three impacts are sometimes referred to as people, planet and profits.

(Network for Business Sustainability)

Even if the three aspects of sustainability (economic, social, and environmental) existed already, the development of the TBL and its connection with stakeholders were basic for the development of tools such as the GRI (Global Reporting Initiative), the AccountAbility Standards, or the DJSI (Dow Jones Sustainability Index) (see Annex A.1.3 to this chapter to find more information). The TBL had also some critics (Norman and MacDonald, 2004) about measurement, aggregation, or transparency of data. The most important claim about limitations for this approach was that the dimensions of people, planet, and profits do not have a common unit of measure or something that helps with the integration; instead, they work as three separate accounts that cannot easily be added up. This aggregation has not come up with a logic to arrive at a net sum. However, despite this fact, the TBL was a kind of revolution to favor the measurement and track by companies, in addition to its economic data, of the social and environmental performance in a meaningful, consistent and comparable way (Pava, 2007). To sum up, the TBL has been associated with the creation of economic, ecological, and social value, although it was still unclear how to address the complex challenges provided by the internal/external trade-offs among different areas of sustainability.

Business sustainability 3.0: truly sustainable business

With the entrance into the new century, we start to recognize the issue of "planetary boundaries" as well as "being green as a need" (see Chapter 1 and Section 2.3.1 (page 9) for further explanations). Under the umbrella of all these aspects, a call for a better global governance was sent; the business sustainability 3.0 was thought to follow these demands. Business sustainability 3.0 recognized that business not only can but must be part of the solution to the Anthropocene's challenges. As proposed by the United Nations, companies can be effective agents of change and can contribute to the needed transformation, maintaining and/or enlarging the resilience of dynamic socioecological systems worldwide.

Dyllick and Muff (2015) defines this type of business sustainability as:

Truly sustainable business shifts its perspective from seeking to minimize its negative impacts to understanding how it can create a significant positive impact in critical and relevant areas for society and the planet. A Business Sustainability 3.0 firm looks first at the external environment within which it operates and then asks itself what it can do to help overcome critical challenges that demand the resources and competencies it has at its disposal.

(Dyllick and Muff, 2015: 10-11)

Within the present situation of planetary challenges, many distinguished opinion leaders and scholars were clamoring for a change – a change in the role of corporations.

- Yvo de Boer, just after resigning from his executive position at the UNFCCC (United Nations Framework Convention on Climate Change), stated: "I have always maintained that, while governments provide the necessary policy framework, the real solutions must come from business" (UNFCCC, 2010).
- Peter Drucker commented on business challenges, "Every single social and global issue of our day is a business opportunity in disguise" (as cited in Cooperrider, 2008, and Dyllick and Muff, 2015).
- Ban Ki-Moon regularly underscored the crucial role the business plays in the realization of Sustainable Development Goals. In a recent speech at the United Nation he stated: "We will also need to activate business as never before, and quickly. We are spreading the word far and wide that every business has a responsibility to improve our world" (UN, 2016, 16 September).

This type of business sustainability requires companies to engage in a novel way, moving from just seeking to minimize their negative impacts (mostly the negative externalities) to being able to create a significant positive effect in different areas of society and the natural environment. Dyllick and Muff specifically comment that:

> The potential for contributing positively will vary largely between companies, their resources, strategies and purposes, and it will vary between different industry sectors and societal contexts. Making a positive contribution to overcome sustainability issues and thus serving the common good becomes the main purpose of a truly sustainable business.
>
> (Dyllick and Muff, 2015: 11)

Linking the macro to the micro: the IPAT fundamental equation

In the early 70s, the American ecologists Paul Ehrlich and John Holdren devised a simple equation, the IPAT equation, in dialogue with another American ecologist, Barry Commoner, identifying three factors that modulated human environmental impact in the world (Ehrlich and Holdren, 1971, 1972a, 1972b; Commoner et al., 1971). A revision on the origins of this so-called fundamental equation can be found in Chertow (2000). In the equation, the human environmental impact (I) was expressed as the product of (1) population (P); (2) affluence (A); and (3) technology (T).

The IPAT equation has been very influential to drive thoughts in the Academia about the impact caused by man on the earth. Although the equation was first used to determine which single variable was the most damaging to the environment, over the years the utilization was modified, recognizing that increases in population and affluence could, in many cases, be balanced by improvements to the environment offered by technological systems. Moreover, the equation was brought into the discussions with the purpose of being used as a sustainability call – for example, it was utilized as a support reasoning for the Rio Conference in 1992.

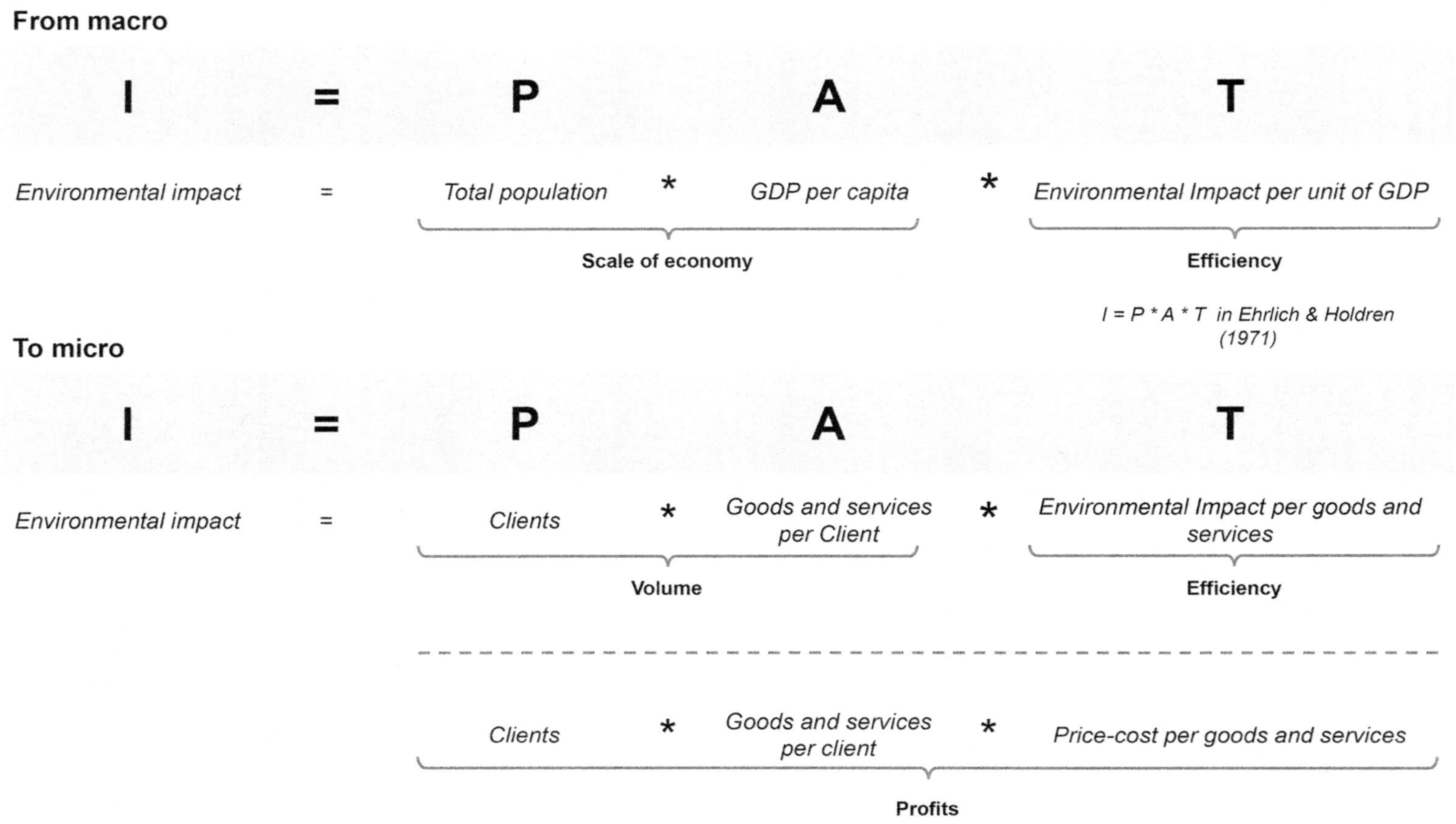

From macro

| **I** | **=** | **P** | ***** | **A** | ***** | **T** |

Environmental impact = Total population * GDP per capita * Environmental Impact per unit of GDP

Scale of economy | Efficiency

I = P * A * T in Ehrlich & Holdren (1971)

To micro

| **I** | **=** | **P** | ***** | **A** | ***** | **T** |

Environmental impact = Clients * Goods and services per Client * Environmental Impact per goods and services

Volume | Efficiency

- - - - - - - -

Clients * Goods and services per client * Price-cost per goods and services

Profits

Figure 2.2 The IPAT fundamental equation: from macro to micro level.
Source: authors' elaboration.

However, how can the equation help us with the understanding of our impacts? Let us provide an example that was used by the time of the Rio Conference in the early 1990s. On that date, world population had doubled in the period 1950–1990. Under new assumptions of another doubling world population for the coming 40 years and with an affluence factor estimated as rising by five for the same amount of time, in order to equalize the impact of our actions for such a period of time technological developments should deflect previous increases by a factor 10 reduction with huge improvements in efficiency. In a nutshell, besides the multitude of assertions given in this call, the understanding of this equation in a simple way emphasized at that time the idea of equalizing impact as one of the main issues of sustainability (Figure 2.2), something that, as we will see in the next chapter, is not enough.

The IPAT equation can also be applied at the business level, having as a target companies. The environmental harm (the impact of a company) can be expressed as a function of other three variables: the number of clients the company has (that stands for P), multiplied by the amount of units of goods/services each client is buying (equivalent to A), and then factored by a measure of efficiency (the impact caused by each unit of good/services, represented by T). In this case, it can be seen as the overall environmental footprint of the company where T should be seen as the environmental footprint of each product. As the environmental footprint of a company can be measured through different units (carbon footprint, water footprint, ecological footprint; see Chapter 6 for further explanation), by using the equation we have a simplified way of analyzing the impact of a company in absolute and relative terms. Moreover, the equation helps us investigate how sustainability strategies at the corporate level can contribute to the effective reduction of the firm's environmental footprint. An example of the application of the IPAT equation at the business level is illustrated in Box 2.2. We use the example of the well-known French cosmetics firm L'Oréal to discuss what the corporate strategical and operational implications of some sustainability-based commitments are.

Box 2.2 L'Oréal and the IPAT equation

L'Oréal, the French world leader in beauty companies, in 2005 defined the commitment to reduce its absolute carbon emissions related to its plants and distribution centers by 50% until 2020, while cutting water consumption and reducing waste by 50% per unit of finished product by the same time and finally shrinking the transport footprint. Later, the company increased the commitment progressively (60%) until its final announcement in 2015: "Reinforcing its commitment to fighting climate change, L'Oréal announces its ambition to become a 'carbon-balanced' company by 2020. Through its sustainable sourcing projects, the Group aims to generate carbon gains corresponding to the amount of greenhouse gas emissions linked to its activities" (Jean-Paul Agon, L'Oréal president, L'Oréal press release 3 September 2015).

The commitment of 50% reduction came together with the ambition for these years (2005 to 2020) to win another billion consumers around the globe. What these commitments means for the company products to be sold?

We can use the IPAT fundamental equation to do a simple analysis based on a hypothetical and fictitious single assumption (you can make other assumptions). Let's imagine that we use the assumption that the new billion forecasted consumers will buy the same quantity of products that the previous one [just to make the term (A) equal in the equation]. Then, we can use the IPAT equation to make a preliminary assessment of what these commitments stand for new product environmental requirements. Then, in the IPAT formulation, filling it with forecasted numbers for 2020:

(P) will double (double clients) and,
(A) will remain equal (similar equal buying power), then:

	I	=	P	*	A	*	T
CO_2 emissions	**0.5**		2		1		0.25
water/wastes	1		2		1		**0.5**

In terms of carbon emissions under these assumptions, an absolute reduction of fifty percent (in bold) translate into a further reduction of 75% per unit of product in its carbon footprint in its plants and distribution centers. The same commitment for other environmental stressors such as water and wastes worked out for L'Oréal in relative terms, the commitment for those stressors is to reduce the use of water or the production of wastes in 50% by unit of finished product (in bold); then, in 2020, if (P) is multiplied by two, (A) remains equal and (T) is reduced by half, the absolute water and waste footprint (I) remains equally to one so the absolute impact for water and wastes would remain identical to what it was in 2005.

Following with similar reasoning, the equation can be used in a deeper context: we can decompose the volume of sales between different products and transform the volume of selling's generated (P*A) into profits for the company by multiplying this volume by the price minus cost of each unit of stuff. This exercise can be used to understand in absolute and relative terms what commitments mean for firms.

Source: authors' elaboration.

Why do companies engage?

Corporate sustainability as a market transformation

Corporate sustainability has become a key factor in the agenda of business leaders and multinational companies. Environmental and social issues are having and will continue to have a material impact on how companies think and act. The evidence grows stronger every day, as witnessed by several new surveys and reports. Still, as recently reported in a large survey realized by MIT Sloan Management Review in collaboration with The Boston Consulting Group (Kiron et al., 2017), the challenge for companies lies in their ability to capture these opportunities and transform them into value for their shareholders and stakeholders.

Our view is that sustainability acts as a major force guiding a profound market trans-formation capable of reshaping the way and the logic of doing business. In the years to come, global corporations as well as small- and medium-size firms, will be asked to: integrate strategy with sustainability; develop sustainability strategies; innovate their operations, their supply chain and their products; and implement new ways of disclosing and reporting sus-tainability outcomes. However, the present pace of change is still not addressing the plan-etary challenges of our century, and a transformational change is required. Only the seeds of this new approach are starting to be seen today. Companies should go ahead, but what are the driving forces that can shape this transformation? In the next section, we will illus-trate some of the external and internal drivers that push sustainability at the business level. We start by analyzing the so-called megatrends, then we study some selected stakeholders. Finally we will look at the role of business leaders and CEOs.

Business megatrends at the beginning of the 21st century

Megatrends are global, sustained and macro-economic forces of development that impact business, economy, society, cultures and personal lives, thereby defining our future world and its pace of change. These global forces, unlikely to go away soon, interact with each other and affect everything, from individuals, to companies, to societies, and all sectors of the economy and all type of organizations, creating changes that shape the way for moving into the future. These forces will create opportunities and risks in the way people do business and business attracts consumers. These will be the forces that have the potential to create disruption for good, challenging innovation and creativity.

The present environmental crisis has mostly been recognized as one of these global meg-atrends (EEA, 2015, HayGroup, 2014; Ernst and Young, 2016). This environmental crisis fol-lows three main assumptions: (1) natural capital is depleting because natural systems are deteriorating; (2) we have only one planet, the earth, so our carrying capacity is limited; and (3) our ecological footprint is increasing. A major assessment of this global environmental crisis was published in 1999 in the UNEP Global Environment Outlook 2000 report (UNEP, 1999). This paramount study drew attention to two critical, integrated, recurring phenomena, as outlined below.

- **Non-resilient social systems.** Societies are threatened by grave imbalances in produc-tivity and in the distribution of goods and services between them. A large proportion of the human population lives in poverty, and a widening gap exists between those who benefit from economic and technological development and those who do not.
- **Non-resilient ecological systems.** Accelerating changes are occurring at the global scale, with rates of economic and social development outstripping progress in achieving internationally coordinated environmental stewardship. Improvements in environmental protection due to new technologies are being cancelled out by the magnitude and pace of human population growth and economic development.

Non-effective resilience together with large global problems of inequalities and welfare redistribution are two sides of the same coin. The environmental crisis megatrend can be analyzed in its relationship with other identified megatrends. Figure 2.3 shows nine large identified megatrends that shapes our future and that have been grouped under the three factors of human activity related to the IPAT formulation. Putting together the work of several agencies and consulting firms (Frost and Sullivan, 2014; HayGroup, 2014; KPMG, 2014; European Environmental Agency, 2015; PricewaterhouseCoopers, 2016; Ernst and Young, 2017), we came up with nine global megatrends, listed below.

- **Growing population.** The world population is growing steadily at around 1% annually, and this comes with a set of social demographic changes. The increase in average incomes and the fall in levels of absolute poverty, particularly over the last decade, suggest that an increasing proportion of the world's population is neither rich nor poor by national standards but finds itself in the middle class of the income distribution. On the other hand, some of this population could face unemployment because some future economic growth could come without the creation of proportionate employment opportunities. In addition, 2 billion persons are included as part of the environmental crisis with large problems of poverty, famine, sanitation and other derived problems.
- **Aging.** The world is undergoing a major population shift with far reaching implications. The elderly population in the Western world is increasing, migration patterns are changing and there are major population shifts in countries. As the population gets older, healthcare issues become increasingly important and, at the same time, the risk of western pandemics like obesity and mental illness are likely to spread.
- **Increasing urbanization.** Three main trends in urbanization are the development of mega-cities, mega-regions, and mega-city corridors and networks. As the built environment increases, new urban lifestyles will need to be accommodated and this will bring about important societal changes. People will discover a wider range of life and career options as well as personal lifestyles. They will have the freedom to make decisions based on values, not economics, that ultimately could transform labor forces.
- **Shifting economic powers.** Global economic shifts will have a significant influence over international politics and governance. A multipolar world is expected, and this will affect other global trends such as the rise of the middle class, poverty aspects, and the role of business in society. Geostrategic competition will increase, and countries other than the BRIC group will have an influence on economic activity. This growth in influence will also bring sweeping cultural changes to the global business and economic landscape.
- **Raising global connectivity.** Interpersonal connections using the latest discoveries in communication and information technology will be a major force of the future. This factor will allow a much more powerful flow of ideas and materials. Connectivity will bring many social changes such as the issue of geo-socialization that will modify how we think, move, interact, communicate, do business, and capitalize on opportunities in the future.
- **Diversifying governance.** Governments will face a mismatch between the increasingly long-term, global and systemic challenges facing social-ecological systems and their more national and short-term focus and powers. Although we expect to see more actions for coordinated governance at the global scale, the possibility of rising international

conflicts would be high. In addition, demands for better leadership, integrity and transparency will also increase.

- **Accelerating technological transitions.** Technology will increasingly impact our lives. Technological planetary breakthroughs on artificial intelligence, robotics, nanotechnology, Fab-labs, bioengineering, machine-to-machine, information technologies, satellites, etc. will transform societies. The growing demand for food, water and energy will also serve as a basis of large moves like de-carbonization and electrification when it comes to energy.

- **Living in a smart world.** The new world of big data allowed us to see how smart concepts (products and services) are predicted to have an eminent position in the years to come. The coalition of technology, innovation and zero defects can produce smart solutions able to deal with future requirements. When you work on these issues related to the environmental crisis you may get into a new green era of development. Increased connectivity will lead to the convergence of products, technologies and industries moving into an interconnected world.

- **Disrupting innovation.** Growing opportunities coming from disruption implies the creation of markets where they do not currently exist or the creation of good products/services for new consumers. This type of innovation should lead to the development of new business models for the future.

Why do companies engage? External drivers

In the previous section, we have seen how ten large megatrends are expected to have important impacts on the coming decades in the business arena. Nine of these trends are shaping the way business will act regarding the three variables of population, affluence, and technology. The other trend links to the environmental crisis and focuses on how the instability and loss of resilience of social-ecological systems will pressure companies to reconsider their strategy and operations and the way in which they create value. The increasing scarcity of ecosystem goods and services is becoming more important for business success and long-term performance.

Management and organization studies can count on more than two decades of literature where scholars have attempted to provide legitimate explanations of "why" companies engage in corporate sustainability. Among these approaches, stakeholder theory (Freeman, 1984) provides a powerful framework to understand why companies address sustainability issues. According to this perspective, stakeholders can affect—and are affected by—the business (Freeman, 1984: 25). Affected stakeholders can exercise pressures by engaging with the firms in multiple ways and through different strategies and tools, therefore inducing firms to change their behavior and acknowledge their instances. Meeting stakeholder expectations is therefore essential for preserving the capacity to create long-term value. Among the many stakeholder groups discussed in literature, there are some that, in our view, today play a particularly important role in driving companies towards environmental sustainability. Accomplishing and aligning with their demands is therefore essential to maintain the firm's reputation and legitimacy. At the same time, when these stakeholders decide to remove their license to operate, companies might find themselves

I   =   P   A   T

I

Non Resilient
Ecological Systems

ENVIRONMENTAL
CRISIS

Non Resilient
Social Systems

P

GROWING
POPULATION

AGING

INCREASING
URBANIZATION

A

SHIFTING ECONOMIC
POWERS

RAISING GLOBAL
CONNECTIVITY

DIVERSIFYING
GOVERNANCE

T

ACCELERATING
TECHNOLOGICAL
TRANSITIONS

LIVING IN A
SMART ERA

DISRUPTING
INNOVATION

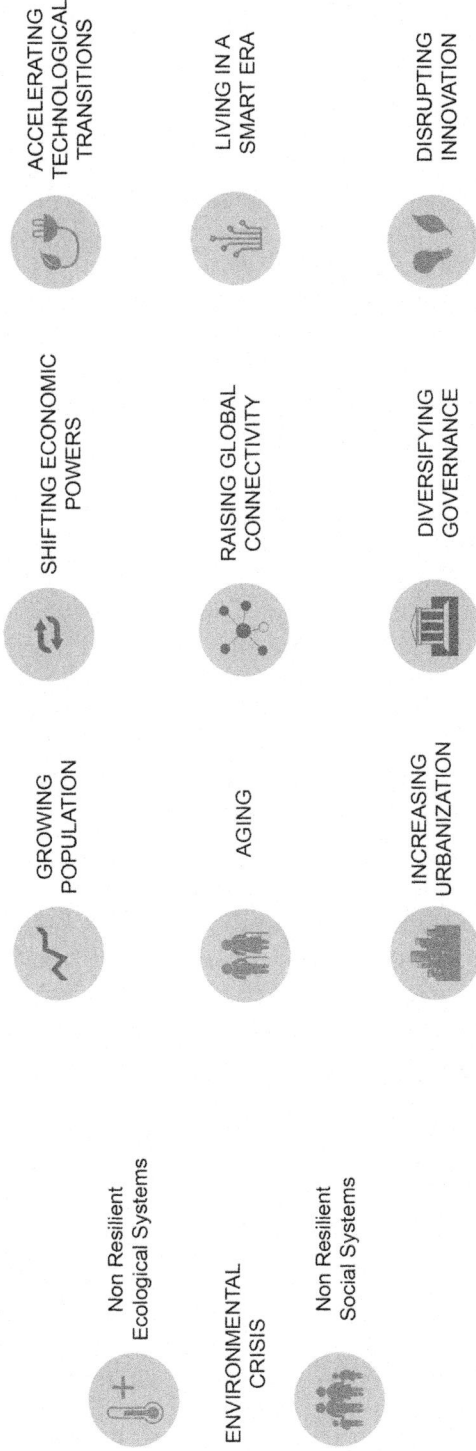

Figure 2.3 Megatrends grouped according to the IPAT factors.
Source: authors' elaboration.

in very complex situations with effects on short-term and long-term profitability. The four external drivers identified are the policy makers, financial markets, consumers, and social movements/NGOs.

- **Policy makers.** When we consider environmental and social themes, a fundamental driver of change is represented by policies and regulations that foster firms, organizations and individuals to modify and adapt their behaviors to the new rules of the game. It is beyond the aim of the present work to explore the broad set of mechanisms and tools in the hand of policy makers, ranging from overarching international conventions (e.g., the case of Paris Agreement on climate change) to long-term policy commitments (e.g., the UN Sustainable Development Goals); or from market mechanisms (e.g., the EU Emission Trading Scheme for carbon emissions) to international principles and declarations (e.g., the ILO Declaration on Fundamental Principles and Rights at Work). Pressures from policy makers can have multiple spheres of influence: they can focus at the global level, at the regional level, or at the local one. Companies, especially multinational ones, are therefore exposed to multiple regulatory and policy risks and they need to align with multiple requests. For example, when we look at environmental problems, standards (e.g., environmental level of emissions) and regulations (e.g., waste management) in different part of the world might be profoundly different. Another aspect that is gaining importance when we analyze policy makers is a growing emphasis on the development of effect-related policy. This is the case of the Paris Agreement that identifies the desired state for our atmosphere, corresponding to the 450ppm scenario. Similarly, when we consider the issue of marine activities and ocean protection, the European Union Marine Strategy Framework Directive has introduced the concept of "Good Environmental Status" – meaning that the different uses of marine resources can be conducted at a level that preserves the health of a clean and productive marine environment, thus ensuring its continuity for future generations. These new approaches to policy and regulations are diffusing at multiple levels with enormous implications for companies' behavior and strategy. An example (see Chapter 8) is science-based targets for climate change that links corporate actions to fight climate change and reduce GHG emissions with the level of decarbonization required to keep the global temperature increase below 2° C, as described in the IPPC Fifth Assessment Report (IPCC, 2015).
- **Financial markets.** Does the financial community care about climate change, biodiversity loss, and human rights? Is Wall Street interested in sustainability? For many years the answer was "No." While sustainability was becoming a material issue for many companies, investors were not a driver pushing firms to reduce carbon emission or address social issues along the supply chains. However, more recently, things have changed dramatically. In his 2018 letter to S&P 500 CEOs, Larry Fink, CEO of BlackRock, probably the largest investment corporation worldwide, offered a strong defense of long-term value creation, advising CEOs that he would not support companies that are not engaging in positively contributing to society and not investing in sustainability (Winston, 2018). Even more importantly, this was not the first time. At least since 2016, Fink has

addressed the question of the short-term markets' hysteria, underlying the importance of environmental, social, and governance challenges in shaping corporate financial performance.

BlackRock's approach to sustainability is part of a broad wave of change that is spreading inside the financial community driven by the consolidation of the Socially Responsible Investing (SRI) movement, the growth of green bonds, the increase of shareholder activism, the divestment from fossil fuels and the emergence of new standards and frameworks for accounting and reporting.

Box 2.3   The Socially Responsible Investing movement

Terms such as responsible investing, sustainable investing or Socially Responsible Investing (SRI) have become more common in the language of investors and asset owners. Each term might have specific nuances and different treats, but all of them share the idea of an investment approach that considers the relevance of environmental, social, and governance (ESG) factors in portfolio selection and management. Moreover, they share a long-term orientation regarding returns. According to the Global Sustainable Investment Review (GSIR) (2017), in 2016 there were about US$22.89 trillion of assets worldwide (about US$12 trillion in Europe and US$8.7 trillion in the US) managed under this approach, with an increase of 25% from 2015. Therefore, the movement of SRI is rapidly taking off and becoming a fast-growing segment of the financial landscape, acting as a catalyzer for engaging companies into sustainability. There are several types of strategy within responsible investing. Based on the logic of portfolio selection, we can identify the following.

- **Negative screening.** Companies are excluded from investment based on criteria relating to their sector, products, activities, policies or performance.
- **Integrated analysis screening.** In this case, the selection is based on an active analysis of the ESG factors as part of the investment research required for the asset allocation.
- **Positive/best-in-class screening.** Here the criteria are to invest in sectors, companies or projects selected for positive ESG performance relative to industry peers.
- **Thematic investment screening.** The selection is based on sustainability oriented themes such as climate change, renewables, clean tech and water.
- **Corporate engagement or active ownership.** The focus is on the utilization of power (shareholder power or the ability to directly engage corporations) to push companies to improve their sustainability performance.
- **Impact investing.** This is a relatively new approach in private markets aimed at addressing social or environmental problems, including community investing, where the investments are specifically directed to traditionally underserved individuals or communities.

Negative screening with US$15.02 trillion and ESG integration with US$10.37 are the largest strategy, while impact investing is still a niche but promising segment (GIIN, 2017).

Source: Authors' elaboration using multiple sources.

This is a new "financial ecosystem" that is fast growing thanks to a new generation of actors and new norms, institution, and initiatives. Several specialized organizations with specific technical competencies have emerged in the last decade. Specialized organizations and data providers such as Bloomberg ESG and Thomson Reuters are feeding asset managers and asset owners with qualitative and quantitative information on environmental, social and governance issues to assess risks and opportunities. Research and rating agencies (e.g., RobecoSAM, Vigeo Eris and Sustainalytics) are offering investors decision-making tools to select their portfolios on the basis of ethical and sustainability factors.

Other initiatives such as the CDP, formerly the Carbon Disclosure Project, provides disclosure systems and pushes companies (as well as cities and regions) to measure and manage their sustainability impacts with regard to climate change, water, ecosystems, forests and supply chains. The mechanism through which these stakeholders operate is that of inducing firms to adopt higher sustainability standards and to ask for accountability and non-financial information disclosure. The effects can drive a pivotal change in business since companies are required to introduce a new set of key performance indicators (KPIs) capable of intercepting social and environmental impacts and embed these metrics into management systems and decision-making process to align with the financial community demand. This kind of non-financial information is needed to integrate and complement the traditional financial measures that are clearly unable to assess sustainability, and that provide incentives that are not in line with social and environmental targets, therefore shifting the corporate focus from the shareholder value idea to a multidimensional approach based on the stakeholder value. Of course, these dynamics are still in the early stages (according to GSIR in 2016 SRI counts for 26.3% of the global managed assets) and need to be sustained and further strengthened in order to broadly diffuse in the markets, affect investor behavior and contribute to orienting corporate strategies towards sustainability.

- **Consumers.** In today's business world, consumers are central to every company strategy and understanding their preferences and purchasing behavior is key for developing a successful brand positioning. Consumer concerns about sustainability and their attitude towards social, environmental and ethical products is therefore paramount for every company that intends to embed sustainability into its strategy. The debate about sustainable consumption and green consumerism is long. It started back in the 80s and has gone through multiple waves of "rise and stumble," alternating optimistic and pessimistic visions, but with very little progress in terms of green product purchasing for many years (Peattie and Crane, 2005: 358). This has increased skepticism about the real interest in sustainable products, especially regarding the willingness to pay for social and environmental benefits. Even if companies and marketing managers must wait for a sustainability call from their consumers, currently, in many markets and for many product categories, individuals are probably not prepared to take these aspects into account when making their shopping decisions. Therefore, what has changed today? Has sustainable consumerism finally taken off? Looking at several market surveys, the sharp increase of sustainability labels (Darnall and Aragon-Correa, 2014) and the diffusion of sustainable brands in some product categories (food, cosmetics and even textiles), it seems that things have finally evolved and matured to the point where we can say we have entered a new era of "conscious consumerism." The shift in the mind of consumers becomes more evident every year. They expect more than price and quality for the things they are buying, and issues such as transparency, honesty, responsibility, carbon impact and reputation are gaining increasing importance in shopping decisions. Moreover, when we look at Millennials and Generation Z (Nielsen, 2015), the interest is even stronger since they really care that their favorite brands are good and sustainable. For these market segments, information about sustainability has a positive impact on their evaluation of a company and their intention to buy. Another important element that seems to have changed is the perception that sustainable products are associated with a higher level of functional performance, while in the past, green was often at odds with quality (Whelan and Fink, 2016). Finally, marketing researchers have identified a specific group of consumers – named the Lifestyle of Health and Sustainability segment (LOHAS) – who act as an active environmental steward group. Personal health, eco-tourism, natural lifestyles, alternative transportation, renewable energy and socially responsible investing are just some of the topics this segment is looking at for personal development and sustainable living. This segment can be up to 20% of the western world's consumers, and they can set the sustainability bar to promote a change in the way many companies act. This new wave of conscious consumers is voting with their wallets, and this is going to affect companies, opening opportunities for those brands that demonstrate commitment to sustainability.

- **Civil society and NGOs.** Advocacy groups and non-governmental organizations have become key forces in society, capable of affecting a firm's behavior both directly and indirectly. On the one hand, they are able to act with a specific campaign to pressure companies to change their strategies and actions. We can go back to the 1990s when Nike was targeted by labor activists and anti-globalization movements for abuse and exploitation of the workers in its supply chain. The result was that for many years, Nike became a synonym for misconduct until a dramatic change towards responsibility and transparency was initiated. Today, Nike is a well-known champion in sustainability, with responsible practices along the supply chain that have become admired by competitors and appreciated by NGOs. On the other hand, these organizations can activate other stakeholders to persuade companies to act: they can exercise pressures on consumers to boycott specific brands, collaborate with media to give visibility to their actions, and lobby with policy makers to obtain favorable regulations or policies. At the same time, thanks to their specialized knowledge, skills and capabilities, these organizations offer firms critical resources to tackle social and environmental issues, providing advice and guidance. This is the case, for example, of WWF, which has started multiple programs engaging and partnering with companies in order to establish management standards and certification systems for several commodities (e.g., timber and paper, tuna and cotton). Other examples are the Rainforest Alliance or Fairtrade, which work hand in hand with business to conserve nature and guarantee sustainable livelihoods to farmers and smallholders. Even Greenpeace, which usually maintains a more independent and watchdog position towards companies, has started to collaborate on specific issues such as climate change, rainforest preservation and pollution. In sum, cross-sector forms of collaboration between firms and civil society represent a growing phenomenon which, according to many observers, might support business transformation towards more sustainable approaches (Seitanidi and Crane, 2014).

Why do companies engage? Internal drivers

We have analyzed what we consider to be the most important external forces that exercise pressure and orient companies towards sustainability; we now turn to internal drivers and particularly look at the role of business leaders. Inputs to strategic decision-making ultimately include the Board of Directors and top managers; the success and the failure of a company in reaching its goals can be usually traced back to the responsibilities of top managers and the Board. In today's competitive context, characterized by high complexity, high volatility and fast dynamics strategic management processes tend to be governed by the Chief Executive Officer (CEO).

CEOs are ultimately responsible for the implementation of the company's strategy regarding long and short-term plans, they are the architects of the mission and vision, and they contribute to shaping the values and the organizational culture. They also position at the interface between the external stakeholders (e.g., financial community and investors, governments and policy makers, customers) and the internal ones (e.g., the Board, shareholders, employees, top managers) and, in doing so, they have a unique visual perspective that allows them to coordinate and align internal resources with external opportunities and risks (Bertels et al., 2016).

Recent studies have underlined that the CEO's leading role and capabilities acquire even more importance when large and impactful strategic and organizational changes are required, like in the case of sustainability challenges. Sitting at the top of the organizational pyramid, CEOs have the responsibility of leading the company towards the transformation required to make it meaningful for the company and its stakeholders. Personal commitment and involvement; capacity to scan information and build strategic momentum; ability to build trust, motivate, and engage top managers and employees on the new journey; these are some of the distinctive features of exceptional leaders, that support effective strategizing and success (Birshan et al., 2017).

When it comes to the journey towards sustainability, the CEO's role is therefore paramount. Let us start with a couple of examples. In June 2012, Paul Polman, the CEO of Unilever since 2009, stated in the opening of a famous interview with *Harvard Business Review*:

We thought about some of the megatrends in the world, like the shift east in terms of population growth and the growing demand for the world's resources. And we said, "Why don't we develop a business model aimed at contributing to society and the environment instead of taking from them?"

(Ignatius, 2012a: 112)

In the same interview, he also commented on the relation between building value for society and company success: "We think that, increasingly, businesses that are responsible and actually make a contribution to society in its positive sense, make it part of their overall business model, will be very successful" (Ignatius, 2012b). Both statements were - and are - courageous and disruptive. They not only contributed to introducing managers and stakeholders to new Unilever's sustainability strategy and the Sustainable Living Plan (see Chapter 8) but also contrasted the conventional idea of CEOs, who usually talk "only" to investors and shareholders, with a mainstream financial jargon.

Similarly, Karl-Johan Persson, CEO of H&M, another company oriented towards sustainability, in an interview included in the H&M Sustainability Report 2013 commented:

Of course, I hope for H&M to continue to grow and contribute to jobs and development around the world. But to continue growing, we need to consider our planet's boundaries . . . I hope that we will be able to produce fashion in a closed-loop, using less of our planet's resources and reducing waste instead. For the resources that we will still need, we must share them fairly between today and future generations.

(H&M, 2104: 3)

Like in the case of Polman, Persson introduced concepts such as planetary boundaries, closed-loop systems and resource scarcity as limit conditions to corporate development and as external factors that shape strategy and decision-making. The examples of Paul Polman and Karl-Johan Persson are not isolated. Other examples of business leaders strongly committing their companies to an inspirational purpose and who want to contribute to making a better world are: Ivon Chouinard, founder of Patagonia; Elon Musk, CEO and Chairman of Tesla; Jeffry Immelt, CEO of GE; and Jochen Zeitz, former CEO of Puma.

So, what makes some CEOs more capable of understanding these challenges and starting a process of change? Looking at literature on the role of CEOs and at prior studies in business sustainability, we can identify and group some distinctive features - a mix of personal moral and cognitive elements that help a CEO shift the organization towards sustainability - as listed below.

- **Beliefs and worldview.** The systems of beliefs and values of the CEO, his/her view of the world and his/her capacity to commit to sustainability as a personal purpose are distinctive traits often associated with this type of leadership.
- **Scanning and sensing.** Understanding the risks and opportunities related to the threats to the resilience of social-ecological systems, the question of planetary boundaries and the implications for the specific business activity is another key feature. Excellent CEOs,

in fact, are able to detect and interpret important strategic information, performing strategic reviews and transforming these inputs into meaningful decisions.

- **Engaging the organization.** Sustainability requires that the whole organization commits to the new social and environmental challenges, undertaking radical and often painful changes. CEOs need to explain to their managers and employees the reasons for these transformations, involve them through symbolic actions, walk the talk, spread responsibilities and introduce novel organizational solutions (e.g., Chief Sustainability Officer, sustainability committee, incentive systems based on social and environmental targets).
- **Engaging stakeholders.** Showing evidence of a clear business case for sustainability, linking the creation of value for the company to the generation of positive societal impact, is still a critical task in order to have approval from investors and shareholders. Similarly, to involve clients, suppliers and NGOs into the new strategy, CEOs need to build reputation and trust through clear commitments and meeting expectations with coherent performance.

Another interesting perspective in analyzing distinctive traits of CEOs and leaders committed to sustainability is offered by Freya Williams. In a book published in 2015 titled *Green giants: How smart companies turn sustainability into billion-dollar businesses*, she analyzes the stories of nine companies with more than US$1 billion in revenues (Unilever, IKEA, Toyota, Natura, GE, Nike, Tesla, Whole Foods and Chipotle) that are committed to a sustainable purpose. A common features of these journeys is the existence of "iconoclastic leaders" – the individuals who started it, who share some particular traits: *conviction*, or the ability to orient the organization in risky and uncertain territories and make people trust them, like in the case of disruptive transformations; *courage*, or the ability to be bold, to stand up and change things; *commitment*, in order to overcome obstacles and guide the company towards the new goals; and *contrarian*, in the sense that sustainability may require an upside-down view of the things, a novel perspective, a different look (Williams, 2015). The author confirms the importance of being at the top of the organization to drive effective and pervasive business transformations. Seven out of nine of these iconoclast leaders are CEOs, and the two remaining are Chief Sustainability Officers with outstanding skills and leadership qualities, like in the case of Steve Howards of IKEA and Hannah Jones of Nike.

To conclude, in this section we have tried to illustrate "why companies engage in sustainability" looking inside the organization. We have not analyzed the role of managers and employees. It is obvious that a company cannot be sustainable if the workforce is not motivated and aligned on this goal, but in the majority of the cases we have studied the change agent is the CEO, and the transformations must start from the head, with passion, commitment and purpose.

The business case for sustainability

Searching for win-win opportunities

A "business case" is an argument, usually documented, that is intended to convince a decision maker to approve some kind of action. There can be many motivations to develop a business case, and every motivation should be focused on increasing the value of the company, either in financial numbers or in its long-term competitive positioning. The argument prepared for the business case should examine benefits and risks involved in taking or not taking the specific action(s). The business case can conclude with a statement for its implementation, a kind of roadmap to facilitate the execution.

As we have seen previously, sustainability today grounds the debate about how to deal with the present environmental crisis. If we move this idea into business, companies have a complex and challenging task: they should develop actions to contribute to minimize such negative impacts as well as improve the company's economic performance. Not only large multinational enterprises but also smaller firms should find the path to develop activities to ensure a sustainable global economy. In the move to this overarching goal, companies and managers should try to find their business case for sustainability.

A large body of management literature in the field of corporate sustainability, both academic and practitioner oriented, has dedicated attention to this topic, seeking whether and how firms can create synergies between managing environmental and social issues and increasing the economic performance (Margolis and Walsh, 2001; Hart and Milstein, 2003; Schaltegger and Wagner, 2006). These studies have covered a broad range of topics from attempting to define what a business case for sustainability is, to investigating the relations between the sustainability performance of a firm and its economic or financial performance. Some works have dedicated attention to understand whether a business case can be designed on purpose, and how it can be managed. Other research has analyzed the trade-offs and conflicting situations between the different sustainability targets (Hahn et al., 2010). Finally, scholars have investigated the direct and indirect variables that can help explaining the business case, analyzing the mechanisms that can transform the environmental and social benefits into economic or financial value (Ambec and Lanoie, 2008).

In this body of literature, one major research stream has been empirical and quantitative, focusing on the attempt to identify a statistical relation between the sustainability performance (environmental and social) and the financial performance. Over the years, hundreds of studies have been carried out following three main methodological approaches:

- **portfolio analysis**, which compares the performance of a specific portfolio of selected companies (e.g., sustainable companies, green companies, socially responsible companies) with benchmark indexes,
- **event studies**, which focus on the effect of environmental or social events (e.g., an oil spill) on the variation of the company shares prices, and
- **multivariate analysis**, which verifies the relation between some environmental and social performance variables on measures of the economic and financial performance (e.g., ROI, ROE, Tobin's q).

The results of these studies, including some meta-analysis works have been used to try to summarize the empirical results of vast research in a comprehensive manner (Orlitzky et al., 2003; Albertini, 2013), show still controversial results. We can try to simplify the debate on the relation between sustainability and economic performance using three frameworks, as listed below.

- **Studies establishing a positive relation.** Several works have found a positive, win-win, relation between the two variables across industries and study context. Theoretically, this relation has been explained using different approaches, like the natural resource-based view theory (Hart, 1995) or the slack resources theory (Waddock and Graves, 1997).
- **Studies establishing a negative relation.** In this case, the theoretical explanation of the link reflects the traditional neo-classical paradigm, by which increasing sustainability reduces profits. This relation is also justified by the existence of trade-offs that cannot always be addressed and solved with win-win solutions.
- **Studies establishing no relation.** Finally, a group of papers have shown inconclusive statistical correlations, demonstrating that the link cannot be proved. This is due to the difficulties in measuring the variables and other methodological research limits.

Based on this broad literature stream, what shall we conclude? Is there any win-win relation between sustainability performance and economic success? Do we have a business case that can motivate companies so that they automatically integrate environmental and social issues into the management activity? Today, the debate on the relation between the sustainability and economic performance has evolved and scholars have acknowledged the necessity to better investigate the mechanisms that drive the performance. Studies have demonstrated that the relation is significantly moderated by factors such as the way environmental, social and economic variables are measured, the length of the period of observation, the type of regulatory context and other contingent factors. At the same time, these studies show that companies that have developed specific capabilities to address sustainability obtain a competitive advantage over other companies.

To conclude, the academic and practitioner public now embrace the view that the relation exists and that firms can simultaneously obtain financial performances while reducing their environmental footprints. What is more interesting is to better investigate "how" this relation is determined, "what" the drivers are that explains the win-win opportunities, and "when" sustainability and business are at odds, like in the case of trade-offs.

Understanding the concept of business case for sustainability

In order to better understand the notion of the business case and sustainability, and before we move into investigating the drivers of the business case to unravel what type of environmental and social initiatives can generate economic success, we introduce a framework (Figure 2.4) that helps us with the explanation of the different business approaches and the actions companies must, or voluntarily can, adopt to address sustainability issues.

Compliance

Actions or activities that a company must do because there is a legal imperative, like in the case of regulations and environmental standards, in corporate sustainability jargon are referred to as compliance. A compliance approach is not a business case, in the sense that companies do not search for an economic benefit and are guided by the necessity to accomplish with the regulatory requests. Sometimes this is not easy as legal requirements can vary by country, state, region or local conditions. Nevertheless, based on empirical evidence, we know that a compliance approach can also generate economic opportunities for companies. For example, a firm that operates worldwide with factories in different regions can adopt different approaches to manage environmental emissions, adapting to the local emission limits. On the other hand, the firm can adopt the most advanced practices to manage environmental issues (e.g., air pollution, water or waste) and enforce them in all worldwide facilities, undertaking higher costs for compliance, but benefiting from economies of scale and optimizing supply chain operations, thanks to the application of a unique environmental practice. These approaches are associated with different short- and long-term costs and benefits that must be assessed and measured before making decisions. For example, choosing to enforce the stricter environmental standards can generate higher cost in the short-term, but in the future when regulations changes and higher standards are enforced, the company might win a competitive advantage over competitors that will be obliged to adapt to the new regulations. The ability to anticipate future regulations and proactively manage compliance is a capability that companies can develop and that can lead to innovative behavior, also driving better financial results (Porter and van der Linde, 1995). However, the starting point for these actions is related to minimizing the negative effects of policies and regulations, and we assume that the company is not looking at possible benefits for making profits out of managing sustainability.

Business case of sustainability

The second type of behavior refers to companies that are seeking what has been called the "business case of sustainability" (Schaltegger and Burritt, 2015; Schaltegger et al., 2017).

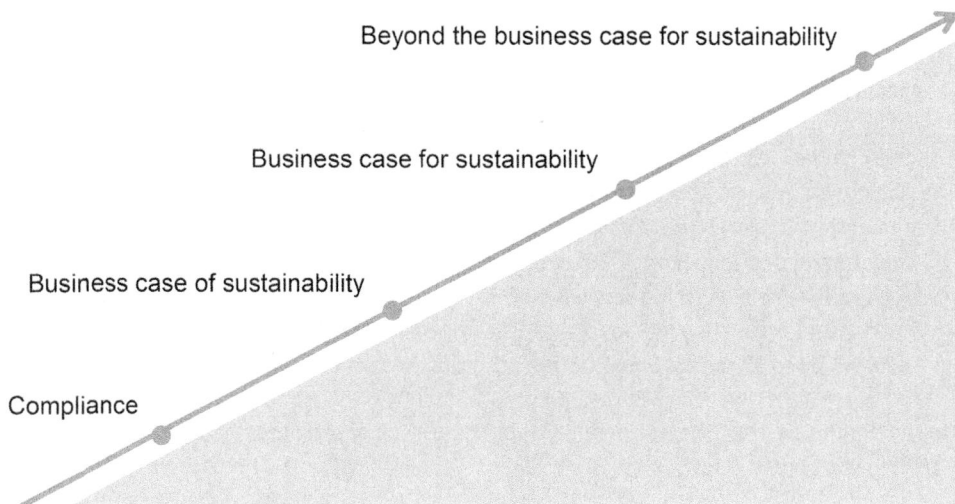

Beyond the business case for sustainability

Business case for sustainability

Business case of sustainability

Compliance

Figure 2.4 From compliance to the business case for sustainability and beyond.
Source: authors' elaboration.

This is a situation in which companies obtain financial performance "while" considering social and environmental issues resulting from the actions implemented. In these cases, economic success and sustainability benefits are not at odds (Hahn et al., 2010), and environmental and social problems can be easily used to increase profitability. In other words, it is about conventional business and the traditional profit maximization logic, while environmental opportunities can be instrumentally used to create more profit. This is the case, for example, of eco-efficiency actions with a very short-term payback period that are not impacting the business practices. Often, business language identifies these actions with the expression "low-hanging fruits," a concept that captures the idea of benefiting the potentialities of money savings from improved water, energy, materials and waste management. In this situation the company can reduce costs and diminish the environmental impact of the operations and can obtain appreciation from stakeholders. For example, in the Pinkton Hotel case (see Annex 3 to this chapter) substituting existing lights bulbs with LED lights or installing low-flow shower heads and toilets, generating immediate benefits with low investments. It is therefore an easy practice to implement.

Business case for sustainability

The concept of business case for sustainability proposes a substantially different view in terms of ethical and operational approach when compared to the business case of sustainability. In this case, the economic success is generated through, not just with, a careful identification, assessment, and selection of the environmental and social actions (Schaltegger et al., 2012: 97). In other words, the initial motivation of the company is to address environmental and social problems. It is a deliberate action to improve sustainability through which a positive economic effect is created. This approach calls for a different evaluation of business success that is not measured only with economic and financial indicators, but also on the basis of the capacity to reach sustainability goals and targets. At the same time, this approach emphasizes the search for synergies among different stakeholders' requests, and among different strategic and operational options. To conclude, Schaltegger et al. (2012: 98) help us with the framing of the business case for sustainability identifying three basic requirements:

- the company must realize a voluntary activity with the deliberate intention to address societal or environmental problems,
- the activity must create a positive business effect, or a positive economic contribution (e.g., cost reduction, increased market share, risk reduction), which can be measured or argued for in a convincing way, and
- a clear and convincing link must exist, showing that the sustainability activities have led or will lead to an improvement in the economic success.

On the pragmatic level, the business case for sustainability requires the company to perform a few tasks, such as:

- examine the strategic opportunities that these social and ecological activities may generate,

- identify, assess, measure and monitor the effect on revenues, costs, and margins and on the competitive positioning related to these actions, and
- identify, assess, measure and monitor the positive social and environmental impact associated to these actions.

To conclude, the business case for sustainability can be seen as an approach that guides companies in the pragmatic search for bundled solutions with combination of the economic, social, and environmental benefits. At the same time, the existence of a business case for sustainability automatically aligns corporate behavior with environmental and social issues. The question that emerges is whether business cases are enough to address the challenges of the Anthropocene epoch. On the one hand, as illustrated by several research and studies (Kiron et al., 2017), companies and managers have still limited capabilities to identify and implement innovative strategies and business models that capture these opportunities. On the other hand, the idea that there are always ways to address trade-offs or conflicting situations where sustainability goals are at odds with economic performance is naïve and simplistic. This will bring us to the introduction of the final approach, which looks at management sustainability beyond the business case.

Beyond the business case

We have acknowledged that the business case for sustainability brings companies into a much more efficient use of natural and social capital. It is therefore a powerful driver for reducing the environmental and social impact, while generating economic value. Since decades of debate on sustainability have acknowledged the fact that the properties of natural resources and ecosystems (e.g., the non-substitutability of some forms of natural capital or the non-linearity of their dynamics) might require radical transformation in the way business is carried out and in the market logics. The planetary boundaries and the degradation of the social-ecological system resilience point at paramount challenges for business and show that in many cases win-win options are simply not available. In other words, there are many situations where the business case "of" or "for" sustainability simply does not exist, while trade-offs between the environmental, social and economic objectives exist (Hahn et al., 2010). For example, if we aim to maintain our temperature increases below the target of 1,5°-2°, current technologies do not allow us to burn the fossil fuel reserves of many companies listed in the stock exchange. This concept, referred to as unburnable carbon (Jakob and Jérôme, 2015) should condition the strategies of many oil, energy and mining firms since it looks evident that environmental constraints are at odds with the possibility of these companies continuing their activities in the long term without dramatic transformations in their business models.

Another example refers to the possibility to find valuable alternatives to the iron cage of consumerism (Jackson, 2013). In many cases, in fact, the improvements that companies can reach through the increase in the resource productivity or through environmental product stewardship can be offset by the sharp growth of the market demand. Moreover, even if technological solution to drastically reduce our environmental impact can be available, consumer can make purchasing decisions that do not go in line with solving the problems we have. In a nutshell, trade-offs in corporate sustainability (Hahn et al., 2010) can refer to different temporal dimensions (short-term vs long-term decisions and orientation), different outcomes (protection of the environment vs development of industrial activities), or different processes (alternative strategies or practices). Moreover, trade-offs can impact multiple scales, from the individual, to the organizational to the industrial.

As a consequence, companies may be required to move beyond the business case. They are asked to radically transform their core strategy and integrate sustainability profoundly into their mission, vision or even changing into purpose driven companies, where the ultimate goal becomes that of having a positive impact on society and nature. These elements will be further analyzed in the next chapter, where we develop the Business In Nature concept, linking companies back to social-ecological systems where they ultimately belong.

Analyzing the drivers of the business case

Our analysis of the relations between sustainability and the business case anyway is not complete if we do not try to examine the drivers and the mechanisms that explain the existence of win-win-win opportunities. Additionally, in this case, we can identify a large stream of literature that has investigated these relations looking inside the corporate "black box" (Ambec and Lanoie, 2008; Whelan and Fink, 2016). This research has analyzed how corporate actions that respond to environmental and social challenges can generate value, impacting on the levers that organizations use to increase revenues, cut costs, or reduce the risk profile.

In order to illustrate the value drivers, we introduce a two-by-two matrix. On the vertical axis we distinguish between two forms of value creation: increase the revenues (e.g., through growth in sales volumes, new products, higher prices) and reduce the costs (e.g., increasing resource productivity). On the horizontal axis, we position the temporal variable, separating between short-term value creation and long-term value creation processes. Different levels of uncertainty are associated to different timeframes. The matrix generates four distinct dimensions through which sustainability drives the value creation process (see Figure 2.5).

The lower-left quadrant focuses on cost reduction. This is the first value driver and probably the most direct link with economic performance through operational efficiency. The improvement in energy and resource productivity and the reduction of pollution and waste are synonymous of savings and allows the firm to reduce the costs associated with fines and taxes. All these aspects can contribute to short-term economic benefits, in line with the Porter Hypothesis that we have illustrated in the previous sections (Porter and van der Linde, 1995).

The lower-right quadrant refers to the risk management dimension. Addressing environmental and social issues can reduce different typologies of risks (e.g., operational, technological, social, regulatory). These risks reveal themselves over a long-term period and can affect many business areas along the entire value chain. For example, the effects of climate change and water shortages on agro-food resources or the impact of extreme weather events on firm assets (e.g., infrastructures, facilities, warehouses) can interrupt the business activity or sharply increase the price of the resources. These events that can have dramatic impacts on the firm's bottom line happen outside of the typical organizational control and require the

REVENUE GROWTH

REPUTATION AND LEGITIMACY

DIFFERENTIATION

- New line of business
- Increase prices
- Increase market share
- New markets

LICENSE TO OPERATE

- Brand value
- Employer engagement and recruitment

Increase revenues

Short-term

Long-term

Value Creation

OPERATIONAL EFFICIENCY

- Increase resource productivity
- Cost reduction

RISK MITIGATION

- Lower cost of capital
- Increase attractiveness for investors

Reduce negative

COST REDUCTION

RISK MANAGEMENT

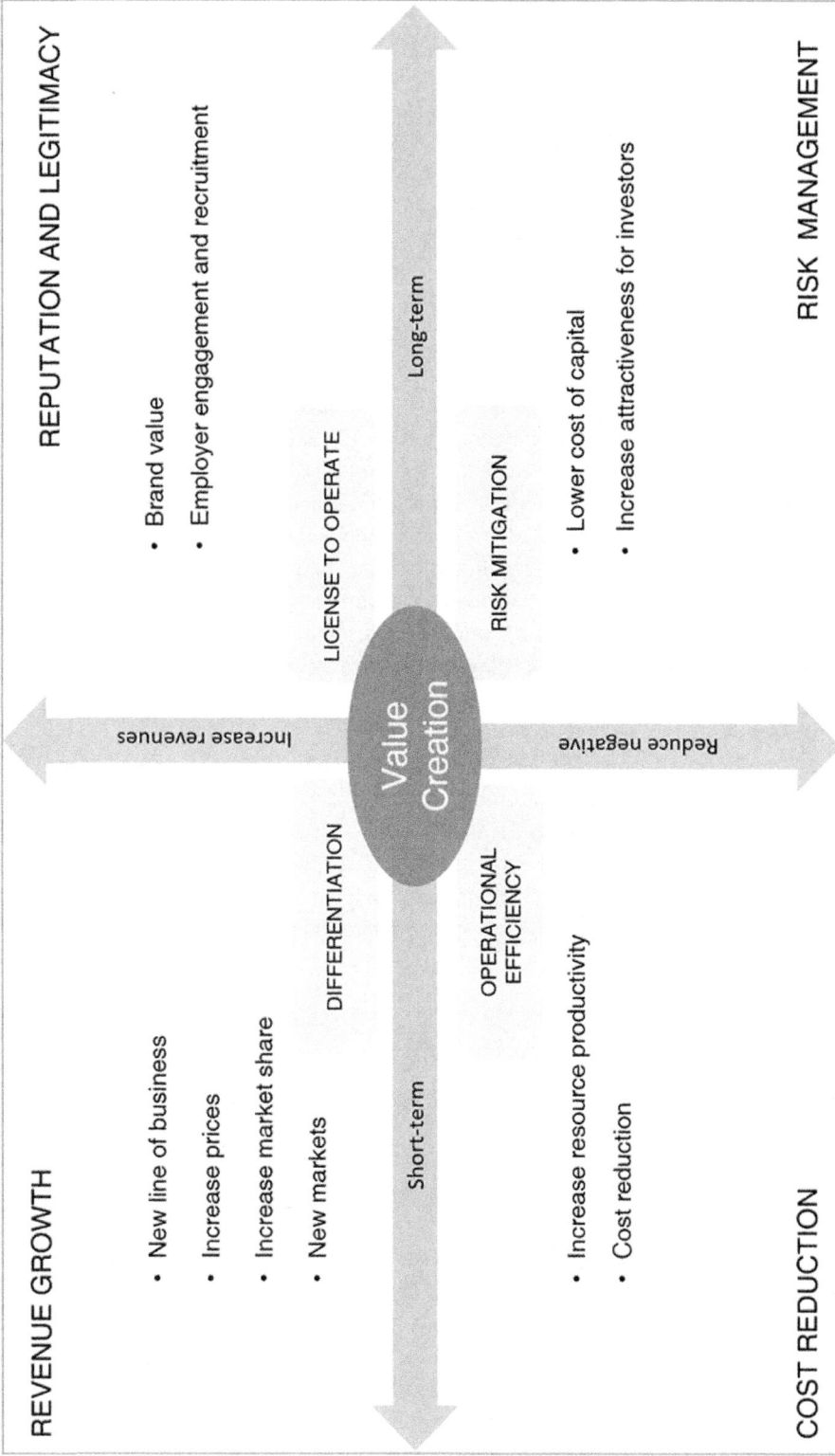

Figure 2.5 Drivers for sustainable value creation.
Source: authors' elaboration.

implementation of specific risk management strategies (Whelan and Fink, 2016). Managing these risks through adequate techniques and procedures can generate possibilities for better future prospects that finally will also increase the firm's value. Furthermore, if we look at the financial community, the focus on sustainability and risks reduction can produce multiple benefits: lowering the cost of capital for the firm, reducing the insurance costs and increasing the attractiveness for investors interested in ESG performance with ultimate effects on the market value (e.g., SRI or pension funds).

The upper-left quadrant addresses the issue of revenues resulting from opportunity-oriented sustainability drivers. Considering environmental and social issues as product attributes allow firms to access certain markets like in the case of green public procurement, or in the case of distribution channels that require products with high sustainability standards and eco-labels. Sustainable products and services can represent potential sources of new revenues accessing novel market segments interested by these types of offer. A growing number of consumers today appear attracted by brands with sustainability, purpose and integrity at their core. Moreover, greener products or services can allow companies to use a differentiation strategy, applying higher prices to those that are willing to pay more, or increasing the market share in the segments where the company is already operating. Customer loyalty and trust can be also associated with these actions. Finally, sustainability can be a driver for innovation, fostering companies to search for novel solutions to respond to the demand needs while radically reducing their environmental footprint and their social impact. New markets for cleaner technologies can be another source of revenues encouraged by sustainability.

Finally, the upper-left quadrant focuses on corporate reputation and legitimacy. The inclusion of different stakeholder's interests into firms' decision-making may increase the legitimacy and the license to operate, opening to the firm new opportunities for growth. Moreover, companies with high sustainable reputation can attract talent, facilitate employer engagement and reduce employee rotation, thus providing a better organizational climate and motivation, with effect on the economic performance and long-term success. All these aspects can further contribute to an improvement in the shareholder value and to increased productivity.

To conclude, we have provided evidence that there are multiple drivers that influence the business case "of" and "for" sustainability leading companies to win-win-win results. These drivers can generate direct or indirect business success and increase the economic performance. Moreover, these variables are often interlinked, with multiple effects that are difficult to unravel. For example, long-term oriented actions that increase reputation or reduce the likelihood of environmental incidents can influence the brand value increasing the revenues. Similarly, a focus on supply chain traceability and transparency reduces the risks at the supply chain level, improves corporate legitimacy and may positively affect the sales contributing to a better market positioning and product differentiation.

As previously underlined, building a solid business case based on clear measures and indicators of environmental, social and economic performance will help guiding companies towards sustainability. Anyway, we must also remember that sustainability frequently requires moving beyond the business case in order to find innovative ways to address the complexity of the social-ecological challenges of our new century.

Summary

This chapter has analyzed the evolution of the concept of corporate sustainability. We first illustrated the process of globalization of environmental issues and the mainstreaming of environmental values. We identified four main paradigms characterized by a different focus and we examined each one in details: the dominant social paradigm, the environmental paradigm, the Sustainable Development paradigm and the social-ecological paradigm. We positioned the debate on corporate social responsibility and stakeholder theory within this timeline, briefly introducing these two theoretical perspectives. We concluded the section providing a brief exam of the social-ecological paradigm, connecting corporate sustainability to the challenges of the Anthropocene.

In the following part of the chapter we have built on Dyllick and Muff (2015) analysis of "business sustainability" practices. We have examined the transformation of this concept over the years, from refined shareholder value management (business sustainability 1.0), to managing for the triple bottom line (business sustainability 2.0), to truly sustainable business (business sustainability 3.0). We also provided a first analysis of the "Business In Nature" (BInN) concept (that will be further discussed in Chapter 3). Then we illustrated the use of the IPAT fundamental equation as a new way to construct a narrative about sustainability practices at corporate level.

Another section of the chapter investigated why companies engage in corporate sustainability. First, we analyzed several different documents (e.g., reports and surveys) from international organizations and specialized consulting companies. We identified ten large megatrends that are going to shape the next decades of this new century and we associated these megatrends with the variables of the IPAT equation. Second, we focused on the exam of different typologies of external and internal drivers. We started with a review of the role of policy makers, the financial markets, consumers, civil society and NGOs in pressuring companies to adopt more sustainable behavior and strategies. Then, we introduced and investigated the growing importance of executives in driving change towards sustainability, with a strong emphasis in leadership aspects.

The last topic discussed in this chapter is the business case for sustainability. A "business case" is an argument, usually documented, intended to convince a decision maker to approve some kind of action. When we look at sustainability, a large body of literature has focused on unraveling the relationship between the environmental (and social) performance and the financial performance of a firm. In order to better understand this relation, we introduced a framework that helped us with the explanation of the different business approaches: compliance, the business case "of" sustainability, the business case "for" sustainability, and beyond the business case. We closed the chapter with the analysis of main drivers that shapes the business case: operational efficiency, mitigation of risks, increase of revenues, improved reputation and legitimacy.

Chapter 2 annexes
Annex 1

A.1.1: Questions for discussion

- Explain the meaning of the dominant sustainability paradigm and that of social-ecological paradigm. How do these paradigms differentiate each other?
- What are Dyllick and Muff's (2015) typologies of business sustainability? Analyze the three different approaches proposed by these scholars and discuss their distinctive features.
- What is the IPAT equation? Illustrate how it can be used to analyze business impact on the environment and corporate sustainability commitment and strategies.
- Why do companies engage in sustainability? Briefly analyze the main theories that can be used to explain corporate responses to the environmental and social challenges.
- What are business megatrends? Why are they important in the corporate sustainability debate?
- What does it mean that sustainability is a market transformation?
- Identify and discuss the role of the financial community and NGOs in driving corporate sustainability.
- Explain why leadership matters in transforming corporate behavior towards sustainability.
- Illustrate the main differences between the business case of sustainability and the business case for sustainability.
- Analyze the business case drivers and illustrate how they can contribute to generate short-term and long-term value.

A.1.2: Assignments

Assignment 1: Sustainability and strategy

Consider a specific set of companies (you may decide a specific sector, a bunch of indexed companies, a national group). Enter in the web URL of the companies selected and move into the sustainability/responsibility section. Using all materials you can obtain from the information given on this website (web pages, sustainability reports, data and commitments, etc.) try to classify selected companies on the basis of the corporate sustainability approaches analyzed in Chapter 2 (1.0, 2.0 and 3.0).

What do you think particular companies should do to move ahead to the Corporate Sustainability 3.0?

Assignment 2: Videos on sustainability and leadership

When we analyze the topic of sustainable leadership and the role of CEOs in guiding companies towards environmental and social challenges, several videos available on the web can be used to activate class discussion and support the lecture. Moreover, videos can be used as openers to introduce a case study, and they can help discussing the topic of why do companies engage: internal drivers.

Here are examples of the types of videos we can show:

(1)   Yvon Chouinard (Patagonia): www.youtube.com/watch?v=O3TwULu-Wjw

This video can be useful to explain the mission of the company, the organizational culture and the product positioning based on the solid focus to "doing less harm possible."

(2)   Ray Anderson (Interface): www.youtube.com/watch?v=F2LOUBme8rw; www.youtube.com/watch?v=iP9QF_IBOyA

This video shows the personal experience of how he became a convinced environmentalist, how he developed a vision and how he proposed to diminish the impact of its products as much as possible.

(3)   Paul Polman (Unilever): www.youtube.com/watch?v=AbDOOVAtvfQ

In this video the CEO of Unilever introduces the Sustainable Living Plan as a new paradigm for sustainable growth. The video addresses multiple aspects of strategy and sustainability, including the relation between stakeholders and strategy, the business case for sustainability and the relation with the financial community.

(4)   Francesco Starace (Enel): www.youtube.com/watch?v=ai7m2ilanZc

In this video the Enel CEO and General Manager Francesco Starace, who participated as speaker to the Opening Session of UN Global Compact Leaders Summit 2016 (UN General Assembly, New York). explains the importance of placing environmental, social and economic sustainability at the heart of the corporate strategy and culture.

A.1.3: Further readings and additional resources

- The World Business Council for Sustainable Development (WBCSD) (www.wbcsd.org/) is a global, CEO-led organization of over 200 leading businesses working together to accelerate the transition to a sustainable world. It has been a key organization, uniquely positioned, to work with member companies along and across value chains to deliver high-impact business solutions to the most challenging sustainability issues. In its web it is easy to find many reports and documents for corporate sustainability aspects. In its Vision 2050 report, the vision for the organization is stated as "to create a world where more than 9 billion people are all living well and within the boundaries of our planet, by 2050."

Annex 2

A.2.1: References

Accenture & United Nations Global Compact. (2013). The UN Global Compact-Accenture CEO Study on Sustainability. Retrieved from www.unglobalcompact.org/docs/news_events/8.1/UNGC_Accenture_CEO_Study_2013.pdf.

Albertini, E. (2013). Does environmental management improve financial performance? A meta-analytical review. *Organization & Environment*, 26(4), 431-457.

Ambec, S. and Barla, P. (2002). A theoretical foundation of the Porter hypothesis. *Economic letters*, 75(3), 335-360.

Ambec, S. and Lanoie, P. (2008). Does it pay to be green? A systematic overview. *Academy of Management Perspectives*, 22(4), 45-62.

Barrena Martinez, J., López Fernández, M. and Romero Fernández, P. M. (2016). Corporate social responsibility: Evolution through institutional and stakeholder perspectives. *European Journal of Management and Business Economics*, 25(1), 8-14.

Berchicci, L. and King, A. (2007). Postcards from the edge: A review of the business and environment literature. *Academy of Management Annals*, 1(1), 513-547.

Berkes, F., Colding, J. and Folke, C. (2003). *Navigating social ecological systems: Building resilience for complexity and change*. Cambridge, United Kingdom: Cambridge University Press.

Bertels, S., Schulschenk, J., Ferry, A., Otto-Menz, V. and Speck, E. (2016, 8 September). Main Report: CEO Decision-Making for Sustainability, Retrieved from Network for Business Sustainability South Africa: https://nbs.net/p/main-report-ceo-decision-making-for-sustainability-bdf21f34-0c7c-47f6-9522-3ac4d826022e.

Bieker, T. and Dyllick, T. (2006) Nachhaltiges wirtschaften aus managementorientierter sicht. In E. Tiemeyer and K. Wilbers (Eds.), *Berufliche Bildung für nachhaltiges Wirtschaften* (pp. 87-106). Bielefeld, Germany: Bertelsmann.

Birshan, M., Meakin, T. and Strovink, K. (2017, April). What makes a CEO "exceptional"?, *McKinsey Quarterly*. Retrieved from www.mckinsey.com/business-functions/strategy-and-corporate-finance/our-insights/what-makes-a-ceo-exceptional.

Bowen, H. R. (1953). *Social responsibilities of the businessman*. New York, NY: Harper and Brothers.

Carbon Tracker, The Grantham Research Institute and LSE. (2013). Carbon Unburnable Carbon 2013: Wasted capital and stranded assets. Retrieved from http://carbontracker.live.kiln.digital/Unburnable-Carbon-2-Web-Version.pdf.

Carson, R. (1962). *Silent spring*. Cambridge, MA: Houghton Mifflin Company.

Chertow, M. R. (2000). The IPAT equation and its variants. *Journal of Industrial Ecology*, 4(4), 13-29.

Clinton, L. and Whisnant, R. (2014, February). Model Behavior 20 Business Model Innovations for Sustainability. Retrieved from Sustainability website http://sustainability.com/our-work/reports/model-behavior/.

Cole, H. S. D. (1973). *Models of Doom: A critique of the limits to growth*. New York, NY: Universe Publishing.

Committee for Economic Development, CED. (1971, 1 June). Social responsibilities of business corporations. Retrieved from www.ced.org/pdf/Social_Responsibilities_of_Business_Corporations.pdf.

Commoner, B., Corr, M. and Stamler, P. J. (1971). *The closing circle: Nature, man, and technology*. New York, NY: Knopf.

Cooperrider, D. (2008, July-August). Sustainable innovation. *BizEd*, 32-38.

Darnall, N. and Aragon-Correa, J. A. (2014). Can Ecolabels influence firms' sustainability strategy and stakeholder behavior? *Organization & Environment*, 27(4), 319-327.

Davis, K. (1960). Can Business afford to ignore social responsibilities? *California Management Review*, 2(3), 70-76.

Drucker, P. (1954). *The practice of management*. New York, NY: Harper & Row.

Dyllick, T. and Hockerts, M. (2002). Beyond the business case for Corporate Sustainability. *Business, Strategy and the Environment*, 11(2), 130-141.

Dyllick, T. and Muff, K. (2015). Clarifying the meaning of sustainable business. *Organization & Environment*, 29(2), 156-174.

Dyllick, T. and Muff, K. (2015). Clarifying the meaning of sustainable business: Introducing a typology from business-as-usual to true business sustainability. *Organization & Environment*, 29(2), 156-174.

Ehrlich, P. and Holdren, J. (1971). Impact of population growth. *Science,* 171(3977), 1212-1217.

Ehrlich, P. and Holdren, J. (1972a). Impact of population growth. In Riker, R. G. (Ed.), *Population, resources, and the environment* (pp. 365-377). Washington, DC: Government Printing Office.

Ehrlich, P. and Holdren, J. (1972b). A bulletin dialogue on the "Closing Circle": Critique: One dimensional ecology. *Bulletin of the Atomic Scientists*, 28(5), 16-27.

Elkington, J. (1994). Towards the sustainable corporation: win-win-win business strategies for sustainable development. *California Management Review*, 36(2), 90-100.

Elkington, J. (1997). *Cannibals with forks: The triple bottom line of the 21st century business*. Oxford: Capstone.

Elkington, J., Hailes, J. and Malkower, J. (1990). *The green consumer*. New York, NY: Penguin Books.

Ernst & Young. (2017). The upside of disruption Megatrends shaping 2016 and beyond. Retrieved from www.ey.com/gl/en/issues/business-environment/ey-megatrends.

Environmental Agency, EEA. (2015). The European environment – state and outlook 2015: Assessment of global megatrends. Retrieved from www.eea.europa.eu/soer-2015/global/action-download-pdf.

Frederick, W. C. (1960). The growing concern over business responsibility. *California Management Review*, 2, 54-61.

Freeman, R. (1984). *Strategic management: A stakeholder approach*. Boston, MA: Pitman Press.

Freeman, R. E., Harrison, J. S., Wicks, A. C., Parmar, B. and de Colle, S. (2010). *Stakeholder theory: The state of the art*. New York, NY: Cambridge University Press.

Friedman, M. (1962). *Capitalism and freedom*. Chicago, IL: University of Chicago Press.

Frost and Sullivan. (2014, May). World's top global mega trends to 2025 and implications to business, society, and cultures (2014 Edition). Retrieved from www.smeportal.sg/content/dam/smeportal/resources/Business-Intelligence/Trends/Global%20Mega%20Trends_Executive%20Summary_FROST%20%26%20SULLIVAN.pdf.

Global Impact Investing Network, GIIN. (2016). 2016 Annual Impact Investor Survey. Retrieved from thegiin.org/assets/2016%20GIIN%20Annual%20Impact%20Investor%20Survey_Web.pdf.

Global Sustainable Investment alliance, GSIA. (2017). The 2016 Global Sustainable Investment Review. Retrieved from www.gsi-alliance.org/wp-content/uploads/2017/03/GSIR_Review2016.F.pdf.

Gore, A. (1992). *Earth in the balance: Ecology and the human spirit*. New York, NY: Penguin Books.

Groucutt, H. S., Petraglia, M. D., Balley, G., Scerri, E. M., Parton, A., Clark-Balzan, L., Jennings, R. P., Lewis, L., Blinkhorn, J., Drake, N. A., Breeze, P. S., Inglis, R. H., Devès, M. H., Meredith-Williams, M., Bolvin, N., Thomas, M. G. and Scally, A. (2015). Rethinking the dispersal of *Homo sapiens* out of Africa. *Evolutionary Anthropology*, 24(4), 149-164.

Hahn, T., Figge, F., Pinkse, J. and Preuss, L. (2010). Trade-offs in corporate sustainability: You can't have your cake and eat it. *Business Strategy and the Environment*, 19(4), 217-229.

Hardin, G. (1968). The tragedy of the commons. *Science*, 162(1968), 1243-1248.

Hart, S. (1995). A natural-resource-based view of the firm. *Academy of Management Review*, 20(4), 986-1014.

Hart, S. L. and Milstein, M. B. (2003). Creating sustainable value. *Academy of Management Executive*, 17(2), 56-67.

HayGroup. (2014). Building the new leader – Leadership challenges of the future revealed. Retrieved from www.haygroup.com/downloads/MicroSites/L2030/Hay_Group_Leadership_2030%20whitepaper_2014.pdf.

H&M. (2014). H&M conscious actions – Sustainability – Report 2013. Retrieved from http://sustain ability.hm.com/content/dam/hm/about/documents/en/CSR/reports/Conscious%20Actions%20 Sustainability%20Report%202013_en.pdf.

Ignatius A. (2012a, June). The Harvard Business Review, an interview with Paul Polman: Captain Planet [Audio file]. Retrieved from https://hbr.org/2012/06/captain-planet.

Ignatius A. (2012b). The Harvard Business Review, an interview with Paul Polman: Unilever's CEO on Making Responsible Business Work [Audio file]. Retrieved from https://hbr.org/2012/05/unilevers-ceo-on-making-respon.

Jackson, K. T. (2012). *Virtuosity in business: Invisible law guiding the invisible hand*. Philadelphia, PA: University of Pennsylvania Press.

Jackson, T. (2013). "Angst essen Seele auf" – 53 Escaping the "iron cage" of consumerism. In U. Schneidewind, T. Santarius and A. Humburg (Eds.) *Economy of Sufficiency*. Wüpertal Institute. 100pp.

Jakob, M. and Jérôme, H. (2015). Unburnable fossil-fuel reserves. *Nature*, 517: 150–152

Jamor, M. and Jeróme, H. (2015). Unburnable fossil-fuel reserves. *Nature*, 517: 150–152.

Johnson, H. L. (1971). *Business in contemporary society: Framework and issues*. Belmont, CA: Wadsworth Publishing Co.

Kiron, D., Unruh, G., Kruschwitz, N., Reeves, M., Rubel, H. and MeyeSr Zum Felde, A. (2017, 23 May). *Corporate Sustainability at a crossroads*. MIT Sloan Management Review and The Boston Consulting Group. Retrieved from https://sloanreview.mit.edu/projects/corporate-sustainability-at-a-crossroads/.

KPMG. (2013, December 23). Future State 2030: The global megatrends shaping governments. Retrieved from www.kpmg-institutes.com/institutes/government-institute/articles/2013/12/future-state-2030–the-global-megatrends-shaping-governments.html.

Lamarck, J. B. P. (1984). *Zoological philosophy – an exposition with regard to the natural history of animals*. Chicago, IL: University of Chicago Press.

Margolis, J. D. and Walsh, J. P. (2001). People and profits? The search for a link between a company's social and financial performance. Mahwah, NJ: Erlbaum.

Maxwell, J. and Briscoe, F. (1997). There's money in the air: The CFC ban and DuPont's regulatory strategy. *Business Strategy and the Environment*, 6(5), 276–286.

Meadows, D. H., Meadows, D. L., Randers, J. and Behrens, W. W. (1972). *The limits to growth*. New York, NY: Universe Books.

Meyer, J. W. and Rowan, B. (1991). Institutionalized organizations: Formal structure as myth and ceremony. In W. W. Powell and P. DiMaggio (Eds.), *The New institutionalism in organizational analysis* (pp. 41–62). Chicago, IL: University of Chicago Press.

Millennium Ecosystem Assessment, MA. (2005a). *Ecosystems and human well-being: Synthesis*. Washington, DC: Island Press.

Millennium Ecosystem Assessment, MA. (2005b). *Ecosystems and human well-being: Current state and trends*. Washington, DC: Island Press.

Millennium Ecosystem Assessment, MA. (2005c). *Ecosystems and human well-being: Scenarios*. Washington, DC: Island Press.

Millennium Ecosystem Assessment, MA. (2005d). *Ecosystems and human well-being: Policy responses*. Washington, DC: Island Press.

Millennium Ecosystem Assessment, MA. (2005e). *Ecosystems and human well-being: Multiscale assessments*. Washington, DC: Island Press.

Mitchell, D. and Coles, C. (2003). The ultimate competitive advantage of continuing business model innovation. *Journal of Business Strategy*, 24(5), 15–21.

Nielsen. (2015, October). The sustainability imperative. New insights on consumer expectations. Retrieved from www.nielsen.com/content/dam/nielsenglobal/co/docs/Reports/2015/global-sustain ability-report.pdf.

Nørgård, J. S., Peet, J. and Ragnarsdóttir, K. V. (2010). The history of the limits to growth. *Solutions*, 2(1), 59–63.

Norman, W. and MacDonald, C. (2004). Getting to the bottom of "triple bottom line". *Business Ethics Quarterly*, 14(2), 243–262.

North, D. C. (1990). *Institutions, institutional change and economic performance*. Cambridge: Cambridge University Press.

Orlitzky, M., Schmidt, F. L. and Rynes, S. L. (2003). Corporate social and financial performance: A meta-analysis. *Organization Studies*, 24(3), 403–441.

Orsato, R. J. (2006). Competitive environmental strategies: When does it pay to be green? *California Management Review, 48*(2), 127-143.

Pava, M. L. (2007). A response to "Getting to the bottom of triple bottom line". *Business Ethic Quarterly,* 17(1), 105-110.

Peattie, K. and Crane, A. (2005). Green marketing: Legend, myth, farce or prophesy? *Qualitative Market Research: An International Journal,* 8(4), 357-370.

Porter, M. (1991). America's green strategy. *Scientific American* 264(4), 168.

Porter, M. and C. van der Linde, C. (1995). Toward a new conception of the environment-competitiveness relationship. *Journal of Economic Perspectives,* 9(4), 97-118.

Post, J. E. (1991). Managing as if the earth mattered. *Business Horizons,* 34(4), 32-38.

PricewaterhouseCoopers and SAM Group. (2006). The Sustainability Yearbook 2006. Insights from SAM's Sustainability Research and PwC's approach to Governance, Risk and Compliance. New York.

PricewaterhouseCoopers. (2016). Megatrends. Retrieved from www.pwc.co.uk/issues/megatrends.html.

Reinhardt, F. (1998). Environmental product differentiation: Implications for corporate strategy. *California Management Review,* 40(4), 43-73.

Ricardo, D. (1817). The works and correspondence of David Ricardo. In *On the principles of political economy and taxation* (Vol 1). London: John Murray.

Roome N. (1992). Developing environmental management strategies. *Business Strategy and the Environment,* 1(1), 1-24.

Salzmann, O., Ionescu-Sommers, A. and Steger, U. (2005) The business case for corporate sustainability: Literature review and research options. *European Management Journal,* 23(1), 27-36.

Sandmo, A. (2015). The early history of environmental economics. *Review of Environmental Economics and Policy,* 9(1), 43-63.

Schaltegger, S. and Wagner, M. (2006). *Managing the Business Case for Sustainability.* London: Routledge.

Schaltegger, S., Lüdeke-Freund, F. and Hansen, E. G. (2012). Business cases for sustainability: the role of business model innovation for corporate sustainability. *International Journal of Innovation and Sustainable Development,* 6(2), 95-119.

Schaltegger, S. and Burritt, R. (2015). Business cases and corporate engagement with sustainability: Differentiating ethical motivations. *Journal of Business Ethics,* 147(2), 241-259.

Schaltegger, S., Hörisch, J. and Freeman, R. E. (2017). Business cases for sustainability: A stakeholder theory perspective. *Organization & Environment,* 1-22.

Schmidheiny, S. (1992). *Changing course: A global business perspective on development and the environment.* Cambridge, MA: MIT Press.

Scott, W. R. (2007). *Institutions and organizations: Ideas and interests.* London: Sage Publications.

Seitanidy, M. M. and Crane, A. (2014). *Social partnerships and responsible business: A research handbook.* New York, NY: Routledge.

Shrivastava, P. (1995). The role of corporations in achieving environmental sustainability. *Academy of Management Review,* 20(4), 936-960.

Shrivastava, P. and Hart, S. (1992). Greening organizations. *Academy of Management Best Paper Proceedings,* 52, 185-189.

Smith, A. (1776). *An inquiry into the nature and causes of the wealth of nations.* London: W. Strahan & T. Cadell.

Stokols, D., Perez-Lejano, R. and Hipp, J. (2013). Enhancing the resilience of human-environment systems: A social ecological perspective. *Ecology & Society,* 18(1), 7.

United Nations, UN. (2016, 19 September). At forum, Ban stresses key role of business in securing way forward on SDGs, UN Sustainable Development Goals. Retrieved from www.un.org/sustainabledevelopment/blog/2016/09/at-forum-ban-stresses-key-role-of-business-in-securing-way-forward-on-sdgs/.

United Nations Environment Programme-UNEP. (1999). Global Environmental Outlook. Retrieved from www.unenvironment.org/resources/global-environment-outlook.

Waddock, S. A. and Graves, S. B. (1997). The corporate social performance-financial performance link. *Strategic Management Journal,* 18(4), 303-319.

Wagner M. (2003). The Porter Hypothesis revisited: A literature review of theoretical models and empirical tests. Retrieved from the Centre for Sustainability Management, University of Lüneburg, website www2.leuphana.de/umanagement/csm/content/nama/downloads/download_publikationen/38-2downloadversion.pdf.

Whelan, T. and Fink, C. (2016, 21 October). The comprehensive business case for sustainability. *Harvard Business Review*. Retrieved from https://hbr.org/2016/10/the-comprehensive-business-case-for-sustainability.

Williams F. E. (2015). *Green giants: How smart companies turn sustainability into billion-dollar businesses*. New York, NY: AMACOM, American Management Association, New York.

Winston, A. (2018, 19 January). Does Wall Street finally care about sustainability? *Harvard Business Review*. Retrieved from https://hbr.org/2018/01/does-wall-street-finally-care-about-sustainability.

World Business Council Sustainable Development (WBCSD). (2010, April 2). Vision 2050: The new agenda for business. Retrieved from www.wbcsd.org/Overview/About-us/Vision2050/Resources/Vision-2050-The-new-agenda-for-business.

Word Commission on Environment and Development. (1987). The Bruntland Report: Our common future. Retrieved from www.un-documents.net/our-common-future.pdf.

Annex 3

A.3.1: Case study resources

In order to improve the learning experience, we suggest the following case studies that address the topic of strategy and sustainability. The first one is about Royal Dutch Shell and it environmental, social and political responsibilities in Nigeria, where the regulatory vacuum calls for a new role for corporations. The second case describes the story of Chipotle Mexican Grill and its rapid growth strategy as a "fast-casual restaurant" based on sustainability, social responsibility and integrity. In particular the case offers the opportunity to analyze the links between investment/divestment decisions and sustainability at corporate level.

- Hennchen, E. and Lozano J. M. (2012). *Mind the gap: Royal Dutch Shell's sustainability agenda in Nigeria*. Case Writing Competition 2012, Corporate Sustainability Track. Available at Oikos International Homepage website. Published by: ESADE Business School, Spain. Prize winner.

- Vijaya (Narapareddy). Zinnoury (2017), *A Burrito without integrity: Is this Chipotle for me?* Case Writing Competition 2017, Sustainable Finance Track. Available at Oikos International Homepage website. Published by: Daniels College of Business. University of Denver, USA. Prize Winner.

Another interesting case that links sustainability strategy, circularity with the business case of and for sustainability is "Umicore's transformation and the monetizing of sustainability":

- Leleux, B. and van der Kaaij, J. (2016). *Umicore's transformation and the monetizing of sustainability*. Reference n. IMD-7-1708. Available at The Case Center. Published by: IMD, Switzerland.

Finally, we suggest a new case that addresses the issue of divesting from fossil fuels and helps with illustrating the role of sustainable finance in combating climate change and driving companies toward more responsible behavior.

- Gödker, K., Oll, J., Sump, F. and Julia Frech, J. (2016). *The Case for divestment: Rockefellers' fortune?* Case Writing Competition 2016, Sustainable Finance Track. Available at Oikos International Homepage website. Published by University of Hamburg, Germany. Prize Winner.

A.3.2: Case study from this book

We suggest the case of Pinkton Hotels as a way to explore through a sustainability program how to green strategy. The Pinkton Hotels story allows to analyze the different typologies of business case and sustainability. In the box below, we provide a short version followed by a list of questions for discussion.

Pinkton Hotels goes green: case study

As Mrs. Lyte was drinking her coffee one morning, she thought about the huge responsibility she had. She was sure it was the right way to go but she couldn't avoid feeling the pressure; it was her first big decision since she had been appointed CEO. She had decided to transform the Pinkton Group into the first 100% eco-friendly hotel chain and it wasn't an easy task. Fortunately, she knew she had the right people to start implementing her plan. She had thought about it too much already; it was time to do something. She finished her coffee and headed out the door, more determined than ever.

The Pinkton Group

The Pinkton Group is one of the most renowned hospitality chains in the world, with more than 50 hotels in 30 countries and eight seafront resorts. Founded by a family of Italian immigrants in the 1960s, it started off as a small hotel in Orange County, California, and grew into a multinational business. The founding family had started a business in a town that survived on agriculture and had been able to bring new life and money to its people. However, the fast and sudden growth of the business had inevitably caused some changes and the values of environmental and social sustainability on which everything had been built had faded away.

When Mrs. Lyte was appointed as the new CEO of the Pinkton Group, her first decision was to bring the hotel chain back to its founding values with the aim of transforming it into the eco-friendliest hospitality chain on the planet. She knew it was a very ambitious mission but, at the same time, she was certain it was possible. She knew exactly who could initiate the project and set the example for the rest of the hotels: Mr. Golman had the determination, ambition and passion to successfully take action.

As a matter of fact, Mr. Golman was the general manager of the Red Maple, one of the Pinkton Group's hotels. It was a medium-sized hotel located in the Swiss Alps. The structure was relatively new, it had been built 25 years ago and could host up to 250 guests at full capacity. Last years' RevPAR (revenues per available room) were €67, almost €2 higher than industry average[1] and perfectly on target.

The hotel industry

Mrs. Lyte was extremely proud of these results; the hotel industry had become very competitive in the past years. Traditionally, the players could be divided into two large groups: branded and independent hotels. Depending on which of the two

---

categories a hotel belongs to, it can decide between four business models: franchised, managed, owned, and leased. However, borders are not so clear anymore: three big trends are changing the rules of the game and business model innovation is necessary to survive[2].

All stakeholders are being affected by technology-based transformation, evolving customer needs and a more dynamic competitive environment. With the advent of Airbnb as the first big player in the sharing economy, for example, competition has been radically transformed.

Among the evolving customer needs category, one of the strongest trends is that of increasingly responsible individuals. More specifically, as more consumers are choosing to engage in eco-conscious lifestyles at home, it is not surprising that this is reflected in their travel choices as well. As a matter of fact, eco-friendly travel options were projected to increase up to 36% in the past year and are estimated to continue growing.[3]

Pinkton and the greening trend

It was time for Pinkton to honor its original values and provide a higher sense of purpose to its employees, guests and communities. Mrs. Lyte lived eco-consciously herself and was well aware of the benefits going green would bring. Starting from the financial side of the story, making the hotels more sustainable would allow for costs savings in different areas (e.g. energy consumption). Reputationally, the opportunity was huge. Riding the wave of the eco-friendly travel trend would put Pinkton one step ahead of its competitors and would build a strong reputation for the chain, and also increase customer loyalty. Finally, the benefit to the natural environment and communities would be important, Pinkton could take an active role in saving the planet.

Obviously, there were costs to be considered as well. The initial conversion costs of new resources, the training costs for employees, consultancy fees . . . they would have to set up a new management and organizational system, cultural change had to be implemented, resistance would have to be overcome. However, Mrs. Lyte was sure a business case for sustainability existed and she was determined to prove it.

Road to sustainability

The goal was clear: making Pinkton Hotels the first hotel chain to be 100% eco-friendly. But the road was winding, a detailed program was needed.

The DearPlanet program was planned to be carried out in three phases in order to gradually introduce sustainability practices aimed at 100% eco-friendliness.

- **Phase 1, Simple:** The pilot plan phase would be carried out in one hotel (the Red Maple Hotel) through simple changes with a short payback period. This was where Mr. Golman took the lead.

- **Phase 2, Complex:** This phase involved other types of difficulties, the business case would have a longer payback period and sustainability would require greater effort and higher investments.
- **Phase 3, Integrated:** In this phase, sustainability is at the very core of strategy and every investment is valued from this perspective.

Phase 1

It can be argued that this was the most important phase of all: the chance to prove that a business case for sustainability existed and the benefits of introducing eco-friendly practices would outweigh the costs.

The idea was to develop a managerial plan to reduce the Red Maple Hotel's environmental footprint, which, in turn, could have bottom-line savings. Mrs. Lyte and Mr. Golman collaborated in defining the non-disruptive and cost-reducing operational practices they would introduce. In this initial phase, they decided to concentrate on improving the guest rooms with five main changes:

1. Substitute existing lights with LED lights.
2. Install low-flow shower heads and toilets.
3. Substitute disposable body soaps with bulk dispensers.
4. Swap plastic cups used for coffee with reusable mugs.
5. Introduce the Project Planet Program[4] (encourage guests to reuse towels and linens by providing them with pillow and towel cards).

After the initial investment was absorbed, a decrease in costs was registered. The largest cost decrease came from the utility bills, especially thanks to the changes in lighting, water flow, and linen washing. The impact of bulk dispensers and reusable mugs was harder to measure but cut back room waste by almost 20%. In addition, guest reviews were enthusiastic about the DearPlanet program and this fostered an improvement in loyalty and reputation, especially online.

The success of this first phase served as a template for the introduction of the DearPlanet program to the rest of Pinkton's hotels. It was a clear example of a Business Case of Sustainability where performing well in environmental and social issues led to a situation of economic success itself.

Phase 2

Mrs. Lyte knew Mr. Golman would not have let her down. She was proud of him and decided to involve him in the next phase of the program. Red Maple Hotel would serve as the pilot location again but, this time, deeper action and personnel involvement were needed.

Mr. Golman put together a small sustainability team that would work on defining and implementing change. They decided to work on three main pillars: energy saving, raw material sustainability, and staff training.

1. **Energy saving:** The hotel switched to a 100% renewable energy provider, sky-lights were installed in common areas in order to have more natural light during the day, new energy-saving air conditioners were placed in the rooms, presence detector sensors were installed in all bathrooms, and a solar heating system was introduced for the swimming pool.
2. **Raw material sustainability:** New organic-cotton bed sheets, curtains and towels were introduced; eco-friendly soaps and detergents were given to the staff; and all paperwork was printed on recycled paper (the amount of paper was also dras-tically reduced with the introduction of digital booking solutions). In addition, all wooden furniture was replaced with new furniture coming from certified suppliers and walls were painted with a new, 100% natural paint.
3. **Staff training:** Employees went through special training sessions in which they were instructed to follow specific directions. Amongst these, they learned how to recycle properly, how to check for leaks or running faucets/toilets/showers and how to turn off any unnecessary lights they found on. In addition, they were given the chance to suggest further improvements they deemed suitable.

Staff involvement and training proved fundamental for the implementation of all the changes. A lot of time and resources were invested in getting them on board and on explaining why change was necessary and how, in the long term, it would bring pos-itive results also from a financial point of view. Mr. Golman knew that involving them from the very beginning and making them feel part of the decision and transformation would help them collaborate and avoid resistance. With a few exceptions, this is exactly what happened.

The implementation of the other two pillars asked for large financial investments as well as a temporary closure to the public. Identifying appropriate and reliable sup-pliers was the first big obstacle to overcome, the sustainability team had to write out a criterion on which to rate the different options and, finally, make a decision. In order to keep the hotel closure as short as possible large restorations were made first such as furniture change, wall painting, air conditioners, and presence detectors. Once these were installed, the rooms were ready to host guests again and the rest of the changes could be done while the hotel was operating.

After almost one year, the Red Maple Hotel had reduced its carbon footprint by 36%, its energy consumption by 28%, and it was 100% eco-friendly in terms of energy, materials, and recycling. It had achieved the LEED and Green Seal certification, and had been admitted to TripAdvisor's green leaders program, boosting its reputation and cus-tomer loyalty. In addition, its employees were enthusiastic about the changes, retention rate improved and a monthly sustainability meeting was introduced for the proposal of new improvements and the evaluation of current practices. Furthermore, a network of "eco-champions" was created and best-practices were shared with employees of all other Pinkton Group hotels. Phase 2 is a clear example of a Business Case for Sustainability.

Phase 3

Mrs. Lyte had been successful in proving that a solid business case for sustainability existed. The best practices developed and implemented in the Red Maple Hotel had gradually been introduced in all the other hotels of the Pinkton chain. In addition, sustainability had been integrated into the groups' culture. A formal statement of values had been laid out and a formal management system with clear and relevant goals and targets had been introduced both at the hotel and at the chain level. The small sustainability team which had been created at the beginning of Phase 2 had been expanded and a whole sustainability department now existed and was responsible for progress measurement and employee training and rewarding.

The DearPlanet program had been a success up to now and Mrs. Lyte felt it was time to start phase 3 and build the first 100% eco-friendly hotel from scratch. Sustainability was now at the very core of strategy and each decision would have to be valued in this perspective. Fortunately, Pinkton Group already had a strong and trusted network of suppliers on which it could rely on for guaranteed sustainable materials, but a lot of hard decisions had to be taken. The goal was to optimize the use of the environment rather than working against it.

Which was the best location? Would they have to rely on expert project designers or could they do it themselves? How many solar panels were needed to be self-sufficient in terms of energy? Was that even possible? Could they partner with local farmers to make the hotel's restaurant 100% sustainable and serve only local food? Could they make the hotel completely paperless? Which type of green roof should they build to insulate the structure best?

Making the right decisions was fundamental: green buildings use on average 26% less energy, emit 33% less carbon dioxide, use 30% less indoor water and send 50%–75% less solid waste to landfills and incinerators[5].

Next steps

The road to sustainability was still long, but Pinkton was on the right track. With the construction of this new hotel they had increased chances to win the National Geographic tourism award and to tap into a niche market which was still not crowded. Pinkton was already the market leader in terms of eco-conscious tourism with all of its hotels holding gold level LEED certifications and it had the potential to do even better. The business case for sustainability existed and was slowly paying off, giving it the chance to make new investments such as the amazing new hotel it was building.

Mrs. Lyte was already thinking ahead. Once the new hotel was ready, why not build electric car charging units in the hotels' parking lots? And why not start offering sustainability classes to its customers? She knew customer education was fundamental; maybe she could find a way of contributing to this . . . or maybe she could start offering eco-conscious travel packages by partnering with local transport companies?

Opportunities were endless; she had to make sure everyone stayed committed and enthusiastic. Fortunately, she was not in this alone.

1 Average Global industry RevPAR = €65.39 according to IHG annual report 2016.
2 IHG trends report 2017.
3 Booking.com report 2017.
4 www.greensuites.com/Environmentally-Friendly-Hotel-Programs/Project-Planet-Program.
5 US Green Building Council.

Source: Author's elaboration on the basis of the Kimpton case and access to different company sources – Kimpton Hotels Balancing Strategy and Environmental Sustainability. Murray Silverman and Tom Thomas. Case studies on sustainability management and Strategy. The OIKOS collection (edited by J. Hamschmidt) Greenleaf Publishing Book (2007)

Questions for discussion

(1) How important is the "business case for sustainability?" Is it necessary to justify the implementation of the DearPlanet program? What costs and benefits exist?
(2) In your opinion, does Pinkton's DearPlanet program involve any potential risks to their business model?
(3) To what extent does the DearPlanet program have marketing value? Would you actively promote it and how?
(4) How would you measure the results of the DearPlanet program?
(5) Would you require each potential product to stand on its own, meeting that criteria that it cost no more than the existing products? Or would you treat it as a whole (entire program must be cost neutral)? What are the pros and cons of each approach?
(6) How would you institutionalize eco-incentives such as this one? How can Mrs. Lyte make sure the program continues and enthusiasm is not lost?

3 The "Business In Nature" concept

Learning objectives

- Examine the interdependence between businesses and social-ecological systems.
- Understand how firms operate in social-ecological systems.
- Analyze the difference between impacts and effects.
- Build the theoretical foundations and principles for the "Business In Nature" (BInN) concept.
- Explain the relation between strategy and sustainability.
- Define a set of corporate sustainability strategies.
- Frame four basic approaches that can guide managers in rethinking the value chain.
- Provide a common language to bridge ecology and business disciplines.

Chapter in brief

This chapter introduces and discusses a new conceptual framework, "Business In Nature" (BInN), a new perspective in tackling the Anthropocene challenges and dealing with corporate sustainability today. This framework transfers concepts previously examined, such as impact/effects, complexity, footprint, resilience and adaptability into the business perspective. Moreover, it provides the new baseline to guide companies in embedding sustainability into strategy.

Interpreting the relation between sustainability and corporate strategy is not an easy task. In the real world of business, though the BInN concept shows compelling conditions for companies, there are several different positions where few firms have fully incorporated sustainability, and still many others do not acknowledge the relevance of environmental and social issues in strategic thinking. In order to better understand this complex mosaic, in this chapter, we discuss how companies embed sustainability. Further, we elaborately consider the competitive dimension of strategy, and we introduce a set of sustainability strategies that companies can develop to manage environmental and social issues. Finally, we briefly introduce a framework that helps identify and understand the multiple options available at the value chain level to tackle sustainability. Essentially, BInN calls for enlarging the boundaries of action along the entire value chain, challenging companies to rethink the way they do business from the sourcing of raw materials to the end of life of the product. We propose the following four basic approaches: (1) making production systems sustainable; (2) managing sustainable supply chains; (3) designing sustainable products and services; and (4) innovating business models for sustainability. These four approaches will be analyzed in-depth in the forthcoming chapters of the book.

Introduction

Back in the 1960s, Garrett Hardin wrote his famous essay "The tragedy of the commons" (Hardin, 1968) in which he stated that the degradation of the environment had to be expected whenever many individuals use a scarce resource. To make his point, he used the example of a pasture that is open to all who want to bring their animals to graze on the grass available there. His work was one of the first calls for sustainable resource use and integrated thinking in the management of the commons. He started to see the initial signals that years later would develop into the globalization of environmental challenges and the mainstreaming of environmental values. Today, the globalization of environmental problems in the form of an unforeseen environmental crisis (both non-resilient social and ecological systems) forces us to reformulate the tragedy of the commons. If you use the atmosphere in place of "pasture land" and carbon dioxide (CO_2) in place of "grass," with the result comes another type of a prisoner dilemma – one that explains why completely "rational" individuals may not cooperate even if it appears to be in their best interests to do so (Dawes, 1975; Soroos, 1994). It is argued that all countries and/or companies would benefit from a stable climate, but there would be winners and losers depending on the response and then, countries and/or companies could often be hesitant to curb their emissions. In addition, uncertainties could arise that may delay actions to address this and other environmental questions. The problems would remain and the environment would be degraded gradually.

Several years after Hardin's essay, Elinor Ostrom, the only woman to win a Nobel Prize for Economics (2009), wrote her highly influential book *Governing the commons. The evolution of institutions for collective action*. In this book, she presented an analysis of economic governance for the commons, and the points she made are still very valid today.

> What one can observe in the world, however, is that neither the state nor the market is uniformly successful in enabling individuals to sustain long-term, productive use of natural resource systems.
>
> (Ostrom, 1990: 1)

We need to change this assumption. Ostrom popularized the need for collective action (that we can frame as collaboration in business jargon) as a way of dealing with the problem of the commons and the management of common-pool resources in a general and/or particular manner. In the move for governing these common pool resources in the best possible manner, Ostrom asked for a distinction between the resource system itself and the flow of resource units produced by the system, recognizing the clear interdependence between these two as a carrying capacity for the entire system. Today, we recognize that the resource system itself is degrading, similar to its flow of resources (see Rockström et al., 2009 in Chapter 1). The large interdependence that this degradation has for the prosperity of the social capital of the earth puts this ecological condition at the risk of potential collapse and other unprecedented consequences. Hence, managing this natural capital, our common-pool natural resources (atmosphere, oceans, aquifer systems, forests, etc.) are critical for human development and survival. As we saw in the prologue, a general claim, a warning to humanity, has been recently issued by thousands of scientists. Firms cannot escape this. They must integrate the protection of the commons as part of their corporate responsible behavior. Future strategies in these social-ecological landscapes must acknowledge the interdependence of businesses and nature moving into desired positive impacts. As Paul Polman (CEO of Unilever) stated in his interview with the *Harvard Business Review* in 2012, "Why don't we develop a business model aimed at contributing to society and the environment instead of taking from them?" (Ignatius, 2012: 12).

Welcome to the 21st century.

Companies in social-ecological systems

Framework for analysis

As we saw in Chapter 1, we have entered the Anthropocene era. This era is characterized by a global environmental change, which is a consequence of a myriad of interactions and interdependences between human societal developments and the earth's natural capital. Global environmental change is not restricted to climate change and greenhouse gas emissions, and cannot be merely understood in terms of a simple cause–effect paradigm. Recent studies of the earth's land surface, oceans, coasts and atmosphere, biological diversity, the water cycle, and biogeochemical cycles make it clear that human activity is generating changes that extend well beyond natural variability. In some cases, alarmingly so, and at rates that continue to accelerate. These changes can cause the earth's system dynamics to reach critical thresholds and as a consequence, we could observe abrupt changes. Global environmental change research over the last decade shows that the earth's system is currently operating well outside the normal state exhibited over the past 500,000 years (ESSP-Future Earth, see Annex 1 to this chapter; Steffen et al., 2015). This has multiple implications for humans' well-being; basic goods and services supplied by the planetary life support system, such as food, water, clean air and an environment conducive to human health are increasingly affected by it. Managing the planet within its resource limits and planetary boundaries, and moving toward Sustainable Development is a very urgent mandate.

What if? What if – in the end – we no longer have a planet to live on, as Al Gore (45th Vice President of the United States of America) suggests in his stunning lesson on global warming in the 2006 documentary "An inconvenient truth" or in his book *Earth in the balance* (Al Gore, 1992)? What if we have a planet that is not able to adjust to human population and/or its diversified societies? What about these societies? What about businesses in these societies? There are too many questions and choices that can vary enormously between people with different values and cultures who are also living in different realities under the same global umbrella, our planet earth. The IPAT formulation (see Chapter 2) initially aimed to equalize, through time, three main factors related to human activities – population (P), affluence (A) and technology (T) – with the impact (I) caused by humans in order to maintain a functional nature. Although we can view this as a positive fact, today's planetary trends are showing us that an equation that globally only equalizes the impact of human activities on the planet is not sufficient. It is not acceptable any longer, not even in the near future. Today, we need to drastically reduce such a global impact. What type of a planet do we want? What type of a planet can we get? What about stability and future? What about peace and prosperity? What about people? The IPAT formulation must be related to an impact (I) targeted as acceptable (within functional limits) by the state of nature in which we want to live our future and it should become a part of our collective decision, the acceptable natural capital. Defining the state of this natural capital by effective targets and adjusting our activities to these targets is a human mission for decades to come; something far ahead of what we have today.

In Chapter 1, we recognized that a framework for analysis that separates the human and the remaining natural environment is not adequate for dealing with our present environmental problems. We opted for using the term social-ecological system in order to find the correct approach to manage these relations and interdependences. On the one side of these systems is the social sub-system, which is formed by us humans (human capital) and what we do (human-derived capital), and on the other side is the ecological sub-system, its structural units and their functions (altogether completing the natural capital). Between the two sub-systems are their interactions. Figure 3.1, from left to right, shows all the pressures and pulses we produce on the ecological sub-system. From right to left are all the flows of natural resources we use as humans in a tangible and intangible manner, the flow of ecosystem goods and ecosystem services. Within this particular diagram, a company must find its position inside the social system and must acknowledge its relation and dependences with the other components of the system.

Social-ecological system accountability: impacts vs effects

Social-ecological systems can be analyzed through different accounting frameworks. From a socio-economic point of view (see Figure 3.1), human activities can produce either negative or positive <impacts> on the ecological sub-system that can translate into negative or positive <effects> observed in the ecological world that can alter its structure and functionality. These effects can produce <consequences> (secondary effects) that change the way in which people are benefiting from the ecological sub-system, forcing us to <respond> and change the way we do things. Social sub-systems are normally monitored through a multitude of economic indicators, such as gross domestic product or household and/or personal disposable income. A human being's impacts coming from these social sub-systems are normally analyzed through a large set of impact indicators commonly used by policy makers in environmental, strategic, and/or regulatory impact assessments, and environmental standards. The issue regarding the development of indicators for the ecological sub-system

SOCIAL-ECOLOGICAL SYSTEM

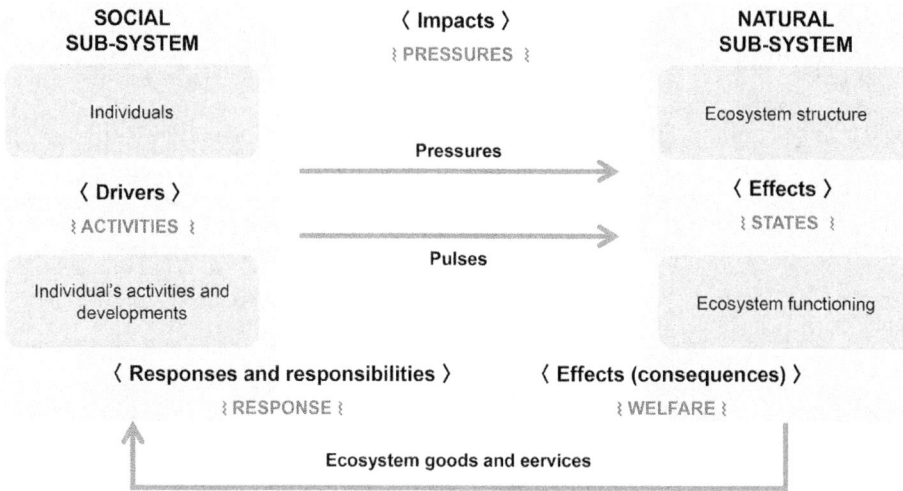

SOCIAL SUB-SYSTEM	⟨ Impacts ⟩ ⟨ PRESSURES ⟩	NATURAL SUB-SYSTEM
Individuals		Ecosystem structure
	Pressures →	
⟨ Drivers ⟩ ⟨ ACTIVITIES ⟩		⟨ Effects ⟩ ⟨ STATES ⟩
	Pulses →	
Individual's activities and developments		Ecosystem functioning
⟨ Responses and responsibilities ⟩ ⟨ RESPONSE ⟩	⟨ Effects (consequences) ⟩ ⟨ WELFARE ⟩	

Ecosystem goods and eervices

Figure 3.1 Social-ecological system diagram showing interactions between sub-systems. Source: authors' elaboration.

is much less advanced, but it considers the definition of the good environmental/ecological status that is desired for a particular "common resource" (atmosphere, oceans, biodiversity, soil, freshwaters, etc.). Finally, the benefits obtained by humans from natural systems can be assessed using market indicators for marketed natural goods or assessing ecosystem service valuation accounting for household and/or personal disposable natural value (Sardá, 2013).

Recently, Cooper (2013) proposed another accounting indicator platform to deal with social-ecological systems from a significantly more natural-base perspective, the Driver-Pressure-State-Welfare-Response (DPSWR) framework, an evolution of the previous Driver-Pressure-State-Impact-Response (DPSIR) framework (European Environmental Agency-EEA, 1999). These platforms were developed to adequately organize information for each environmental issue of the interrelation between the human and ecological sub-systems (Figure 3.1). In this framework, social systems (people's capabilities and their activities) become ⟨drivers⟩ of change (D). They ⟨pressure⟩ constantly or in pulses the natural related systems (P). Those ecological sub-systems (structural units and functions they made) can alter their ⟨states⟩ (S) that in turn can translate into the degradation of fundamental natural resources used by humans (natural goods and ecosystem services), thus diminishing human ⟨welfare⟩ (W). The recognition of such degradation should induce humans to make adequate policy ⟨responses⟩ (R) to solve the pattern of accelerated degradation. The information generated through the DPSWR framework expressed issues in a highly inter-related form, which is something that we cannot observe if we use other sets of indicators that regard different pieces of the social-ecological system puzzle in isolation (see Table 3.1).

Table 3.1 Terminology and definitions to describe interdependencies between social and ecological sub-systems.

Policy (impact/effect)	Socio-Ecology (DPSWR)	Policy - Socio-Ecology Explanation
Activities/ associated aspects	Drivers	An activity or process intended to enhance human welfare.
Impacts	Pressures	A means by which at least one driver causes or contributes to a change in a state.
Effects	States	An attribute or set of attributes of the natural environment that reflects its integrity as regards a specified issue (or change therein).
Consequences	Welfare	A change in human welfare attributable to a change in state.
Response	Response	An initiative intended to reduce at least one impact (state or welfare change).

Source: authors' elaboration adapting Cooper, 2013.

From the previous illustrations emerge the necessity to understand both languages describing the same phenomena and its affected components with different terms. The BInN concept aims to clarify different words used for the same term coming from natural or social sciences. Combing these two different languages of analysis is basic for a common understanding of the terminology used to express the interactions and interdependences between the two sub-systems of a social-ecological system (Figure 3.1; Table 3.1). Our activities in the social sub-system (drivers of change) are impacting the ecological sub-system (pressuring it) and producing several effects (changes in the state of nature) and consequences, modifications by which we benefit from natural resource flows (our welfare) that ultimately will require social responses or will translate into responsibilities, but also into large depletions of these granted resources. We still need to find a magic formula to solve the problem of the degradation of our natural capital, our "commons." It is necessary to gain a clear understanding of the importance of this capital and to find adequate ways to maintain it functional, in order to ensure the future provision of goods and services. A clear vision is needed. In addition, all relevant actors, including companies, are responsible and should act accordingly.

A second clarification that needs to be undertaken to correctly understand all this jargon is to provide clarity on the distinction between what are a company's environmental impacts and their associated effects. In general, these two words, used for many applications, are often used interchangeably. However, they are rather different. When used in the context of business and environment, there may be clear differences, but there is significant confusion in the literature. For the purpose of this book, environmental impact is the action we (as a company) undertake physically on the environment (e.g., it can be observed as the amount of physically destroyed forest, the amount of extraction of something from nature, the amount of wastes dumped into the environment and the amount of gasses emitted into the atmosphere or effluents into the waters). On the other side, an environmental effect is a change or result of the previous impact (e.g., loss of biodiversity in a forest, loss of hectares of a natural environment and alteration of the concentration of pollutants in a particular environment). The effects, understood as changes in the states of nature, can produce some environmental consequences (secondary effects of those changes) to refer to the issues resulting from these changes. Box 3.1 provides an example of climate change and relates it to the functioning of the environment and its natural processes.

Box 3.1   Climate change as an example for environmental jargon

In its Global Risk Report 2018 (an annual base survey on global risk perception for the World Economic Forum of Davos in 2018), climate change (risks of failure in adaptation and mitigation, extreme weather events, and natural disasters) escalated to the top global risk for the coming decade (see Chapter 1 Annex 1 for links). Climate change is the result of the change in the composition of the global atmosphere and natural climate variability is having strong consequences due to its interconnection with many other issues. Scientific evidence for warming of the climate system is unequivocal: "The current warming trend is of particular significance because most of it is extremely likely (greater than 95 percent probability) to be the result of human activity since the mid-20th century and proceeding at a rate that is unprecedented over decades to millenniums" (IPCC, 2014: 4). Directly or indirectly, climate change is the consequence of a large economic failure, one of the largest negative externalities produced by human actions; its causes and consequences are global, its impacts and effects will be persistent, and its derived uncertainties and risks are large.

Climate change is any long-term significant change in the "average weather" of a region or the earth as a whole. It is produced by the process of global warming, an increase in the average temperature of the earth's near-surface air and oceans that has occurred since the mid-20th century and is projected to continue. Most climate scientists agree that the main cause of the current global warming trend is human expansion of the "greenhouse effect," the change in the thermal equilibrium temperature of the planet by the presence of an atmosphere containing gas that absorbs and emits infrared radiation. A different set of natural and artificial gases produce this effect, the most important contributors being CO_2 and methane.

For all the above reasons, we will make use of the case of climate change to explain various terminological aspects discussed previously. See Figure B3.1 in this box.

Today, the engine of the "production/consumption machine" that runs human activity on the planet is 80 percent fueled by energy that is obtained by burning fossils fuels. From these burning activities, in 2017, global emissions from fossil fuels and industry were computed as 36.8 $GtCO_2$ equivalents. If you add other human activities such as agriculture, forestry and livestock that emit other greenhouse gasses, the result is a number between 45 to even 50 $GtCO_2$ equivalents. This is considered the total annual accumulated **impact** of our man activities.

These gases move into the atmosphere. Simultaneously, some of these gases are trapped by natural sinks, such as forests and phytoplankton, but unfortunately today, the rate at which they are trapped are no longer the same as the rate at which we put them into the atmosphere. Once upon a time, before our industrial revolution, both processes (gas emissions and gas absorptions) were in balance, but now CO_2 from burning fuels and forests is rapidly increasing almost twice as fast as it is being absorbed by plants and the ocean (approximately 54% and 46%, respectively) and it could be worse as scientists are starting to believe that some of those "sinks" are in fact getting saturated. This is why the arrow in the figure is going up (more percentage of emitted annual gasses remain in the atmosphere every year). The **natural process** created through thousands of years of the planet's evolution has been altered.

The "greenhouse effect" produced by this alteration (change in the thermal equilibrium temperature of the planet due to the increased presence of gases in the atmosphere that absorb and emit infrared radiation) has different effects as there are changes in the state of the atmosphere and other physical related aspects. There are three very important **effects**:

(1) The concentration of CO_2 equivalent in the atmosphere. The most important heat-trapping (greenhouse) gas has moved into 407.62 ppm (January 4, 2018) [the recommendation from scientists was that it should not go beyond 350pp (Rockström et al., 2009)] and is increasing (see the arrow in the figure).
(2) The global average temperature for the earth is currently increasing at a rate of 0.3ºC Celsius every 10 years and will continue to increase (see the arrow in the figure). Global temperatures have increased by 0.97ºC since the start of the industrial revolution and 17 of the 18 warmest years on record have been in the last 20 years.
(3) Radiative forcing, the difference between insolation absorbed by the earth and energy radiated back to space, increased positively to 2.29 watts m^{-2} since 1750. This is increasing at a rate of 0.33 watts m^{-2} a decade (see the arrow in Figure B3.1).

    Other effects not analyzed here include sea level rise, 3.2 mm per year; arctic ice minimum, 13.2 percent decrease per decade; land ice, 286 gigatons per year. See NASA Vital Signs for details (Chapter 1 Annex 1).

The changes in the state of nature translate into different secondary effects – the consequences that we and the rest of species of the planet need to deal with. These are the **consequences** of climate change. There is a large list of consequences such as droughts, floods, extreme weather events, wildfires, wildlife exposure, human food security and appearance of past illness.

Finally, understanding climate change as a serious threat for humanity, United Nations made a call through the United Nations Framework Convention of Climate Change (UNFCCC) to deal with this issue. The long-term series of meetings, conferences, and other agreements such as the Kyoto Protocol (1997), has finally resulted in a new policy **response**, the "Paris Agreement" document (2015) that will enter in force in 2020. In this document, a global response is provided in its article 1 ("holding the increase in the global average temperature to well below 2°C above pre-industrial

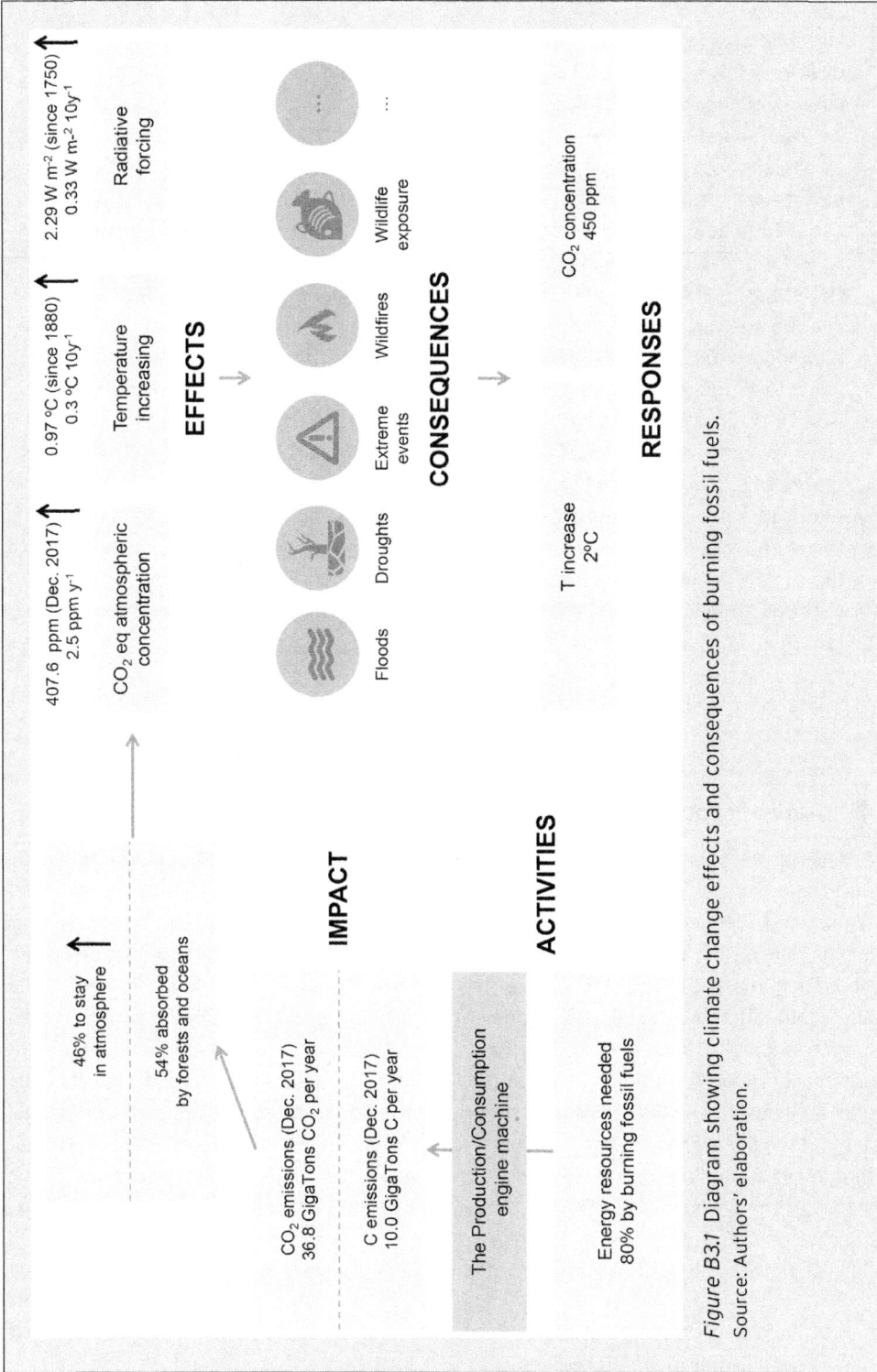

Figure B3.1 Diagram showing climate change effects and consequences of burning fossil fuels.
Source: Authors' elaboration.

levels and pursuing efforts to limit the temperature increase to 1.5°C above pre-industrial levels . . ." (UNFCCC, 2015: 2) to no more than 450 ppm of atmospheric CO_2 equivalent concentration.

The framework (DPSWR) organized the information related to climate change in a clear manner showing the interdependences between both sub-systems. It also shows us how numbers increase. If CO_2 equivalents increase by 2.5 ppm per year in an exponential manner and we do not reach the peak of emissions and reduce them drastically during the coming years, we will cross the agreed 450 ppm in less than 15 years. However, sometimes, even if numbers seem clear, the natural process does not follow our human logics. In fact, social-ecological systems are complex adaptive systems, and they follow non-linear and uncertain evolutionary patterns. As such, we need to be ready for future uncertainties. In an excellent paper (John Sterman comment for the *New York Times* on 29 January 2009 on Solomon et al., 2009 paper), the author has given us some thought for reflection: "Our mental models suggest that if we stop the growth of emissions, we will stop global warming, and if we cut emissions, we'll quickly return to a cooler climate. We tend to think that the output of a process should be correlated with (look like) its input. If greenhouse gas emissions are growing, we think, the climate will become warm, and if we cut emissions, we imagine that the climate will cool down. In systems with significant accumulations, however, such correlational reasoning does not hold." We may need to change our way of thinking too.

Source: authors' elaboration from different sources.

Companies in social-ecological systems

Businesses develop within social systems, which are becoming increasingly more volatile, uncertain, complex and ambiguous (VUCA) as they have been defined in the VUCA world (Wolf, 2007; Bennett and Lemoins, 2014). Social systems, as illustrated in Chapter 1, are intrinsically nested at different scales (a city is evolving inside a region, the region inside a watershed, the watershed inside a country, or more, the country inside a continent and so on) within different ecological environments. Social and ecological systems are linked and interconnected through a variety of flows (energy, materials, information, and knowledge) and feedback loops. Businesses are an intrinsic part of these social systems, but they also play a pivotal role within ecological systems. As businesses operate and grow, they interact and modify the ecological functioning at different scales in a nested manner. The fact that the social and ecological systems are intertwined and nested at different scales, further increases the system complexity and uncertainty, limiting our capacity to foresee the consequences of business actions on nature (see Figure 3.2 for a diagram).

In all these scales, businesses interact with the ecological sub-systems in two ways:

- companies pressure those systems in various forms (e.g., pollutants and waste), therefore modifying their natural "states",

- they receive and consume the flow of the natural resources-ecosystem goods and services-that they use for their operations.

Ultimately, companies can be affected by what happens in ecological systems at the specific scale level where they operate or as a result of the interconnectedness of the scales. The first is the case of a firm that makes intensive use of fertilizers and pesticides in agriculture; in the long-term, this massive use of chemicals can impact soil fertility, reducing the overall land productivity. The latter is the case of a firm affected by climate change. Climate change occurs on a global scale, but extreme weather events can impact business activities at a local level, in regions that are particularly exposed to droughts or heavy rains. Similar to the social context, the ecological context is also becoming more uncertain, more unpredictable, more volatile, more degraded, and the implications of this trend on the global economy can definitively force social conditions to be altered.

Businesses' long-term success depends on the set of resources on which they operate (economic, social and ecological) and today, as previously highlighted, some of these resources can no longer be taken for granted. Natural resources are a very special case. They have specific properties that differentiate them from other types of resources and that often make them irreplaceable or not substitutable. They can be classified as renewable or

NESTED SOCIAL-ECOLOGICAL SYSTEMS

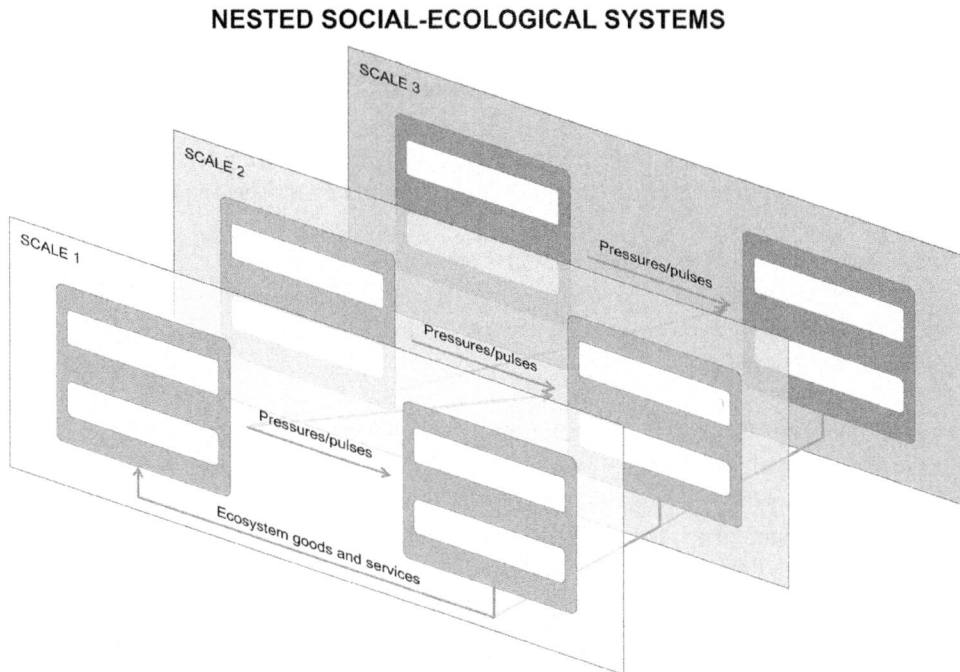

Figure 3.2 Interconnectedness of social-ecological systems at multiple scales.
Source: authors' elaboration.

non-renewable, but when those resources are consumed at a rate faster than the rate of replacement (quantity), or when they are rapidly degraded (quality), the natural resource stock (particular units) can diminish and eventually extinguish with the consequent loss of their ecological functions, and therefore of their services.

If ecological conditions are not maintained owing to our "business as usual" production and consumption patterns, the consequences can be dramatic. Therefore, we need to develop innovative solutions that modify the way we do things today to ensure that social-ecological systems will become resilient again. As Jason Clay, Senior Vice-President of WWF said: "[W]e need to begin to manage this planet as if our life depended on it, because fundamentally, it does" (Elizabeth, 2017). To conclude, no matter which social-ecological system we operate in, we must drastically reduce the impact of our activities (production and consumption patterns) to be able to maintain the resilience of the ecological sub-system (structure and functions) and to ensure the provision of the ecosystem goods and services we need. Understanding the limits and the carrying capacities of the social-ecological system is a basic need for every company that aims at long-term survival.

Business In Nature

Doing business and doing good

In the last decades, despite the efforts that companies have put in "greening" their operations, their overall ecological footprint has increased with a corresponding growth in the human ecological footprint on the planet. Applying the IPAT equation (see Chapter 2) to the business world, even if a company improves the efficiency in its "T-factor" term, this can be offset by an increase in its activities due to their growth (A-factor) or by the fact that other companies may join this company in its particular sector of activity (P-factor). As a result, in absolute terms, the entire impact (I-factor) continues to rise. If you double the efficiency, but simultaneously are able to double your sells, the environmental problem generated is the same, but it can become even worse as the effects accumulate into the natural environment. This process, known as the "rebound effect" (see Grubb, 1990, as an example with regard to energy), is dramatically limiting many of our efforts to reduce our ecological footprint, and hence calls for more radical actions. Therefore, there are two key questions that need to be addressed without delay:

- To what extent are current environmental management practices compatible with ecological sustainability?
- Moreover, to what extent is the reduction intensity in the I-factor of the equation in line with the maintenance of ecological integrity and its resilience?

There are no clear answers to these questions, but large reductions in the I-factor basically become an imperative mandate for the years to come and put pressure for radical changes in the manner that we produce and consume. In other words, we must rethink the way we are doing business, while contributing in a positive manner to the society without damaging our natural capital. Basically, we need to develop a different method to drive our gigantic production and consumption "machinery" so that it ensures the resilience of the social-ecological systems in an acceptable, functional manner.

After years of absolute dominance (Crouch, 2011), the mainstream business model based on a narrow focus of monetary results, strong short-termism, and a disruptive competitive approach which benefits few (particularly financial investors and top managers) at the expense of many (comprising society, local communities, ecosystems and ecosystem services, and future generations) is facing much criticism (Ghoshal, 2005; Stiglitz, 2012; Tencati and Pogutz, 2015). If we want to build a new and sustainable pattern of development in this century, business contribution is fundamental (Schmidheiny, 1992; UN Global Compact and ICC, 2015; Kiron et al., 2013). However, it needs to be aligned with the biophysical processes that are controlling our planet.

John Kotter, one of the leading scholars in the field of leadership on corporate change based at Harvard Business School, help us with understanding how businesses can address these challenges and contribute to a more sustainable society. His popular fable *Our iceberg is melting* (Kotter and Rathberger, 2006) about a penguin colony in Antarctica provides some basic lessons for managers and illustrates the way change needs to be managed globally. In the story, penguins have lived on the same iceberg for years. When a bird discovers that a serious problem is appearing – the ice is starting to melt – no penguin is ready to listen. They are fine with their lives and nobody wants to change. This fable is about acknowledging the urgency for a change. Similar to the case of penguins and the ice-melting, several signs of environmental crises are standing in front of us, calling for urgent changes in the way we do things. John Kotter's approach to change has identified a sequence of actions and factors that normally leads change to success (Kotter, 1995, 1996). Among these factors, creating a sense of "urgency" about the problem and putting people (leaders, managers, and employees) on the same page are fundamental aspects to start the transformation. In other words, change is also a question of time as the "amount of change" that you need to fix the problem does not increase in a linear manner, but is exponential. In a video linked to this fable on changing and succeeding, John Kotter says:

> Change kind of fluctuates, maybe that's true over a millennium, but in our lifetime, when we think ahead one, two, five years, we are on this kind of curve [the exponential one] and it has huge implications because if you get the challenge of change right, the opportunities are unreal; and if you do not, if you go into denial, if you just do not want to deal with [it], if you do not know how to deal with because nobody has shown you their simple formulas, life can be really bad.
>
> (Kotter, 2011: 4.10 to 4.40, see also Annex A.1.2 to this chapter)

Concerning the relationship between business and the natural environment, this lesson from John Kotter shows us that it is not time to "go into denial," but it is time to find the "simple formulas" to make the right changes we need. Moreover, there is the question of urgency. We need to undertake this transformation quickly as our pattern of growth and ecosystem degradation follows an exponential curve (check out the concept of "great acceleration" highlighted in Chapter 1); therefore, the changes that we need to address the planetary boundaries' challenges become larger and need to be quicker as time passes. In Figure 3.3, we represent this challenge. The responses required to address a problem are significantly more demanding in t_9 than in t_4. At the business level, managers must understand urgent transformations in the way firms organize and strategize, starting with a new approach to corporate sustainability.

The mainstream economy pursuit by the business world is still characterized by finding short-term gains and by developing business competitive models where corporate success is practiced at the expense of nature, society, and future generations. This needs to be changed fast enough to preserve the earth's system dynamics. In the previous chapter, building on Dyllick and Muff's work (2015), we have observed that a business's approach to sustainability has mainly focused on the search for win-win solutions that are able to increase shareholder value (business sustainability 1.0) and on the triple bottom line framework (business sustainability 2.0). We also called for a much deeper change in the expected contribution of businesses to Sustainable Development. We need to move beyond the traditional language of economics and consumerism, and evolve businesses into a force that can positively contribute to the challenges we have ahead in relation to our natural environment. Until now, most firms that have approached sustainability have worked on the direct impacts of their activities to minimize the pollutants of production processes and products, improve energy efficiency, and reduce risks. Both the frameworks (corporate sustainability 1.0 or 2.0) have become more popular in management practice, leading many firms to undertake important changes in their organizational routines and simultaneously reducing the harm caused to social-ecological systems. However, the idea that firms can't grow indefinitely but have to respect natural carrying capacity and ecosystem dynamics as a potential limit, has never been

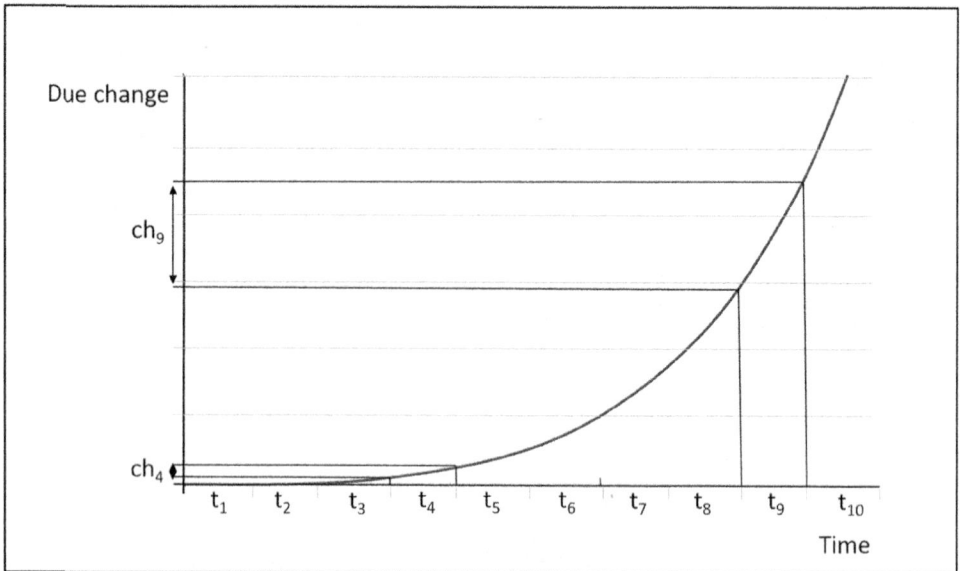

Figure 3.3 Implication of change of an exponential curve in the great acceleration.
Source: authors' elaboration.

seriously taken into account. As previously pointed out, in many firms the strong increases in volumes of production and sales have not been counterbalanced by their improvements in eco-efficiency. Today, we need major changes to address these challenges.

In other words, we have substantially integrated ecological issues into a "business as usual" approach instead of advancing a more radical transformation ensuring that the deep nature of business and markets can be maintained in a resilient social and ecological system. Although such a critique is considered unrealistic and naïve by many, a pragmatic shift in company targets from eco-efficiency to eco-effectiveness (Dyllick and Hockerts, 2002; Braungart et al., 2007) and toward a multiple-bottom-line perspective (Dyllick and Muff, 2015; Tencati and Pogutz, 2015) is necessary and may support the diffusion of more environmentally sustainable business models.

In a nutshell, the present environmental crisis (one of the huge megatrends of this century – see Chapter 2 in this book) has largely not been addressed by companies. Moreover, their responses have been mainly isolated and "egocentric." Instead of focusing on collaborative relationships along the whole supply chain (from raw material suppliers to consumers), many firms have concentrated their efforts on internal solutions aimed at reducing the direct environmental impact generated at the production site level. Furthermore, in many cases environmental risks have been approached as isolated, while pollutants coming from many different sources and with various impacts have different time and spatial scales, impacting the resilience of multiple ecosystems (Folke, 2006). As such, business practices do not work appropriately with the concept of leaving resilient social-ecological systems.

The issue of linking social-ecological systems to business organizations is of central importance for the analysis of almost any action related to Sustainable Development. Corporate sustainability calls for a new holistic approach (Whiteman et al., 2013; Winn and Pogutz, 2013; Tencati and Pogutz, 2015). Business units operate in the social-ecological paradigm though their involvement in nested social-ecological systems and they are forced to learn how to internalize the different needs of these systems at a local, regional, continental or global level to allow them to be resilient while developing their business operations. Considering this purpose, we are advocating for the application of a BInN view in line with the call of Dyllick and Muff (2015) on truly sustainable businesses.

The BInN concept calls for a profound socio-technological transformation. Businesses must engage and contribute to the development of a novel techno-economic cycle based on green economy, clean innovations, and responsible behavior. This systemic transformation also requires firms to create a new pathway with regard to production-consumption mechanisms. Planetary boundaries provide constraints to our current market structures and logics, and we must escape what Tim Jackson has defined as the "iron cage of consumerism" (Jackson, 2009: 9; Jackson and Senker, 2011). Acknowledging the interdependencies of social-ecological systems is paramount. We believe that businesses must play a leading role in activating a virtuous cycle that links the goal of economic profitability, with enhancing social capital (*Prosperity*) and drastically reducing the impact on ecological systems, both to maintain its functionality (*Effectiveness*), and to ensure its resilience and the provision of ecosystem goods and services (Figure 3.4). In other words, we must recognize the idea that not only the global production-consumption machine is fueling the world for development and welfare, but nature and ecosystems also contribute and constitute its ultimate pillar.

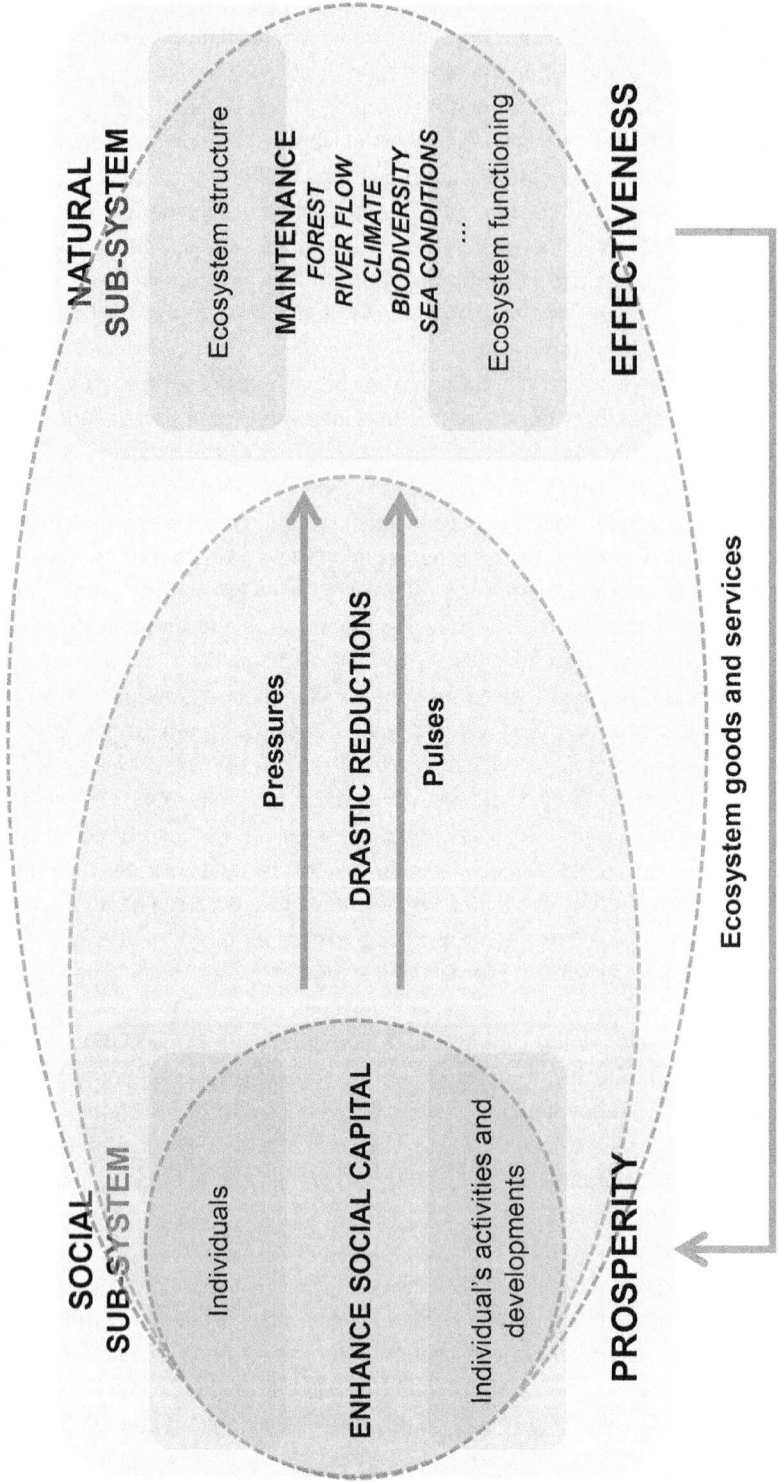

Figure 3.4 Social-ecological system and business transformation: ensuring prosperity and effectiveness.
Source: authors' elaboration.

As has been repeatedly said, it is not sufficient to wait for the prediction of the future and to adapt our institutions and behavior to it; today, we need to shape the future we want, a future that acknowledges the new epoch of the Anthropocene and emphasizes the idea of humans/business in nature.

In our call of action for BInN, corporations should meet their business goals without compromising the ability of natural systems to provide the natural resources and ecosystem services upon which our well-being and that of other living species depend. Building on these initial ideas, we define Sustainable Development in this century as:

> Enhancing social well-being and economic prosperity, while maintaining the integrity of natural systems and its potential for the provision of ecological goods and services.

Moreover, we frame the concept of BInN as:

> Creating a significant positive impact in societies as well as increasing the resilience of social-ecological systems.

In doing so, we provide a definition that suggests a new role for corporations in societies and in nature. We move from the concept of value maximization to that of positive impact, and by including the resilience of social-ecological systems, we provide a framework that connects organizations to the concept of planetary boundaries. To achieve these objectives, we should be able to transform the current prevailing worldview, the social values and norms, and economic logics, in order to favor a transition to a new vision based on resources' conservation, instead of conquering nature; on cohesion and collaboration, instead of harsh competition; on diversity, instead of standardization; on quality, instead of quantity; and on entertainment, instead of haste.

Now that we have provided a definition of our key concept, it appears equally important to highlight the four principles that represent practical implications and offer a path to operationalize BInN, helping companies to "walk the talk."

Extending the boundaries of intervention

The BInN view requires the adoption of an extended approach to analyze the relation between firms and social-ecological systems that recognizes the multiple scales of interactions. Firms should go beyond the traditional corporate boundaries in order to encompass all phases of the value chain (from extraction and cultivation of resources to the end of life of the products) as well as to internalize the interdependences with the social-ecological systems in which they act, either based on the pressures exerted or on the needs of the ecosystem goods (raw materials, water, energy, etc.) and ecosystem services (provisioning, regulating, supporting, and cultural).

Following this view, life cycle thinking is an essential competence for business organizations. Firms need to reduce the resource use and emissions/wastes flows to the environment, while improving their products' performance through the entire life cycle.

Moving from a focus on environmental impacts to a focus on environmental effects

Historically, companies have been controlled and regulated by the impacts they have through the use of environmental standards or cleaner production regulations (see Chapter 4 of this book). This issue has generated an impact-driven way of thinking in firms. The BInN view asks firms to not only sharply reduce the impact they generate on ecosystems, but also to stabilize or, better, reduce their targeted negative effects. In ecology, the expression "states of nature" refers to preserving good states of ecosystems (atmosphere, freshwater system, marine systems, forests, etc.) that allow these systems to maintain their integrity, functionality, and resilience. Maintaining the state of nature is also paramount for societies, individuals, and companies to be resilient and to flourish.

The importance of moving to this new approach with a focus on the effects is also demonstrated by the attention that policy makers are devoting to regulate these states of nature through regulations. During the last two decades, as it is almost impossible to directly link ecological effects with a particular driver through a causal–effect relationship (even a particular company), one of the most important responses we have observed from legislators (clearly expressed in the European context) has been the transition from a regulatory approach based on impact-driven policy instruments to a new one in which effect-driven policy instruments became the most important ones.

By developing a long-term vision (that corresponds to the global state) to be reached for a particular natural system as a whole (the condition of the atmosphere, the good marine environment, the good ecological quality of freshwater, etc.), policy makers affect companies' behavior and actions. Essentially, business organizations should plan accordingly and collaborate each other being aware not only about the impacts generated, but also about the undesirable effects of production processes and products. By introducing the BInN concept, we attempt to move organizations from impact-driven to effect-driven thinking.

Innovating products, services and business models for the future

The third principle refers to innovation. Environmental scientists and ecologists, such as Johan Rockström and Will Steffen (Rockström et al., 2009; Steffen et al., 2015a, b) have highlighted the importance of contrasting the "great acceleration" (see Chapter 1) bending the curve that is reducing the exponential growth that characterize many trends in our society, decoupling prosperity from the material consumption of nature. Achieving true decoupling will mean more than increasing gains in eco-efficiency owing to new clean processes and products, as discussed in the previous sections.

BInN calls for a transformative change in the way companies respond to the market demand, and this can be reached only when radically new generations of sustainable products and services are designed, launched, and diffused into the markets. However, radically new products and services may not be sufficient in a planet of 9 to 10 billion people. The type of major leap that is required is more substantial. Changes need to entail the way businesses approach the markets, how resources are extracted and combined along the value chains, how value is generated for consumers, and how value is produced for the company. This is about the development of new and sustainable business models. Business model innovation allows firms to radically rethink how goods and services taken from nature are utilized to produce products and services, and how the environmental values contained in these products and services are delivered to and reflected in customers' preferences. The business model perspective is, at the same time, cognitive, technological and systemic, and opens new trajectories for re-organizing our business logic. Moreover, it offers great opportunities to innovative and responsible companies that acknowledge operating in social-ecological systems and want to act as a positive force for a change toward a more sustainable economy.

Developing collaborative approaches

Due to the planetary scale of the environmental challenges, it is necessary to understand that these crises cannot be solved with the effort of merely a few actors. It is not about optimizing or maximizing the outcomes in isolation. Hence, the role of companies lies in influencing others and pushing broader transformation at the level of industries. A move from the dominant focus on competition to a focus on collaboration among peers and with stakeholders (e.g., clients, suppliers, NGOs, policy makers, and investors) is the last pillar of the BInN view. To fight environmental problems, innovative solutions need to be developed and diffused along the value chains, possibly on a global scale. This is why technology transfer and cooperation between firms and stakeholders is so important. For businesses, this idea emphasizes the importance of developing new business models that better fit the collaborative approaches, similar to the case of supply chain cooperation schemes, cradle to cradle design, and circular business models. We will illustrate these approaches in the forthcoming chapters.

Conceptual frameworks for analysis

The implementation of the BInN concept in a company should be based on the above-mentioned four principles: (1) a broader view of the firm based on an extended value chain approach with a clear life-cycle thinking view and taking into account ecosystem goods and services; (2) the development of strategic driving goals based on effects; (3) the design of transformative innovations in products, services, and business models; and (4) the development of collaborative schemes with multiple stakeholders. These four pillars should permeate both the strategic thinking and implementation into the value chain operations. In the following sections, we are going to provide two conceptual frameworks in order to facilitate the analysis.

Corporate sustainability strategies

Strategy and sustainability

When we consider the concept of strategy and strategy formulation, an immediate question emerges: What is strategy? Essentially, strategy is a broad concept with many decades of doctrine and theorizing. It can be applied at the level of the entire corporation (corporate strategy), or at a narrower level such as in case of a specific business strategy (also competitive strategy) or a functional strategy (marketing or financial strategy). At its simplest, we can consider strategy as a plan of actions that aims at establishing the conditions for long-term success, and therefore the continuity of the firm. For example, Jay B. Barney, states that strategy is about the theory of how to achieve high-level performances in markets and industries within which the firm operates (Barney, 2011: 1301). Johnson et al. (2008: 9) define strategy as "the direction and scope of an organization over the long-term, which achieves advantage in a changing environment through its configuration of resources and competences with the aim of fulfilling stakeholder expectations." According to Michael Porter (1996), who considers the competitive dimension, strategy is about being different from other competitors, combining a diverse set of activities to generate unique value for customers.

From these definitions, long-term orientation, resources and competences and future expectations are important aspects to be considered in every strategy formulation. Strategy brings us into a desired and unique position for the future. It creates the road map to bring a company's vision into reality. It determines how organizational resources, skills, and competencies should be combined to create competitive advantage, which is the superior performance of the company over its competitors (Porter, 1981; Grant, 2010). Therefore, at the foundation of strategy is the ability to understand the needs and demands of customers, stakeholders, and society at large.

What happens when we relate strategy to sustainability? As discussed in the previous chapters, sustainability has become part of the demand of stakeholders, creates new business opportunities, and can impact a company's long-term survival. When we consider corporate behaviors, there are multiple interpretations of how sustainability can be embedded into a company and what a sustainable company is. Only a few companies are "born sustainable" or with the purpose of pursuing a broad-based commitment to sustainability (Eccles et al., 2012), while for a large majority of firms, sustainability requires a conscious decision about the level of integration with the corporate and business strategy and with the organization culture. In Figure 3.5, we visually illustrate the relation between strategy and sustainability as two spheres that can more or less overlap.

In the first case (A), core strategy and sustainability orientation are not integrated. This may be the case of a company where the strategic goal is not influenced by sustainability objectives. The company addresses environmental and social issues mainly for compliance reasons or to respond to emergencies. Actions attain the level of production processes or can require the adaptation of some product features only to respond to policy makers' regulations (e.g., end of life product recovery, packaging recycling). The mission, vision, decision-making processes, the way through which the company competes, and the organizational culture are not "contaminated" by sustainability elements.

On the other extreme, (C), we find companies where the strategy and sustainability orientation fully overlap. Generally, these are firms that are created with a sustainable purpose and have implemented a strategy and business model that responds to specific environmental and social challenges. In these cases, sustainability is in the DNA of the organization and it fully orients decision-making at the enterprise, business units, functional levels, and day-to-day operations. Moreover, sustainability is fully embedded in their organizational culture. Examples of these companies are Patagonia, Vaude, Seventh Generation, The Body Shop, Nature's

Path, Ethiquable and Freitag. Specialized clean tech companies in areas such as renewables, mobility, energy efficiency, or circular economy (see later in this book, Chapters 0 and 0) can also be included in this group. An interesting case of corporate strategy and sustainability integration is Brazil's Natura Cosméticos acquisition of The Body Shop (see Box 3.2.).

Box 3.2 Natura cosmetics and the Body Shop

In September 2017 Natura, the Brazilian cosmetics group, completed the acquisition of The Body Shop, the make-up brand funded in 1976 by British entrepreneur Anita Roddick. The Body Shop had been suffering from a steady decline in its sales over the last few years. In 2006 L'Oréal bought the UK-based company with grand expansion plans. Over the years, due to a dilution of the company's ethical image and more powerful competitors in the natural beauty segment, The Body Shop financial performance dropped down, with sales declining by 5% and operating profit declining by 38% in 2016.

L'Oréal decided to put the Body Shop for sale, and Natura stepped into the breach, offering about US$1 billion for the deal. Natura is strongly focused on ethical principles such as protection of biodiversity, sustainable production, use of environment friendly ingredients and packaging materials. Sustainability is a key element that unifies the two companies and can help in their market expansion. On the one hand, The Body Shop acquisition allows Natura to become a global cosmetic player: the joint entity will cumulate sales of about US$3.5 billion and more than 3,200 stores around the world. On the other hand, The Body Shop is expected to receive a global push from Natura to further expand its sustainable and ethical brands worldwide.

To conclude, Natura's The Body Shop buyout is an example of corporate strategy where sustainability is deeply integrated in the decision-making process. The Body Shop seems a perfect match for Natura's green approach: a balanced way to grow preserving a strong focus on environmental and social issues.

Source: authors' elaboration from different sources.

In the middle, (B), we have companies where sustainability is partially integrated. In this case, there is a convergence or an overlap between sustainability opportunities and business objectives, but the complexity of the organization and its governance model make it challenging to have sustainability fully embedded. This is the case of many companies where sustainability has become a relevant commitment that is accommodated at various levels with an adaptation of strategy. In these situations, sustainability search for a business case without modifying its identity or its market logic. In these situations, trade-offs in terms of management goals, such as short-term versus the long-term view, represent a barrier to the process of integration. This is the case of many listed companies intensely exposed to the short-termism of financial investors. Another hybrid situation (where sustainability is

(A) NOT INTEGRATED   (B) PARTIALLY INTEGRATED   (C) INTEGRATED

| Core | Sustainability | Core | Sustainability | Core Sustainability | Core strategy |
| strategy | orientation | strategy | orientation | strategy orientation | Sustainability orientation |

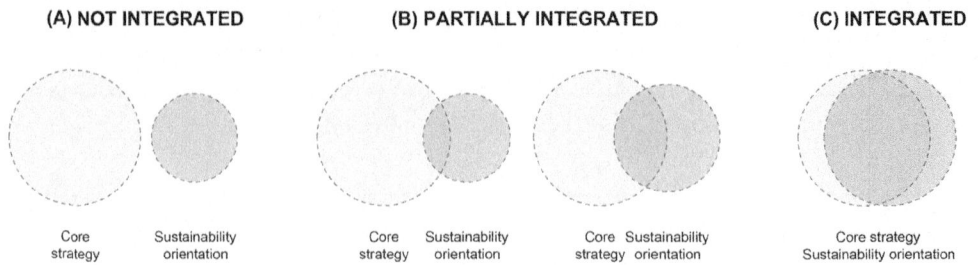

Figure 3.5 Strategy and sustainability level of integration.
Source: authors' elaboration.

only partially embedded) refers to those companies that have a very complex organizational structure. For multi-business firms that manage large portfolios of brands, or for firms that compete in multiple regions, the effort to spread the sustainability strategy through the organization may find resistance with regard to specific business areas or geographies. In these cases, some brands or regional departments can be more advanced in integrating sustainability than others. Finally, in several firms, sustainability has been strongly integrated only in some functional areas, such as operations and supply chain management, but not yet in the marketing or finance departments, where the dominant logics are less open to environmental and social principles.

To conclude, we acknowledge that there are different options when it comes to approaching sustainability and integrating it into strategy. Though environmental issues such as climate change or waste can be the top agenda for many firms, the level of connectedness with strategy differs significantly from one company to another, and from one industry to another. Today, this is a complex mosaic, but we consider that the future may look different. The BInN concept provides the overarching framework for a deeper transformation of the dominant business logics. In a world of resource scarcity, it will make sense for firms to integrate the process of strategy formulation with sustainability and contribute to social-ecological resilience. However, where should a company start? What about the mission, vision, and purpose?

Integrating sustainability into strategy: mission, vision and purpose

All companies need to define a *mission, vision* and a set of *values* (Chandler, 2016; Kenny, 2014). They are the fundamental steps in building the foundation of a successful strategy and they are an important part of a company's identity. The vision and mission are tightly connected and intertwined. The *vision* describes what the organization seeks to do; it presents the desired future and what the organization wants to become. It is aspirational. The *mission* statement illustrates the reason for the being of an organization and it defines how the company intends to respond to the needs of its stakeholders. It can incorporate the company identity, its products and markets, its technologies, and it can also incorporate elements of the uniqueness of the company when it comes to its market positioning. Generally, mission statements are longer than visions, and they reflect the company's values, responsibilities toward stakeholders, and the strategic priorities. The last element of the strategic orientation

Table 3.2 Mission and vision: some examples.

Mission	**Starbucks:** "To inspire and nurture the human spirit – one person, one cup and one neighborhood at a time." (www.starbucks.com/about-us/company-information/mission-statement)
	L'Oréal: "offering all women and men worldwide the best of cosmetics innovation in terms of quality, efficacy and safety. By meeting the infinite diversity of beauty needs and desires all over the world." (www.loreal.com/group/who-we-are/our-mission)
	Google: "Organize the world's information and make it universally accessible and useful." (www.google.com/intl/en/about/our-company/)
Vision	**Unilever:** "Our vision is to grow our business, while decoupling our environmental footprint from our growth and increasing our positive social impact." (www.unileverme.com/about/)
	Toyota: "Toyota will lead the way to the future of mobility, enriching lives around the world with the safest and most responsible ways of moving people. Through our commitment to quality, constant innovation and respect for the planet, we aim to exceed expectations and be rewarded with a smile. We will meet our challenging goals by engaging the talent and passion of people, who believe there is always a better way." (www.toyota-global.com/company/vision_philosophy/)
	IKEA: "At IKEA our vision is to create a better everyday life for the many people. Our business idea supports this vision by offering a wide range of well-designed, functional home furnishing products at prices so low that as many people as possible will be able to afford them."(www.ikea.com/ms/en_SG/about_ikea/)

Source: authors' own elaboration on multiple sources.

of a company is given by the set of values. *Values* are ideals and principles that guide the actions of the organization and contribute to shape its character and desired culture.

Therefore, addressing the sustainability challenges requires understanding the extent to which the sustainability goals and actions will have to be tied to the mission, vision, and values. In Table 3.2, we have reported some examples of missions and visions of international leading corporations. In some cases, sustainability has been embedded in these statements, while this is not true for others. As previously illustrated, each company can decide to integrate sustainability into strategy at different levels. In our view, this decision is first reflected in the changes that the company undertakes with regard to its vision and mission.

Another important concept is the company's *purpose*. In the traditional approach to strategy, the term purpose was used as a substitute of mission and vision. In recent years, with the rise of corporate sustainability and social responsibility, several companies have started to craft their purpose in a completely different way, looking more openly and honestly to their audience, and substituting or integrating the mission and vision. Moreover, the expression "company with a purpose" or "brand with a purpose" has emerged to identify those organizations that aim at a specific goal of "doing good" to their customers, employees, and local communities.

In these cases, the purpose drives the business strategy and the decision-making process, and inspires day-to-day operations. In other words, we can state that these companies use their businesses to have a positive impact on the society and environment, and not just to make a profit (in line with our definition of BInN). In Table 3.3, we have illustrated some examples of companies that have embedded sustainability principles into their mission statement, changing to a purpose.

Table 3.3 Examples of purposes and sustainability.

Nestlé	"Nestlé's purpose is enhancing quality of life and contributing to a healthier future. We want to help shape a better and healthier world. We also want to inspire people to live healthier lives. This is how we contribute to society while ensuring the long-term success of our company." (www.nestle.com/aboutus)
Unilever	"Our Corporate Purpose states that to succeed requires the highest standards of corporate behavior towards everyone we work with, the communities we touch, and the environment on which we have an impact." (www.unileverme.com/about/)
Coca-Cola	"Our Roadmap starts with our mission, which is enduring. It declares our purpose as a company and serves as the standard against which we weigh our actions and decisions. To refresh the world . . . To inspire moments of optimism and happiness . . . To create value and make a difference." (www.coca-colacompany.com/our-company/mission-vision-values)
Triodos' Bank	"Our mission is to use the money entrusted to us by savers and investors to work for positive social, environmental and cultural change. More specifically, we are in business to: help create a society that protects and promotes the quality of life of all its members and that has human dignity at its core; enable individuals, organizations and businesses to use their money in ways that benefit people and the environment, and promote Sustainable Development; offer our customers sustainable financial products and high quality service." (www.triodos.com/en/about-triodos-bank/who-we-are/mission-principles/)
Nature's Path	"Leaving the Earth Better. In everything we do we aim to work in harmony with nature - mirroring its patterns and its ancient wisdom. From water and the soil to tiny pollinators and birds, we're passionate about protecting our planet and leaving this earth better than we found it." (www.naturespath.com/en-us/our-path/)
H&M	"We want to make fashion sustainable and sustainability fashionable. The commitment of our employees is key to our success. We are dedicated to creating a better fashion future and we use our size and scale to drive development towards a more circular, fair and equal fashion industry." (https://about.hm.com/en/about-us/h-m-group-at-a-glance.html)

Source: authors' elaboration on multiple sources.

This analysis shows that the process of integrating strategy and sustainability can be interpreted in different ways. Some firms have incorporated sustainability in the long-term strategic plan. For these organizations, sustainability has become a driver of value creation, and a fundamental asset to build trust for consumers, investors and other stakeholders. In certain cases, these efforts have guided a deep transformation into purpose companies. These companies are still few, but they are leading important changes in their industries and markets. They have become iconic brands, admired by consumers, trusted, and often imitated, by peers. However, we still have a multitude of firms that have not started to change or are very slowly moving. For these firms, social and environmental issues are still peripheral to the vision and mission, and they have not integrated these into the core strategies. As previously stated, we think that today, a strategy without sustainability is simply a bad strategy, and implies many risks for long-term firm survival.

Dyllick framework for sustainability strategies

Independent from the relationship between the core strategy of a firm and its sustainability orientation (see Figure 3.5), companies have the possibility of working with a different set of strategies that address environmental and social issues. For example, a company can attempt to minimize the greenhouse gases (GHG) impact of the production processes focus-ing only on renewable energy, while another company can focus on consumer segments oriented toward sustainability, positioning its offering on the base green product attributes. Partnering with suppliers to develop a sustainable and resilient supply chain can be another option to reduce the risks of business interruption or emergencies related to environmental or social crisis. Each one of these strategies corresponds to different actions, requires differ-ent resources and competences, can provide market opportunities, and generates different costs and revenues to the company. How can we classify these approaches? How do these strategies generate value for the company, society and environment?

The literature on corporate sustainability is rich with contributions from multiple theoret-ical perspectives that have tried to explore and systematize the above mentioned corporate behavior. Stuart L. Hart (1995) with the natural-resource-based view (NRBV) theory has pro-vided one of the most important frameworks to analyze different sustainability strategies, linking these strategies to competitive advantage (see Box 3.3).

Box 3.3   Natural-resource-based view conceptual framework

One of the most important contributions in managerial literature to understanding sustainability strategies is provided by Stu L. Hart in a landmark study published in 1995. This paper extended the resource-based view of firms to address the ques-tion of interdependence between businesses and nature: "It is likely that strategy and competitive advantage in the coming years will be rooted in capabilities that facilitate environmentally sustainable economic activity – a natural-resource-based view of the firm" (Hart, 1995: 991).

Hart suggests three stages of environmental strategies, each associated to differ-ent driving forces and with a different source of competitive advantage, as follows.

- Pollution prevention represents the first level when companies focus on elimi-nating pollution and waste before they are created. This logic builds on the fact that prevention of pollution pays and companies can grab low hanging fruits with regard to eco-efficiency (Porter and van der Linde, 1995). Prevention of pollution is associated with low-cost competitive advantage.
- Product stewardship is the second level and extends pollution minimization from the manufacturing process to the entire life cycle of the product. Hart links sus-tainability strategies to consumption patterns, and introduces green design and innovation. In this case, the competitive advantage relates to "strategic preemp-tion" or establishing product standards (Hart and Dowell, 2011: 1466).
- Sustainable Development is the third strategy and incorporates economic and social concerns. Moreover, Hart moves the focus from "less environmental dam-age" to contributing in a positive manner to society and incorporates a long-term value creation perspective.

Source: authors' elaboration from different sources.

Over the years, several scholars have built on Hart's NRBV, and have extended its approach in light of important developments that have emerged in environmental science, and in the field of business and sustainability (Orsato, 2006, 2009; Nidumolu et al., 2009; Hart and Dowell, 2011). Sustainability strategies have become common jargon in management and in specific fields such as operations, supply chain, and marketing. They have also been used in multiple ways by firms and practitioners.

In order to provide a comprehensive framework on this topic, we build on the work by Thomas Dyllick and Thomas Bieker (Dyllick, 2000; Bieker and Dyllick, 2003), further elaborated by Rupert J. Baumgartner and Daniela Ebner (2010). We identify five types of sustainability strategies that can contribute to generating benefits to the company and society according to two different dimensions. The first dimension captures the distinction between actions at the company level and actions oriented at the market level. In the first case, the attention is on what happens inside the company (internal); in the second, the idea is to consider actions related to consumers and suppliers (external). The second dimension identifies strategic actions with a purpose to increase revenues (offensive) versus actions that contribute to reduce costs and risks (defensive). The five sustainability strategies are illustrated in Figure 3.6.

- Reputation-based strategy that provides benefit in terms of image and reputation.
- Risk-based strategy focuses on risk reduction and risk control.
- Efficiency-based strategy that refers to improvement of productivity and efficiency.
- Innovation-based strategy that builds on product differentiation in the market.
- Transformation-based strategy that generates value through a deeper change and a new business value proposition.

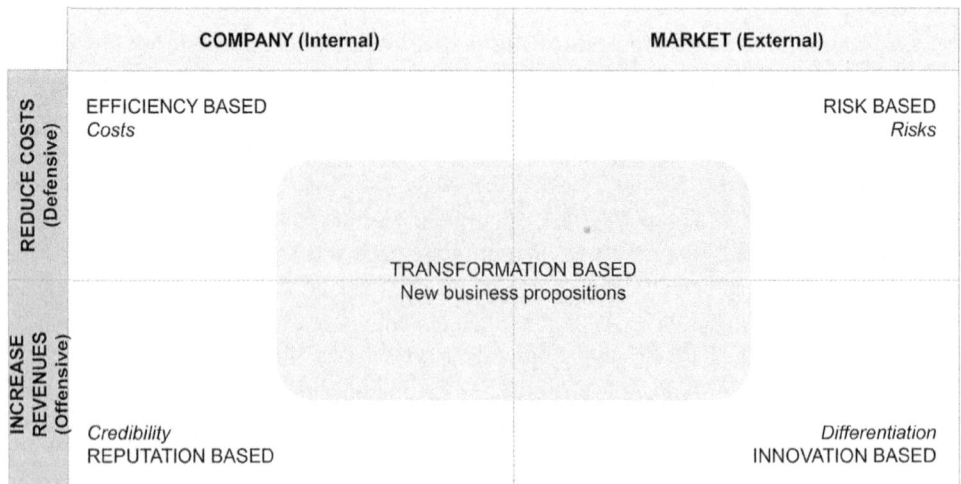

Figure 3.6 Framework for the analysis of sustainability strategies.
Source: authors' elaboration adapted from Dyllick, 2000, and Bieker and Dyllick, 2003.

Reputation-based strategy

This strategy is designed to enhance the credibility of the company against potential nega-tive image or reputational risks. Although this approach can be initiated in a defensive man-ner, or as a reaction to an emergency or crisis, there are many other cases where companies proactively identify and develop actions to build reputation. The value associated with this approach is to improve a firm's credibility, its image, and to reduce transaction costs with key stakeholders (customers, society, NGOs, governments, investors, etc.). Moreover, this strat-egy can be exploited to build legitimacy and acquire the social "license to operate," which helps firms to sustain market growth.

Other sustainability strategies benefit from the credibility and legitimacy issues; there-fore, a reputation-based strategy can strengthen other approaches, such as enhancing trust and customer loyalty or increasing the capacity to attract and engage talented employees.

Risk-based strategy

A risk-based strategy is designed to reduce, minimize, and control corporate risks arising from unresolved ecological or social issues that can escalate into high societal problems (such as climate change, resource scarcity, water issues, biodiversity degradation, and novel enti-ties). These risks, unlike traditional business risks, can manifest themselves over a long-term scale, can impact multiple aspects of the business activity, and in many cases, are outside the organization's ability to exercise control. Table 3.4 provides examples of many typologies of risks that can impact firms using the example of the oil and gas industry.

Managing any environmental related risk requires the application of procedures and frameworks designed first to assess the risks, and then to identify and appraise the options available to reduce, control or avoid potential harmful effects. Assessing a risk involves both the analysis of the likelihood of the hazard, and the consequences in terms of damage and loss of value for the firm, society, and the natural environment. Management strategies to address the risk are typically implemented in the form of internal policies, programs, best management practices, standard operating procedures, monitoring and reporting proce-dures, and even stewardship and training campaigns. Due to all these aspects, the introduc-tion of risk analysis methods dealing with environmental aspects in firms has become an important issue.

Efficiency-based strategy

When sustainability entered the business agenda, "eco-efficiency" became one of the high-est and growing priorities in corporate sustainability. This strategy focuses on resource efficiency and is designed to facilitate productivity improvements at the process level (for example, manufacturing activities and assembling). In a very open sense, efficiency is tradi-tionally considered as doing more with less, and it can positively create cost reduction and superior operative margins (Porter, 1980, 1990). The value here is seen to improve produc-tivity and process yields, and thus reduce costs. In the mid-1990s, some pioneer compa-nies first realized that win-win opportunities could be captured owing to initiatives aiming

Table 3.4 Typology of environmental risks in the oil and gas industry.

Operational risks	Companies encounter significant problems involving ecological externalities and dependences in many stages of their value chains. For example, a company operating in the oil and gas industry will have significant risks in oil exploration and production (see the recent Deepwater Horizon Platform of TransOcean-BP accident-2010), oil transportation (many cases of oil spills, such as the Exxon Valdez-1989), oil processing (explosions in refineries such as the Texaco one in Texas – 2005) and oil distribution (the vandalized oil pipeline explosions in Nigeria, Nigerian National Petroleum Corporation –1998, 2006).
Regulatory risks	Global climate agreements, such as the UNFCCC Paris Agreement (2015), put pressure on governments to raise regulatory tools in order to reach global targets and commitments. This regulatory issue can be of strategic importance for the core business of some industrial sectors, such as the oil and gas industry. This industry, similar to other energy related companies, has the highest direct carbon intensity, and therefore the largest regulatory exposure to emissions compliance schemes.
Technological risks	Technology is a key element in the oil and gas industry. Internal technological advancements as well as research and development can positively impact companies by allowing them to align the operations to the legal and societal demands (carbon capture and storage could be a good example) or by improving the extraction of its needed raw materials. However, technology can become a high-pressure factor when it allows other industries to compete with them in the market for the provision of energy as is the case with renewable energy.
Financial risks	All issues related to climate change risks, along with the impact of carbon pricing, have direct or indirect implications and exposes companies to financial vulnerability. The investment community demands more information from companies about the exposure to climate risks, as well as the prospective cost of their carbon emissions. Based on these requests, companies should disclose information about climate-related risk in their reporting, similar to their disclosure of other material risks to support informed investment and insurance underwriting decisions.
Political risks	Political risk goes beyond the mere effect of regulations. Although oil and gas companies prefer countries with stable political systems, and a history of granting and enforcing long-term leases, many of them do business in developing countries with unstable governments and a history of sudden nationalization, or more subtle, as found in nations that adjust foreign ownership rules to guarantee that domestic corporations gain an interest.
Market/competitive risks	Climate change is affecting market demand, and growing segments of consumers are asking for new cleaner technologies and low-carbon products. These new requests are opening opportunities with regard to emergence of new industries, such as renewables and electric mobility. Oil and gas companies may face competitive and market risks as a consequence of these new expectations that can accelerate the technological obsolescence of carbon-based solutions.
Societal risks	Various stakeholder groups (e.g., social movements, non-government organizations, and local communities) play an important role in providing the social "licence to operate." Managing relations with these groups is today key to maintaining the acceptance of operating activities in specific communities or to obtaining the approval of new projects. A good example on how to manage this type of risks could be the stakeholder consultation and outreach made by the French company Total on its carbon capture and storage project at the Lacq pilot project (Total, 2015).

Source: authors' elaboration on different sources.

at increasing the resource and energy efficiency, producing water savings, and reducing the amount of byproducts, toxics, and waste (Porter and van der Linde, 1995). Investments in eco-efficiency relates to the introduction of cleaner technologies and to the adoption of organizational innovation, such as environmental management systems (see Chapter 4) that can disclose hidden opportunities and eventually transform such investments into sources of competitive advantage.

Innovation-based strategy

The innovation-based strategy considers the development of novel products and services that respond to new customers' needs with regard to environmental and social issues. These days, sustainability has become an important driver of organizational and technological innovations, and companies can decide to differentiate their offerings based on the environmental performance of the product or service they sell. Chapter 6 elaborately explores the opportunities related to this strategy. Two sources of competitive advantage can be generated. First, companies can increase their market share through brand differentiation and market positioning. In this manner, companies can expand revenues, but to be successful, consumers must be willing to pay for the green and social features. Elements such as credibility and trust, supply chain traceability and transparency, and availability of information about the product environmental and social performance, are extremely important for this strategy to be successful. Second, innovative strategies can generate additional revenues as sustainable products can allow firms to gain access to new markets (new geographical market, new distribution channels, and new consumer segments, such as Millennials and Generation Z).

Transformation-based strategy

We define the last typology of strategy as "transformation-based." In this case, the idea is that the company actively contributes to shape the structural change of the market and society toward sustainability, contributing to maintaining or even enhancing the resilience of social-ecological systems. There are cases where novel solutions that need to be spread in the market require the development of new infrastructures, complementary technologies, regulations and policies, and the development of new demand and market needs. If we consider the case of renewable energies, electric vehicles, bioplastics and organic food, companies that are developing these innovative technologies, products and services in order to be successful must contribute to radically transform the market conditions, introduce new market logics, inform and educate consumers, and favor the introduction of adequate policies. This strategy engages the company from inside, calling for a long-term vision, new competencies and new managerial practices; however, it also requires different forms of intense collaborations with multiple stakeholders external to the organization, including customers, governments and competitors along the entire value chain. Simultaneously, this approach can generate disruptive results in terms of value generation and benefit to society and the natural environment.

The five approaches examined are not exclusive of each other. On the contrary, firms can develop multiple sustainability strategies that reinforce the process of integrating sustainability into the core strategy. A focus on risks respond to the question of resource scarcity and ecosystem service degradations, and allow to address raw material price volatility or risks of supply chain interruption. An eco-efficiency approach that reduces resource consumption supports the same strategic objective, decreasing the dependency from goods and services provided by ecosystems. Similarly, an innovation-based approach aiming at developing a sustainable brand needs to be grounded in a reputation-based strategy that can be constructed working on communication initiatives and stakeholder engagement processes in order to build trust, credibility and legitimacy.

To conclude, the BInN concept that we have outlined and discussed in this chapter provides a compelling framework and justification for companies to address sustainability challenges. Companies in the 21st century will be required to integrate their core strategy with sustainability as a result of their interdependence with social-ecological systems. The sustainability strategies that we have sketched in this section show different strategic responses to the challenges of the Anthropocene epoch that companies can perform combining their skills and competencies. Adopting these strategies, companies can mitigate their impacts and their effects, they can adapt to the planet's dynamics and planetary boundaries, and contribute to the resilience of social-ecological systems. Moreover, these strategies underline the competitive opportunities that exist in managing environmental and social issues, linking the actions to the business case that we have illustrated in the previous chapter.

Implementing BInN into the value chain

In this final section, we briefly introduce a framework that builds on a paper by Antonio Tencati and Stefano Pogutz (2015), with the goal to help companies implement the BInN concept. This framework, that will be further elaborated in Chapters 4 to 7, attempts to organize the range of actions that companies can adopt to address the sustainability challenges along the entire value chain: from an attempt to optimize the environmental impact of production units to redesigning products, services and supply chains, to the introduction of radically new business models (Figure 3.7). Indeed, these initiatives have different impacts on the organization, on its capacity to successfully compete, and to create long-term value for its stakeholders (Tencati and Pogutz, 2015). Simultaneously, they require the development of new knowledge and competencies that are not always available inside the firm boundaries; therefore, seeking new collaborations with stakeholders.

Focus on production units and internal processes

The first set of actions focuses on production units and internal processes, and generally requires relatively limited organizational changes and investments (Figure 3.7). This is the case of companies that are managing sustainability mainly around eco-efficiency aspects, searching for the benefits derived from an increase in resource productivity and mainly focusing on cleaner production. Today, the appearance of a complete set of clean technologies aimed at moving to zero impact helps in the development of radical reductions of environmental

Focus on sustainable business models

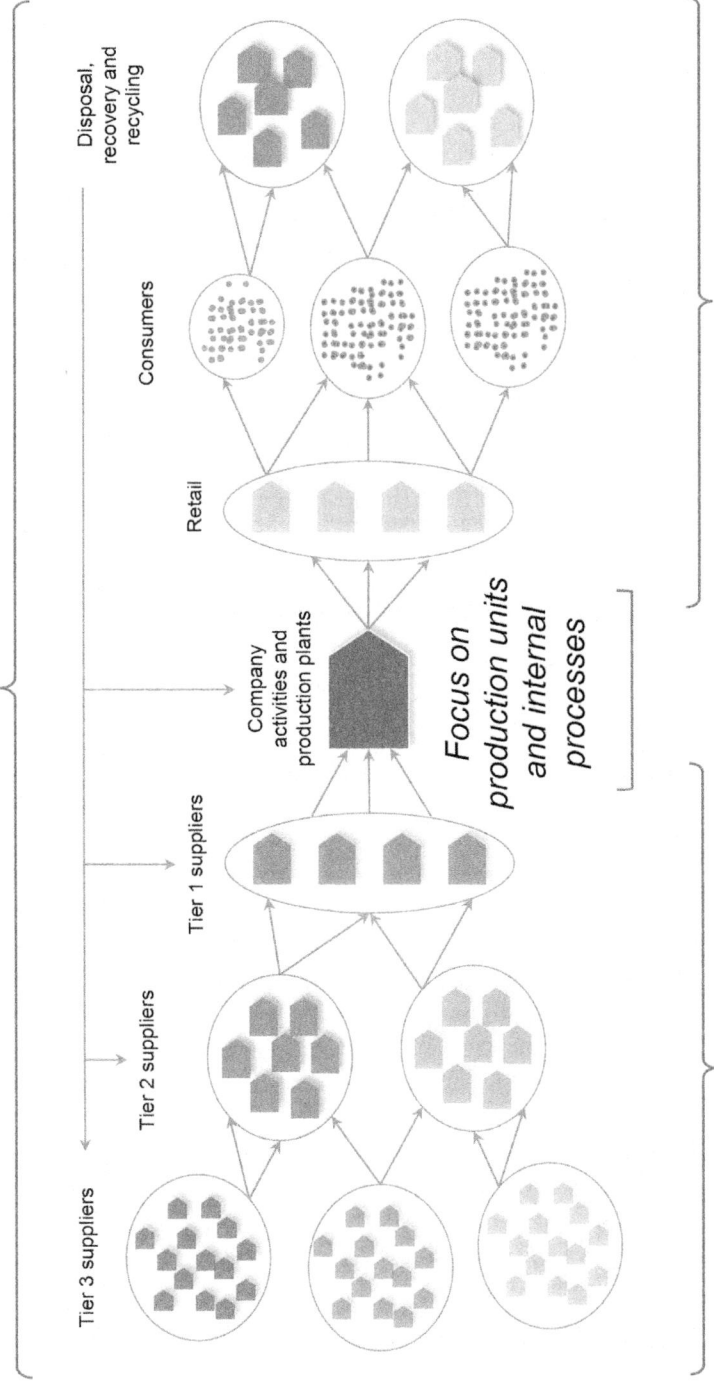

Figure 3.7 Designing a sustainable value chain.

Source: authors' elaboration adapted from Tencati and Pogutz, 2015.

impacts. Environmental management systems generally support this approach, which is driven by technical departments and engineering competencies, and promotes local optimization of environmental factors. All these tools will be explored in Chapter 4.

Focus on sustainable supply chains

The second group of actions is related to the management of sustainable supply chains. A growing number of companies are focusing upstream, to the very early stage of raw material extraction and cultivation. Two main drivers are guiding these responses: on the one side, for some typologies of environmentally intensive commodities, increased resource scarcity, market prices and the risk for supply chain disruption; and on the other side, increased stakeholders' pressures for traceability and transparency of corporate supply chains. For several industries, the ecological footprint significantly depends on the impact generated by second-, third- or even fourth-tier suppliers. The response in these cases involves a thoughtful re-design of the supply chain that goes well beyond putting some "sustainable development" into purchasing or introducing some code of conduct into the relation with first-tier suppliers. Examples of how different sectors, industries, and companies are managing these changes will be explored in Chapter 5.

Focus on designing sustainable products and services

The third set of initiatives refers to designing sustainable products and services. This approach extends the notion of corporate sustainability downstream, toward the final consumers, and the recovery/recycling of products at the end of their life cycles. These actions involve the marketing function, modify the product-system, and the relation with final consumers. The growing market segment of "Lifestyles Of Health And Sustainability" (LOHAS) consumers, the interest in sustainability of Millennials and Generation Z, and overall and increased attention toward green brands, provide the basis for successful green differentiation (Kotler, 2011; Reinhardt, 1998). As a result, we can observe that several companies are trying to win new market segments or consolidate their reputation and brand value by "greening" their offer. Simultaneously, through the design of energy-efficient products and eco-efficient packaging or eliminating potentially harmful substances, firms have learned that they can respond to environmental challenges in a significantly more convincing manner. In fact, among scholars dealing with business sustainability, it is broadly known that a large part of the environmental impact in many product categories (up to 80% in the case of durable goods such as cars, household appliances, and electronics) occurs during the consumption phase or after the products' disposal; therefore, when the goods are transferred to consumers. Focusing downstream allows companies to significantly reduce their environmental footprint. Examples about how to design new products and services required will be explored in Chapter 6.

Focus on new business models

Finally, the fourth level consists of the development of radically new business models grounded in the knowledge that the challenge of sustainability asks for path-breaking solutions and requires profound transformations in the way companies do business. Novel examples are related to the diffusion of organic products in sectors such as food and textile, or the sharp growth of renewable resources that are shaking the electricity industry in several countries. Managers and entrepreneurs are unceasingly required to find new ways to create and deliver value to final consumers. The sustainability challenge is certainly ambitious and risky for a company, but it can offer new opportunities to compete, transforming the industry and leading to the greatest environmental and economic results. When it turns out that resources are not limitless, as was considered until few years ago, and when the idea of endless consumption is no longer affordable, business model innovation may have real system effects, redefining market logics, such as in the case of the circular economy, and changing the relations between ecosystems and companies and between humans and nature. The use of new business models to deal with the environmental crisis will be explored in Chapter 7.

Summary

In this chapter we have introduced and discussed a new conceptual framework – "Business In Nature" (BInN) – in line with today's sustainability needs. This new concept requires a socio-technological transformation that should allow us to escape from past industrial revolutions and to face the future humanity challenges for this century. In order to solve our large social-ecological problems, we must develop another socio-technological revolution based on green economy and responsible behavior, able to change the logics of production-consumption mechanisms and to facilitate large transformative solutions by the incorporation of new radical ideas and sustainable business models.

We define BInN as "creating a significant positive impact in societies as well as increasing the resilience of social-ecological systems." The BInN concept rests on four basic principles: (1) extending traditional corporate boundaries in order to encompass all phases of the corporate value chain; (2) moving from an environmental-driven focus on environmental impacts to a new one on environmental effects; (3) developing a series of products, services and business models; and (4) shifting from a dominant focus on competition to a new focus on collaboration. In the coming chapters we will discuss how this new concept can be implemented following four new areas of influence: internal processes, supply chain aspects, innovative design schemes for product and services and new business models.

Chapter 3 annexes

Annex 1

A.1.1: Questions for discussion

- Analyze social-ecological systems' accountability, and the jargon used in ecology and policy. What is the difference between impact and effect?
- Explain the difference between impacts and effects in the case of climate change.
- What are the implications of nested social-ecological systems scales (spatial and temporal) for businesses?
- What are the implications of an exponential curve in the great acceleration with regard to change? Consider the implications for business organizations and their policies.
- Discuss the concept of "Business In Nature" and its key principles. How can the BInN concept influence strategy?
- What are the relations between strategy and sustainability?
- What are "sustainable born" companies?
- What are purpose companies?
- Why should sustainability integration into strategy influence the corporate mission and vision?
- What is the Natural-Resource-Based View of the firm and what type of sustainability strategies are introduced by Stu L. Hart (1995)?
- Analyze the different sustainability strategies and discuss the business opportunities related to each. Link these strategies with the business case of and for sustainability.
- How could a company implement the BInN concept? Why do we need an expended value chain perspective to address the Anthropocene epoch's environmental and social challenges?

A.1.2. Assignments

Assignment 1: Climate change

Using the web URL https://climate.nasa.gov/, view the latest data on the vital signs of the planet, including arctic sea ice levels, CO_2, sea levels, global temperature and land ice. Promote a discussion to understand (using this example) the impacts, effects, key indicators, consequences, uncertainties and, in general, the global climate issue.

If you want to continue with this exercise, you can move into policy (mitigation and adaptation strategies) for the climate challenge by viewing the web page of the United Nations Framework Convention of Climate Change or the Summary of the International Panel of Climate Change - 5th Assessment Report. The panel's full summary of the policymakers report is online at:

http://ipcc.ch/pdf/assessment-report/ar5/syr/AR5_SYR_FINAL_SPM.pdf.

You can also establish a discussion about the role of corporations in addressing the global climate issue.

Assignment 2: John Kotter videos on "change in organizations"

Enter John Kotter's web page or find a different set of YouTube videos. Here, we provide a list.

- Accelerate! The evolution of the 21st century organization (www.youtube.com/watch?v=Pc7EVXnF2aI)
- John Kotter in succeeding in a changing world - training video (www.youtube.com/watch?v=7ohEBDLPaTE)
- The importance of urgency (www.youtube.com/watch?v=zD8xKv2ur_s)

If possible, encourage students to read the booklet "Our iceberg is melting" (Kotter and Rathberger, 2006). (www.youtube.com/watch?v=tFnsLRp1VUw)

Another interesting booklet to read is "That's not how we do it here!" (Kotter and Rathberger, 2016). (www.youtube.com/watch?v=ewAAK06JrSQ)

After discussing some of these videos and readings, start a discussion on organizational change and urgency connecting the great acceleration of the Anthropocene with Figure 3.2.

Assignment 3: Sustainability and strategy

Select a company (the list below is just illustrative) and critically analyze the integration of sustainability into strategy.

Questions intended to support the analysis are:

- Has the company integrated sustainability into its core strategy? To what extent?

- What about the mission, vision and purpose? Has the company changed these statements to incorporate sustainability? What in your view is missing?
- What barriers do you see that obstacle a solid integration?
- What type of sustainability strategies is the company focusing on? Why?

On this subject, write a 3,500-word individual paper. Each paper must meet the following quality standards:

- **Contents.** Are the ideas and concepts discussed in the chapter well developed and applied to the paper? Are the ideas and concepts sufficiently original? Is the terminology appropriate and clear?
- **Well-reasoned analysis.** Is the analysis logically developed and are the concepts well connected? Is the central argument well-built and well supported by evidences?
- **Organization.** Does the paper have a clear structure? Are there clear topics and sections?
- **Style.** Does the paper use appropriate academic/professional tone? Are references sufficient and well organized?

Examples of companies you can analyze are Nike, Adidas, Coca-Cola, Pepsi, L'Oreal, H&M, Puma, Philips, BMW, Marks & Spencer, Walmart, Carrefour, Siemens, GE, Barilla, Illy, IKEA, Nestlé, Unilever, P&G, Dell, Apple, Intel, Syngenta, Toyota, Enel, E.ON, VF Corporation, JP Morgan and BNP Paribas.

You can also consider the following links to find interesting companies that are integrating sustainability into strategy:

- http://sustainability.com/our-work/reports/2017-sustainability-leaders/
- www.corporateknights.com/reports/2018-global-100/

A.1.3: Further readings and additional resources

- ESSP-Education Science System Partnership (www.essp.org/about-essp/why-essp/). The ESSP is a joint initiative or an integrated study of the earth system, how it is changing, and the implications for global and regional sustainability. It encompass four previous global environmental research programs:
  - Diversitas, an integrated program of biodiversity science,
  - IHDP, International Human Dimensions Programme on Global Environmental Change,
  - IGBP, International Geosphere-Biosphere Programme, and

- WCRP, World Climate Research Programme.

In 2012, these four programs were included under the umbrella of Future Earth.(www.futureearth.org/)

- The "Story of Stuff" is a nonprofit organization aimed at transforming the way we make, use and throw away stuff so that it is better for people and the planet. Annie Leonard, the founder, launched her popular video "The story of stuff" in 2007. Freely available, the video is a good tool to star to talk about the value chain of a particular stuff. The web page contains plenty of other resources being the organization famous by being community minded, solution focused and action oriented.
  (https://storyofstuff.org/movies/story-of-stuff/)

Annex 2

A.2.1: References

Accenture & United Nations Global Compact. (2013). The UN Global Compact-Accenture CEO Study on Sustainability. Retrieved from www.unglobalcompact.org/docs/news_events/8.1/UNGC_Accenture_CEO_Study_2013.pdf.

Banerjee, S. B. (2003). Who sustains whose development? Sustainable development and the reinvention of nature. *Organization Studies*, 24(1), 143-180.

Barney, J. B. (1991). Firm resources and sustained competitive advantage. *Journal of Management*, 17(1), 99-120.

Barney, J. B., Ketchen, D. J. and Wright, M. (2011). The future of resource-based theory: Revitalization or decline? *Journal of Management*, 37(5), 1299-1315.

Baumgartner, R. J. and Ebner, D. (2010). Corporate sustainability strategies: Sustainability profiles and maturity levels. *Sustainable Development*, 18, 76-89.

Bennett, N. and Lemoins, J. (2014). What VUCA really means for you. *Harvard Business Review*, January-February 2014.

Bieker, T. and Dyllick, T. (2003). Nachaltiges wirstchaften aus managementorientierter sight. In E. Tiemeyer and K. Wilbers (Eds.), *Berufliche bildung für nachhaltiges wirtschaften* (pp. 87-106). Bielefeld, Germany: Bertelsmann Verlag.

Birkinshaw, J. and Piramal, G. (2005). *Sumantra Ghoshal on management: A force for good*. Harlow, United Kingdom: Prentice Hall.

Chandler, D. (2016). *Strategic corporate social responsibility: Sustainable value creation* (4th edn). London and Thousand Oaks, CA: SAGE Publications.

Cooper, P. (2013). Social-ecological accounting: DPSWR, a modified DPSIR framework, and its application to marine ecosystems. *Ecological Economics*, 94, 106-115.

Crouch, C. (2011). *The strange non-death of neo-liberalism*. Cambridge: Polity Press.

Dawes, R. M. (1975). Formal models of dilemmas in social decision making. In M. F. Kaplan and S. Schwartz (Eds.), *Human judgement and decision processes: Formal and mathematical approaches* (pp. 87-108). New York: Academic Press.

Dyllick, T. (2000). Strategischer einsatz von umweltmanagementsystemen. *Umweltwirtschaftsforum*, 8(3), 64-68.

Dyllick, T. and Hockerts, K. (2002). Beyond the business case for corporate sustainability. *Business Strategy and the Environment*, 11(2), 130-141.

Dyllick, T. and Muff, K. (2015). Clarifying the meaning of sustainable business: Introducing a typology from business-as-usual to true business sustainability. *Organization & Environment*, 29(2), 156-174.

Eccles, R. G., Perkins, K. M. and Serafeim, G. (2012). How to become a sustainable company. *MIT Sloan Management Review*, 53(4), 42-50.

Elizabeth, L. (2017). What does it take for a large corporation to go green? OZY and JPMorgan Chase & Co. Retrieved from www.ozy.com/acumen/what-does-it-take-for-a-large-corporation-to-go-green/80281.

European Environmental Agency, EEA. (1999). Environmental indicators: Typology and overview (Technical report No. 25/1999). Retrieved from www.eea.europa.eu/publications/TEC25.

Folke, C. (2006). Resilience: The emergence of a perspective for social-ecological systems analyses. *Global Environmental Change*, 16(3), 253-267.

Ghoshal, S. (2005). Bad management theories are destroying good management practices. *Academy of Management Learning & Education*, 4(1), 75-91.

Gore, A. (1992). *Earth in the balance: Ecology and the human spirit*. New York: Penguin Books.

Grant, R. M. (2010). *Contemporary strategy analysis* (7th edn). Oxford: Blackwell Publishers.

Grubb, M. J. (1990). Communication energy efficiency and economic fallacies. *Energy Policy*, 18(8), 783-785.

Hardin, G. (1968). The tragedy of the commons. *Science*, 162(3859), 1243-1248.

Hart, S. L. (1995). A natural-resource-based view of the firm. *The Academy of Management Review*, 20(4), 986-1014.

Hart, S. L. and Dowell, G. (2011). A natural-resource-based view of the firm: Fifteen years after. *Journal of Management*, 37(5), 1464-1479.

Ignatius A. (2012, June). The *Harvard Business Review*, an interview with Paul Polman: Captain Planet [Audio file]. Retrieved from https://hbr.org/2012/06/captain-planet.

Intergovernmental Panel of Climate Change-IPCC. (2014). Climate Change 2014 Synthesis Report. Summary for Policymakers. 35 pp.

Jackson, T. (2009). Prosperity without growth?: The transition to a sustainable economy. Sustainable Development Commission, London.

Jackson, T. and Senker, P. (2011). Prosperity without growth: Economics for a finite planet. *Energy & Environment*, 22(7), 1013-1016.

Johnson, G., Scholes, K. and Whittington, R. (2008). Exploring corporate strategy. Harlow, United Kingdom: Pearson Education Limited.

Kenny, G. (2014). Your company's purpose is not its vision, mission, or values. *Harvard Business Review*, September 2014.

Kiron, D., Kruschwitz, N., Reeves, M. and E. Goh (2013). The benefits of sustainability-driven innovation. *MIT Sloan Management Review*, 54(2), 69-73.

Kotler, P. (2011). Reinventing marketing to manage the environmental imperative. *Journal of Marketing*, 75(4), 132-135.

Kotter, J. P. (1995, January). Leading change: Why transformation efforts fail. *Harvard Business Review*. Retrieved from https://hbr.org/2007/01/leading-change-why-transformation-efforts-fail.

Kotter, J. P. (1996). *Leading change*. Boston, MA: Harvard Business Review Press.

Kotter, J. and Rathberger, H. (2006). *Our iceberg is melting*. New York, NY: St. Martin's Press.

Nidumolu, R., Prahalad, C. K. and Rangaswami, R. M. (2009, September). Why sustainability is now the key driver of innovation. *Harvard Business Review*, 87(9), 56-64.

Orsato, R. (2006). Competitive environmental strategies: When does it pay to be green? *California Management Review*, 48(2), 127-146.

Orsato, R. J. (2009). Sustainability strategies: When does it pay to be green? London and New York: Insead Business Press. Palgrave McMillan.

Orstrom, E. (1990). *Governing the commons. The evolution of institutions for collective action*. Cambridge: Cambridge University Press.

Ostrom, E. (2009). A general framework for analyzing sustainability of Social-Ecological systems. *Science*, 325(5939), 419-422.

Pearce, D., Markandya, A. and Barbier, E. B. (1989). *Blueprint for a green economy*. London: Earthscan Publication Ltd.

Perrini, F. and Tencati, A. (2008). La responsabilità sociale d'impresa: strategia per l'impresa relazionale e innovazione per la sostenibilità. *Sinergie*, 77, 23-43.

Porter, M. (1980). *Competitive strategy: Techniques for analyzing industries and competitors*. New York, NY: The Free Press.

Porter, M. E. (1981). The contribution of industrial organization to strategic management. *Academy of Management Review*, 6(4), 609-620.

Porter, M. (1990). *The competitive advantage of nations*. London: MacMillan Press.

Porter, M. (1996). What is strategy? *Harvard Business Review*, November-December 1996.

Porter, M. and van der Linde, C. (1995). Green and competitive: Ending the stalemate. *Harvard Business Review*, 73(5), 120-134.

Prahalad, C. K. and Hamel, G. (1990). The core competencies of the corporation. *Harvard Business Review*, 68(3), 79–91.

Reinhardt, F. L. (1998). Environmental product differentiation: Implications for corporate strategy. *California Management Review*, 40(4), 43–73.

Rockström, J., Steffen, W., Noone, K. et al. (2009). A safe operating space for humanity. *Nature*, 461, 472–475.

Sardá, R. (2013). Ecosystem services in the Mediterranean Sea: The need for an economic and business-oriented approach. In T. B. Hugues (Ed.), *Mediterranean Sea* (pp. 1–33). New York: Nova Publishers.

Schmidheiny, S. (1992). *Changing course: A global business perspective on development and the environment.* Cambridge, MA: MIT Press.

Solomon, S., Plattnerb, G.-K., Knuttic, R. and P. Friedlingstei. (2009). Irreversible climate change due to carbon dioxide emissions. *Proceedings of the National Academy of Sciences*, 106(6): 1704–1709.

Soroos, M. S. (1994). Global change, environmental security and the prisoner's dilemma. *Journal of Peace Research*, 31(3), 317–332.

Starik, M. (2006). In search of relevance and impact. *Organization and Environment*, 19(4), 431–438.

Steffen, W., Broadgate, W., Deutsch, L., Gaffney, O. and Ludwig, C. (2015a). The trajectory of the Anthropocene: The great acceleration. *The Anthropocene Review*, 2(1), 81–98.

Steffen, W., Richardson, K., Rockström, J., Cornell, S. E., Fetzer, I., Bennett, E. M., Biggs, R., Carpenter, S. R., de Vries, W., de Wit, C., Folke, C., Gerten, D., Heinke, J., Mace, G., Persson, L. M., Ramanathan, V., Reyers, B. and Sörlin, S. (2015b). Planetaryboundaries: Guiding human development on a changing planet. *Science*, 347(6223), 1259855.

Stiglitz, J. E. (2012). The price of inequality: How today's divided society endangers our future. New York: W. W. Norton & Company, New York-London.

Tencati, A. and Zsolnai, L. (2009). The collaborative enterprise. *Journal of Business Ethics*, 85(3), 367–376.

Tencati, A. and Pogutz, S. (2011). Respect for nature: The need for innovative business patterns. In O. D. Jakobsen and L. J. T. Pedersen (Eds.), *Responsibility, deep ecology and the self: Essays in honor of Knut J. Ims on his 60th anniversary*, Oslo: Forlag 1.

Tencati, A. and Pogutz, S. (2015). Recognizing the limits: Sustainable development, corporate sustainability and the need for innovative business paradigms. *Sinergie*, 33(96), 37–55.

TOTAL, S.A. (2015, September 15). Carbon capture and storage: the Lacq pilot. Project and injection period 2006–2013. Retrieved from www.globalccsinstitute.com/publications/carbon-capture-and-storage-lacq-pilot-project-and-injection-period-2006-2013.

UNFCCC, 2015, Adoption of the Paris Agreement. Report No. FCCC/CP/2015/L.9/ Rev.1, Available at http://unfccc.int/resource/docs/2015/cop21/eng/l09r01.pdf (UNFCCC, 2015).

United Nation Global Compact & International Chamber of Commerce. (2015). Scaling Up Sustainability Collaboration: Contributions of Business Associations and Sector Initiatives to Sustainable Development. Retrieved from www.unglobalcompact.org/docs/issues_doc/development/BusinessAssociationsSectorandSD.pdf

Whiteman, G., Walker, B. and Perego, P. (2013). Planetary boundaries: Ecological foundations for Corporate Sustainability. *Journal of Management Studies*, 50(2), 307–336.

Winn, M. and Pogutz, S. (2013). Business, ecosystems, and biodiversity: New horizons for management research. *Organization & Environment*, 26(2), 203–229.

Wirtenberg, J. (2014). *Building a culture for sustainability: People, planet, and profits in a new green economy.* Santa Barbara, CA: Praeger.

Wolf, D. (2007). Prepared and resolved: The strategic agenda for growth, performance and change. dsb Publishing (p. 115).

Annex 3

A.3.1: Case study resources

In September 2017 Natura Cosmeticos, the Brazilian leader in the cosmetic industry, announced its acquisition of The Body Shop from L'Oreal. This new adventure looks full of synergies for both companies and may help the company founded by Dame Anita Roddick to reinvigorate its image and brand positioning. The Body Shop, in fact, has seen its sales slump in recent years. In order to better understand the issue of the purpose company, strategy and integration of sustainability into the corporate and business strategies, we suggest the use of the following case that analyze the acquisition of The Body Shop from L'Oreal, and how its ethical image suffered after the takeover.

- Maseeha Qumer, S. and Purkayastha, D. (2017), *"'Enrich Not Exploit': Can New CSR Strategy Help Body Shop Regain Glory?"*, Oikos Case Writing Competition 2017, Corporate Sustainability Track. Available at Oikos International Homepage website. Published by IBS Hyderabad, IFHE University, India.

Another interesting case on sustainability, strategy integration and a purpose-driven company is Hopworks Urban Brewery, a sustainability-focused brewpub that produces certified organic beer. The case explores how a growth strategy and decision on capital investment can affect all aspects of a sustainability-focused business.

- Pullman, M., Greene, J., Liebmann, D., Ho N. and Pedisich, X (2015), *Hopworks Urban Brewery: A case of sustainable beer*. Oikos Case Writing Competition 2017, Corporate Sustainability Track. Available at Oikos International Homepage website. Published by Portland State University, United States. Prize winner.

Rethinking the corporate value chain

4 Driving production systems sustainable

Learning objectives

- Review the concepts of pollution prevention and cleaner production.
- Analyze the concept of eco-efficiency.
- Illustrate the notion of best available techniques.
- Discuss the main concepts regarding environmental risk management, and environmental management systems).
- Drive the move from eco-efficiency to eco-effectiveness.
- Understanding production and distribution units and how to adopt clean technologies as a first response to environmental challenges.

Chapter in brief

This chapter examines production unit systems and their relationship with the management of sustainability. The most classical view is that of a company in which the borders of intervention are found in the working areas of factories, offices and distribution centers. This chapter starts with a discussion of historical views of pollution prevention and cleaner production with a focus on one of the main concepts that highlights the relationship between businesses and the environment – the concept of eco-efficiency. Then, this chapter focuses on environmental management systems as an evolution of the concept of total quality management extended to relationships with the natural environment. Although the chapter focuses mostly on the ecological dimension of sustainability, we draw readers' attention to the possibility of joining sustainable management frameworks with both environmental and social concerns. The final part of this chapter discusses the development of clean technologies and their relationship with the main megatrends that have been observed at the beginning of the 21st century. These technologies could help society to move from the concept of eco-efficiency to the concept of eco-effectiveness.

Introduction

For many decades, the traditional model of industrial activity followed a linear system model. Manufacturers used raw materials to develop products to be sold in markets in order to enhance people's welfare. In the process of production and at the end of the products' life cycles, waste that was generated needed to be disposed of in the best possible way. However, people soon started to realize that waste materials polluted natural systems and, ultimately, their own societies and themselves. Social awareness rapidly increased, and different regulatory norms evolved. Many waste management requirements and environmental standards were introduced to businesses, forcing companies to include some activities in their industrial linear processes to control the amount of waste and pollution generated. In this way, companies responded to this new set of environmental regulations. Initially, most managers saw compliance as the main focus of attention. However, some companies responded by designing new industrial system processes that could drastically reduce the amount of waste and pollution generated or by more effectively managing their generated waste. This innovation began to pave the road to a cleaner production paradigm and to introduce aspects of eco-efficiency to companies.

We also discuss the development of environmental management systems in the business field, which mirrors the rise of cleaner production. During the last 50 years, the use of management systems in businesses has been a major focus of attention. In the 1950s, many concerns were raised about the necessity of working with the principle of continuous improvement throughout a company, from high level strategic planning and decision-making to the detailed execution of labor in factories and beyond. Following these concerns, a new management approach described as Total Quality Management (TQM) became popular, especially when William E. Deming wrote his popular book *Out of the crisis* (Deming, 1982) and summarized his famous management philosophy. The book promotes the plan-do-check-act (PDCA) approach to process analysis and improvement. TQM describes the culture, attitude, and organization of a company that strives to provide customers with products and services that satisfy their needs while improving quality and eradicating defects. Different quality tools were developed following the popularization of TQM. These tools, such as those created by the American Society of Quality Control, the new industrial practices on quality systems requirements (QS-9000) developed by the three large car manufacturers in the US, and the appearance of different awards such as the Deming Prize in Japan and the Malcolm Baldridge National Quality Award in the US, did much to popularize the quality movement. In 1979, the International Standard Organization (ISO), an organization that provides requirements, specifications, guidelines, and characteristics that can be used consistently to ensure that materials, products, processes and services are fit for their purpose, established a technical committee (TC-176) to work on international standards in the quality field. In 1987, the ISO released the ISO9000 quality management and quality assurance series of standards. Since then, the so-called ISO9001, "a systematic set of activities and procedures that are implemented to ensure that the end product or service meets the quality requirement or specifications expected from a given process" (Fox, 1994; Jeffries, 1999), has been the most recognized worldwide standard for quality management. The timeframe of the TQM movement overlapped with the promotion of the Sustainable Development environmental paradigm and the environmental movement, and, thus, this TQM movement was the basis for the development of environmental management systems.

However, despite the use of cleaner production methodologies and environmental management systems, society remains entrenched in a general behavioral trend towards the degradation of a vast majority of the planet's natural resources. Cleaner production and eco-efficiency are necessary, but they are not sufficient. Almost two decades after the turn of the century, society needs to move toward the use of new, cleaner technologies that can drastically reduce environmental impacts and must create an eco-effective disruptive revolution. In this chapter, we concentrate on the efforts within the traditional boundary of the firm, its production units, the types of environmental tools currently used to reduce the impact of manufacturing activities, and what tools may be available in the future for the same purpose.

Cleaner production

Technology has been always seen as one of the most powerful forces for solving environmental problems. Using the IPAT fundamental equation (see Chapter 2), the "T-term" (technology) always shifts to reduce increases in the "I-term" (impact) when increases in the "P-term" (population) and the "A-term" (affluence) are observed. Although we believe that technology alone will not solve the current acute environmental crisis and that the other two factors (population and affluence) must also contribute to this change, in this chapter, we simply review how technology has been used to reduce environmental problems.

Technology is always improving and advancing, new technologies become more efficient than the previous ones, and even clusters of technologies can develop to create technological phases to address the relationship between business and the environment (e.g., pollution prevention technologies, end-of-pipe technologies, best available technologies, clean technologies, etc.). In each step, the new technology generates less pollution in its life cycle than the one it replaces. However, a basic question still remains: Is this process of technological development sufficient?

Waste management and pollution prevention

In social-ecological systems, the impacts of activities normally produce changes in the states of the physical-chemical and biological properties of these systems. If an impact is caused by the introduction of additional substances or wasted components in quantities that constitute a health risk for humans and ecosystems (in accordance with regulated environmental quality standards), then pollution is relevant to the discussion (Table 4.1 specifies basic definitions). Zero pollution should be a goal for advanced societies, and, therefore, waste management becomes one of the primary actions required. Waste management should aim to deeply reduce or stop the adverse effects of wasted components on the environment, whether these effects take the form of health issues, ecological process alterations, or aesthetic concerns. In this chapter, when not specified, we use the term "waste" comprehensively to refer to solid waste, effluents, air emissions, noise and radiation.

Waste can take a solid, liquid or gas form, and each form involves different methods of disposal and management. Management practices deal with all type of waste, and these practices are not uniform across countries, regions and sectors, but most are highly regulated. We do not specify all of these issues, as they can be found in many manuals and national legislations. Clearly, however, companies need to know how to deal with all regulations applicable to their activities as well as how to deal with non-regulated activities that take place in the worldwide environment. Following a classical view of a firm, companies have often reduced their intervention borders to include factories, offices, and distribution centers. However, a modern company must take care of these issues with a much larger vision, expanding the borders of intervention to include other elements of the company's value chain.

Correct waste management practices are carried out following three main steps: waste minimization, waste valorization, and adequate final treatment and disposal of waste. These steps follow logical guiding principles to ensure an appropriate waste management procedure to basically: (a) reduce the quantity of disposable waste; (b) use the residual value of materials; and, (c) minimize the negative effects on the environment with adequate final ways of disposal.

Waste minimization comes first. The best practice is to reduce the volume of waste generated by companies and the pollution problems that follow. It is clear that avoiding pollution before having to deal with its consequences is the best policy to pursue. Waste minimization involves the establishment of organizational, operational, and technological measures to reduce the amount (quantity and quality) of generated waste that requires special treatment or disposal activities under feasible economic and technical levels. Businesses can often modify their current practices to reduce the amount of waste generated by changing the design, manufacture, purchase, or use of materials and products. In 1975, the 3M company pioneered the concept of pollution prevention with the creation of its well-known Pollution

Table 4.1 Some selected definitions relevant to cleaner production.

Pollution	The introduction of toxic substances into the environment, including the addition of natural substances in unnatural quantities.
Waste	Any substance or object, without economic value, which its holder disposes of or is required to dispose of pursuant to the provisions of a national law in force (according to European regulation).
Waste Management	The collection, transportation, and disposal of garbage, sewage, and other waste products. Waste management encompasses the management of all processes and resources for the proper handling of waste materials, from the maintenance of waste transport trucks and dumping facilities to compliance with health codes and environmental regulations. (according to BusinessDirectory).
Environmental Quality Standard	The degree of concentration of a substance in a body of water (e.g., lake, river, reservoir, etc.) that should not be exceeded if a specified environmental quality objective is to be maintained (according to BusinessDirectory).
Environmental Standard	A policy guideline that regulates the impacts of different human activities on the different sectors of the environment.
Pollution Prevention	Actions businesses and organizations should take to avoid pollution incidents, including permissions needed for waste disposal.

Source: BusinessDictionary, 2017; European regulation, 2017.

Prevention Pays (3P) program (see Box 4.1). The 3P program aimed to prevent pollution at the source (in product design and reformulation and manufacturing processes) rather than removing pollution after it had been created. When 3P was launched, the concept of applying pollution prevention on a company-wide basis and documenting the results was an industry first. Today, this program has been widely recognized, has received numerous awards and has been copied by many companies elsewhere.

Box 4.1 The Pollution Prevention Pays (3P) program of 3M

3M is a science-based technology company committed to improving lives and doing business in the right way. One of its core values is to respect the social and physical environment around the world. For decades, 3M has demonstrated deep care for all aspects of the environment. The company has further extended its reputation as an industry leader in innovation by working with the idea of pollution prevention, that is, avoid polluting before having to deal with its consequences.

In 1975, the 3M company pioneered the concept of pollution prevention with the creation of its well-known Pollution Prevention Pays (3P) program. The basis of this program was that, by reducing the amount of waste generated in the first place, the company could save money on pollution control and raw materials in a clear win-win solution. The 3P program was prepared to deal with different activities to eliminate pollution at the source through product reformulation, process modification, equipment redesign, efficiency improvement, the recycling and reuse of waste materials, and further involvement in the development of new products. In addition, further savings could result from lower operating costs for pollution treatment facilities, decreased raw material requirements, reduced fuel consumption, and increased sales of existing or new products.

3P projects depended on the voluntary participation of employees. These projects had to meet three fundamental criteria: (1) eliminate or reduce a pollutant; (2) benefit the environment through reduced energy use or the more efficient use of manufacturing materials and resources; and (3) save money through the avoidance or deferral of pollution control equipment costs, reduced operating and materials expenses, or increased sales of an existing or new product. Later that century, 3P projects had to meet the previously mentioned criteria plus one of the following special criteria: (1) **Excellence:** use a unique or original design and involve significant technical accomplishment; (2) **Green Step:** illustrate a reduction in emissions during manufacturing relative to a similar family of products or a reduction in emissions by the customer following a life cycle management analysis; (3) **Guardian:** reduce or eliminate toxic emissions during manufacturing, reduce or eliminate toxic emissions by the customer, or introduce a new product that has no toxic emissions; and (4) **Mover:** demonstrate improvement in goods distribution.

From 1975 through 2015, more than 12,700 3P programs prevented over 4.3 billion pounds of pollution and resulted in economic savings of nearly US$2 billion for 3M.

In addition, since 2005, 3M has prevented nearly 1.4 million gallons of water usage and nearly 9.3 million metric tons of greenhouse gas emissions. In the new millennium, 3M deployed a series of environmental metrics related to its environmental and operational performance that can easily be found in its sustainability reports.

The sustainable journey of the 3M company continues to expand. Today, 3M has several commitments through 2020, and these commitments are based on clear metrics that are aligned with the United Nations Sustainable Development Goals. 3M has benefited greatly during the forty years of the 3P program. Some of its important elements are:

- understanding the competitive advantage of a pollution prevention culture that can be recognized as a business strategy,
- setting long-term goal ideologies by considering the incorporation of pollution prevention into a company's mission, vision or value statements,
- generating momentum by engaging employees, a very valuable resource, in activities that differ from their jobs from time to time,
- facilitating networking, as pollution prevention presents a good opportunity to learn and partner with peers, and
- considering measurements of success, goals, expectations, commitments and so on.

The 3P program has been consolidated with the following motto "eliminate pollution at the source and you will generate more economic benefits."

Source: authors' elaboration using multiple sources.

Waste valorization is the second step to be formulated in a waste management program. This step refers to the recovery and reuse of materials, initially without value, to be incorporated in new production processes as new raw materials, to be employed for energy recovering, or to be sold as by-products. The lack of appropriate treatment technologies and the lack of by-product markets need to be scrutinized because, if such technologies and markets are found, they can generate new opportunities for a company. The appearance of internal and external recycling activities also favors the reduction of raw materials and energy consumption and the valorization of waste. The implementation and commercialization of materials through by-product exchange markets (a service that puts companies that produce by-products in touch with industries that can use them as raw materials in new production processes, thereby acting as a catalyst for the market in reusable waste) can also serve to avoid costs and, sometimes, to generate new profits. The use of clean technologies can facilitate the decomposition of waste in less-hazardous products that can be recovered and reused with appropriate management tools. In any of these cases, the advantage of valorization tools is that they can be used to transform the costs of waste management into income. Finally, materials and substances that cannot be either minimized or valorized need to be prepared for proper final disposal. Proper disposal is especially important for hazardous materials, for which waste treatment operations are a necessary first step to modify the physical, chemical and/or biological characteristics of waste to reduce toxicity or to stabilize toxic constituents before handling this waste for final storage in industrial dedicated facilities.

Cleaner production and eco-efficiency

The concept of cleaner production was introduced by the United Nations Environmental Program platform on Industry and Environment in 1989. Cleaner production was defined as "the continuous application of an integrated preventive environmental strategy applied to processes, products, and services to increase eco-efficiency and reduce risks for humans and the environment" (UNEP/IEO, 1990), As introduced (and listed below), the concept applies to production processes, products and services.

- **Production processes:** these include savings in raw materials, water and energy; the reduction of toxic raw materials; and the reduction, both in quantity and toxicity, of all types of emissions.
- **Products:** this involves reducing negative impacts along the life cycle of a product from the extraction of raw materials to the product's ultimate disposal.
- **Services:** this involves incorporating environmental concerns into designing and delivering services.

Cleaner production came about as a step beyond waste management and pollution prevention efforts by formalizing actions and creating national policies that can reduce waste and pollution, by changing attitudes, and by introducing responsible environmental management through evaluations of the introduction of better technological options. Essentially, cleaner production deals with the source of the problem rather than with the symptoms.

Cleaner production was seen as a preventive, integrated and continuous strategy aimed at improving eco-efficiency. Ultimately, this strategy could also translate into better organizational performance through the reduction of costs, the enhancement of productivity, increases in profitability, and the more efficient use of source materials and energy. In western countries, many aspects of cleaner production have been formally regulated, but, in general, implementing these aspects helps companies avoid regulatory costs and exposure to further liabilities, which, in turn, can yield insurance savings and facilitate access to capital from financial institutions and lenders. Cleaner production should facilitate the advancement of corporations to closed loop system operations in which all excess materials are recycled back into the process (see Figure 4.1). These concepts have been further developed under the notion of the circular economy and will also be addressed in Chapter 7.

Cleaner production can be implemented through a hierarchy of measures: (a) continuous improvement through benchmarking actions to discover and implement industrial best practices; (b) process reengineering based on new process know-how and cost effectiveness (measures for improving eco-efficiency); and (c) technological deep changes and the introduction of best-available techniques.

Best practices can be defined as a full set of appropriate personnel and management actions and control of industrial activities that facilitate the minimization of waste. The term refers to a group of actions intended for improvement and pollution prevention that can be implemented with minimal cost and that present a rapid payback of the investment. Table 4.2 shows a list of different possible best practices.

OPEN SYSTEM   **CLOSED SYSTEM**

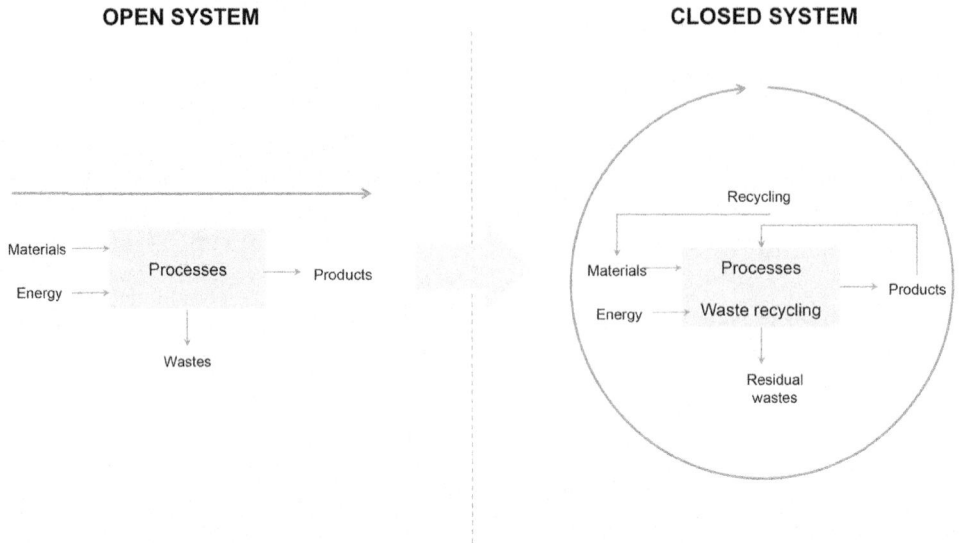

Figure 4.1 From open industrial systems to closed industrial systems.
Source: authors' elaboration.

Table 4.2 A list of industrial best practices

Collect and clean	Taking appropriate managerial and operational actions to prevent pollution and to enforce existing operational instructions.
Everything in its place	Good housekeeping, including the efficient operation of machinery, the monitoring of raw material flows and training staff.
Warehouse order and management	Establishing the necessary actions that guarantee facilities in a perfect state and equipment in good condition.
Material substitution	Phasing out polluting material that can be substituted and switching from solvent-based paints to water-based alternatives.
Installation shut-downs	Performing genuinely planned shut-downs, declassifications, management and reclassifications, and handovers of facilities and reports.
Waste segregation	Allowing the separation of waste and the reuse of materials wasted in processes for other useful applications within the company.
Better use of control procedures	Improving operational procedures, having equipment instructions ready and processing record keeping in order to run processes more efficiently.
Installation maintenance	Following good maintenance practices, which should themselves be considered a resource, to generate substantial savings in source materials and energy.
Process modification	Cutting the number of phases in the processes, switching from chemical to mechanical processes, introducing closed-loop processes, and/or incorporating new clean equipment, among other possibilities.
Communication procedures	Facilitating communication between staff and employees that can result in the introduction of preventive and corrective measures.

Source: authors' elaboration.

A more advanced framework for introducing cleaner production actions involves combining cleaner production with business process reengineering (BPR) through the application of an eco-efficient strategy. BPR frameworks, the analysis and redesign of workflows within and between enterprises for optimization, were popularized by Hammer (1990) and Hammer and Champy (2001), based on the idea that the radical redesign and reorganization of an enterprise are sometimes necessary to lower costs and increase quality. As these authors stated, at the heart of BPR is the notion of recognizing and breaking away from the outdated rules and fundamental assumptions that underscore operations. These outdated rules need to be changed to improve performance. One such outdated rule is that investments in eco-efficient measures are costly. Thus, creating business processes with the goal of maximum eco-efficiency could be a new rule and could sometimes require the application of BPR frameworks that could lead to better and cleaner production. In fact, 3M's sustainability leadership described in Box 4.1 came mainly from its eco-efficiency success, and these practices are increasingly becoming the rule in business.

Mirroring the development of BPR, the World Business Council for Sustainable Development (WBCSD) coined the term "eco-efficiency" in 1992 during the preparatory meetings of the Rio Conference and the origin of the Sustainable Development paradigm. Eco-efficiency was based on the concept of creating more goods and services while using fewer resources and creating less waste and pollution. The eco-efficiency revolution was soon used as a means for companies in the private sector to implement the principles debated in Rio, and, for some time, the term became synonymous with a management philosophy geared towards sustainability. This revolution's vision aimed to facilitate the production of economically valuable goods and services while reducing the ecological impacts of production. Basically, this vision meant introducing innovations to produce more with less and do things right. According to the WBCSD, the critical aspects of eco-efficiency are:

- the reduction of the material intensity of goods or services,
- the reduction of the energy intensity of goods or services,
- the reduced dispersion of toxic materials,
- improved recyclability,
- maximum use of renewable resources,
- greater durability of products, and
- increased service intensity of goods and services.

The WBCSD connected this concept with that of corporate sustainability.

However, as will be discussed, this connection should be considered only partially. Nevertheless, this term is widely used in the triple bottom line strategy (see Chapter 2). Based on Schmidheiny (1992), DeSimone, Popoff and the WBCSD explained that:

> eco-efficiency is achieved by the delivery of competitively priced goods and services that satisfy human needs and bring quality of life, while progressively reducing ecological impacts and resource intensity throughout the life-cycle to a level at least in line with the Earth's estimated carrying capacity.
>
> (DeSimone, Popoff with the WBCSD, 1998: 89)

The alliance between eco-efficiency and cleaner production was noticeable. However, whereas eco-efficiency starts from issues of economic efficiency that have positive environmental benefits, cleaner production starts from issues of environmental efficiency that have positive economic benefits. The combination of these two concepts was soon applied in management practice, leading many corporations to important changes in their organizational routines and BPR activities. Ever since, many firms have worked out the direct impacts of their activities to not only minimize pollution from production processes and products, improve energy efficiency, and reduce risks but also develop innovative strategies for product and service differentiation that strengthen these firms' reputations and competitive advantages.

Eco-efficiency has been driving environmental performance in companies for years. When used in conjunction with economic performance, the main principles of this concept are those behind the notion of "win-win solutions" (Porter and van der Linde, 1995), and if they are used together with socio-efficiency (Dyllick and Hockerts, 2002) the three concepts can be applied to the business case of corporate sustainability 2.0 (Triple Bottom Line) and its "win-win-win" factor. Eco-efficiency programs and actions can be measured, and these measures can serve as indicators of environmental performance. Furthermore, if economic value and environmental impact are related through eco-efficiency and socio-efficiency, these concepts can be used as indicators of organizational performance (Freeman et al., 1973; McIntyre and Thornton, 1978). At the turn of the century, two scholars, Frank Figge and Tobias Hann, came up with the notion of the sustainable value approach, a value-based perspective on the use of environmental resources (Figge, 2001; Figge and Hahn, 2004, 2005, 2013). As companies create shareholder value when they use economic capital more efficiently than their peers do, they can also create sustainable value when they use environmental and social resources more efficiently. The efficient use of capital (and the creation of economic value) can work hand in hand with the efficient use of environmental resources (and the creation of sustainable value) (Porter and van der Linde, 1995; DeSimone and Popoff, 1998; Orsato, 2006; Figge and Hahn, 2013).

The sustainable value approach (Figge and Hahn, 2004) is based on the fact that a company creates sustainable value when it uses its set of economic, environmental, and social resources more efficiently than its peers do. The sustainable value approach measures corporate sustainable performance in monetary terms. The method integrates environmental, social, and economic indicators into a monetary analysis based on opportunity costs, as is normally done in financial systems. The sustainable value approach compares the resource use of a company to a benchmark and, thus, defines the cost of each resource through opportunity costs. This approach is all based on a simple measure of eco-efficiency obtained by dividing a company's economic value by the environmental impact created by the use of some resources. The aim of this approach is to correlate strong eco-efficiency improvements with lower environmental impacts and higher economic returns (Schaltegger and Burritt, 2000). The creators of this approach have developed a guide to sustainable value calculations that can be used and adapted as needed on a case-by-case basis. (Figge et al., 2006). A summarized version of this guide is presented in Box 4.2.

Box 4.2 A guide for sustainable value calculations (adapted from the ADVANCE guide)

Sustainable value (SV) logic

The logic behind the SV approach can be explained in the following example: Company A has an economic return related to the usage of a particular environmental resource (in this case, we use CO_2 emissions as an example) that provides an indication of usage efficiency. By comparing this data with a benchmark, the company can see if it is more or less efficient than the average company (the more euros per use of the environmental resource, the more efficient the company is).

This guide illustrates these calculations in five steps.

Scoping

Definition of the benchmark: The analysis benchmarks the organization against others to define the opportunity costs. The value assessment depends on this comparison, so defining the appropriate benchmark as well as the credibility of the data sources used is critical. A positive SV indicates that the company uses its environmental resources more efficiently than the benchmark. The analysis can compare the company against national economies, companies in a given sector, or other types of benchmarks, but the benchmark needs to be clearly defined from the very beginning.

Definitions of resources to be included in the analysis: The environmental resources to be considered need to be measurable (some of the most recognized and reported environmental resources include carbon dioxide (CO_2), carbon dioxide equivalents (CO_2-equ.), methane (CH_4), energy consumption, ozone emissions, sulfur-dioxide emissions (SO_2), nitrogen oxides emissions (NO_x), waste generation, water use,

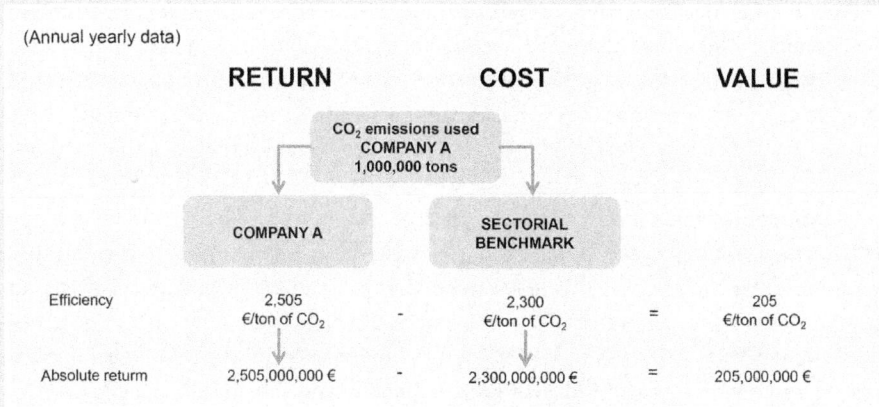

Figure B4.2a

and emissions of volatile organic compounds-VOC). Other social metrics, such as work accidents, the work force, and so on, can be also considered, but all included metrics need to be clearly measurable.

Definition of return: The economic return figure to perform the analysis (e.g., profits, EBIT, net or gross value added, turnover, etc.) needs to be defined. Again, it is important to find comparable data for the benchmark. The guide suggests using the gross domestic product as the return figure a national economy and gross value added at the company level.

Definition of timespan: A yearly time frame is advisable both to collect data and to potentially be able to obtain a data series.

Data Mining

Collecting company data: Environmental and social data can be collected from company reports and websites. The collected data need to be congruent with the scope and time span of the analysis. The economic data used in the return calculations must address the same activity areas and time span.

Collecting benchmark data: Clearly, the benchmark data needs to be comparable with the company data by covering the same scope and time span. Benchmark data depends on the intended comparison (e.g. national data, regional data, sectorial data, competitor data, etc.). This step is probably the most difficult part of the analysis, as there are limited availability of standardized information. In any case, criteria need to be established to ensure comparability.

Calculating Sustainable Value (SV)

Company returns by resource employed: The definition of the return value to be calculated is also critical. The analysis calculates the return to the company resources and compares this value to the return to selected benchmark from similar resources. Once these calculations have been performed, the efficiency of the benchmark can be used to calculate the opportunity costs of the company's resources (the opportunity cost is defined by the return that the benchmark would have created with the company's resources).

Sustainable value creation: Each of the resources used in the analysis ultimately has a value contribution. Then, the analysis compares the company's return with the opportunity costs for each environmental resource. Once this analysis has been performed for each resource, the amount of value created by the entire bundle of resources can be determined. For the final estimation of the SV, the aggregated resources need to be weighted based on the benchmark level of efficiency (e.g., in the original ADVANCE guide of the authors, the creation of the EU15 GDP in 2003 required 5.23 M tons of SO_x emissions and 9.72 M tons of NO_x emissions, meaning that to create GDP in the EU15, each ton of SO_x was bundled with 1.8 tons of NO_x, so SO_x had

(only three resources used)

COMPANY A (Positive Sustainable Value)

	Return Company A	Efficiency Company A	Amount of resources used	Efficiency Sectorial Benchmark (weighted factors)	Opportunity Costs
CO_2 emissions	2,505,000,000 €	2,505 €/ton	1,000,000 tons	2,600 €/ton	2,600,000,000 €/ton
NOx emissions	2,505,000,000 €	3,131,250,000 €/ton	800 tons	2,500,000,000 €/ton	2,000,000,000 €/ton
SO_2 emissions	2,505,000,000 €	5,010,000,000 €/ton	500 tons	4,500,000,000 €/ton	2,250,000,000 €/ton

Value Contribution COMPANY A

	Return Company A		Opportunity Costs		Sustainable Value contribution
CO_2 emissions	2,505,000,000 €	–	2,600,000,000 €	=	-95,000,000 €
NOx emissions	2,505,000,000 €	–	2,000,000,000 €	=	505,000,000 €
SO_2 emissions	2,505,000,000 €	–	2,250,000,000 €	=	255,000,000 €
	2,505,000,000 €	–	2,283,333,334 €	=	221,666,666 €

Figure B4.2b

COMPANY A (Positive Sustainable Value)

Return Company A		Opportunity Costs		Sustainable Value contribution
2,505,000,000 €	-	2,283,333,334 €	=	221,666,666 €

$$\frac{2,505,000,000 \,€}{2,283,333,334 \,€} = 1.097 : 1$$

Figure B4.2c

a weighting factor of 1.8 compared to NO_x). Thus, the sustainable value analysis is finally complete. In the example given (Figure B4.2b) above only three resources were selected in order to keep the analysis brief.

Taking company size into account

Return to cost ratio: Company size needs to be taken into account when comparing different companies. For this purpose, the guide uses the so-called return to cost ratio (RCR). The RCR compares the return of the company to the return the benchmark would have created with the company resources (opportunity costs). A larger RCR indicates that the company yields more return per unit of resource and, thus, is more efficient. The guide also sets up specifications (see the example above) to provide positive values for more or less efficient companies in the analysis.

Interpretation and communication

Interpretation: The RCR value can give the company a measure regarding its efficiency relative a desired benchmark. This value can help the company to guide its future activities.

   Communication: The values obtained in the SV calculations can be used in company reports to communicate environmental and social performance in line with society's transparency demands.

Source: adapted from Figge et al. (2006), ADVANCE Guide to sustainable value calculations.

Best Available Techniques (BATs)

A special application of the integrated preventive environmental strategy behind the aspects related to cleaner production is the use of the so-called best available techniques (BATs), which are defined as the most effective and advanced technological solutions used in the development of activities together with their methods of operation. The term refers to and is based on the following concepts:

- **Best (B)** means the most effective in achieving a high general level of protection of the environment as a whole.
- **Available (A)** refers to techniques developed on a scale that allows implementation in the relevant industrial sector under economically and technically viable conditions and taking into consideration the costs and advantages, regardless of whether the techniques are used or produced within a specific country as long as they are reasonably accessible to the company.
- **Techniques (T)** include both the technology used and the way that the installation is designed, built, maintained, operated, and decommissioned.

The designation of a BAT indicates the practical suitability of a particular technique to provide the means to avoid emissions above regulated limits, or, where that is not practical, the means generally to reduce emissions and the impact on the environment as a whole. In principal, environmental standards (see Table 4.1 for definition) are policy guidelines that regulate the impact of different human activities on different sectors of the environment. However, environmental standards leave waste producers free to comply or not comply with these standards through misbehavior or bad practices; the use of BATs ensures that these environmental standards cannot be exceeded and, thus, facilitates obtaining the desired environmental quality.

Although the BAT concept was first used for all types of industrial installations at the 1992 OSPAR Convention for the protection of the marine environment of the northeast Atlantic, it is now widely applicable in western countries to almost all industrial installations that could possibly have a moderate and/or large pollution impact. In Europe, BATs are regulated by the Industrial Emissions Directive (2010/75/EU on industrial emissions, which replaced the former Integrated Pollution Prevention and Control (IPPC) Directive). In the US, BATs or similar concepts are used, especially in the context of the Clean Air Act (42 U.S.C. 7401 et seq.; 1970) and the Clean Water Act (33 U.S.C. 1251 et seq.; 1972). Other countries and regions of the world are aligned with the use of BATs. In any case, environmental quality standards must be based and accurately reflect the latest scientific knowledge as well as health and technical information about each one.

In the case of Europe, BATs have been clarified through legislation. BATs in Europe are organized by a European Committee – the IPPC Bureau. Technical working groups produced the so-called BAT reference (BREF) documents. A BREF document is fundamentally a technical document dedicated to a special industrial sector (e.g., cement and lime manufacturing) or to a horizontal topic (e.g., cooling systems). See the list in the linked web page associated with Annex A.1.3 to this chapter. A BREF document should contain a number of elements leading to conclusions regarding which techniques are generally considered to be BATs in the sector of focus. A BREF document by itself cannot determine the emission limits and BATs for every installation, but competent authorities are responsible for issuing permits following the advice of BREF documents on BATs. The implemented BATs and their (associated) executive conclusions serve as a reference for setting the permit conditions for installations covered by directives in Europe. In this case, the combination of best technologies and license permits for operation purposes takes a further step toward the evolving aspects of cleaner production.

Environmental management systems

The appearance of environmental management systems (EMS)

Decades ago, the globalization of environmental issues and the further mainstreaming of environmental values reinforced a worldview that required better care for planetary natural resources. Several accidents and environmental catastrophes that occurred in the 1980s created pressure and forced companies and industrial sectors to closely examine the relationship between business and the environment based on the quality aspects of their operations and the impacts of their activities on the environment. Examples include the Bhopal disaster in central India in 1984 (Asia), produced by a gas leak and posterior explosion from a plant owned, managed and operated by Union Carbide India Limited; a fire in a Sandoz warehouse in Switzerland in 1986 (Europe) that resulted in a massive release of chemicals into the atmosphere, the Rhine river, and its surrounding soils; and the oil spill of the Exxon Valdez tanker in Prince William Sound, Alaska in the US in 1989 (North America). Accidents such as these were the beginning of the development of sectorial guiding principles, such as those used by the oil industry (American Petroleum Institute), and the rise of several initiatives regarding guiding principles and codes of conduct.

Several organizations were mobilized to develop environmental codes of conduct so that their signatory members could agree on rules of behavior. In 1988, the American Chemistry Council established its popular Responsible Care® program, a code of management practices that followed a pollution prevention code (see Table 4.3). This code obligates its signatories to improve their performances in terms of health, safety and environmental quality; listen and respond to public concerns; and report their progress to the public. The Responsible Care® program is still valid almost thirty years later (see the web link in Annex A.1.3 to this chapter). The Coalition for Environmentally Responsible Economies (CERES) (see the web link in Annex A.1.3 to this chapter), a non-profit organization based in the US, developed the CERES principles in 1989. This ten-point code of corporate environmental statements is publicly endorsed by companies as an environmental mission statement or ethic (Table 4.3), and, by adopting these principles, corporations publicly affirm the belief that they and their shareholders have a direct responsibility for the environment and must conduct their business as responsible stewards of the environment. They should seek profits only in a manner that leaves earth healthy and safe, believing that corporations must not compromise the ability of future generations to sustain their needs. One year later, in 1990, the International Chamber of Commerce (ICC) launched its Business Charter for Sustainable Development. The Charter (see the web link in Annex 2.2 to this chapter) consists of a short introduction and a set of 16 principles for environmental management that have been compiled (see Table 4.3). All of these codes of conduct were developed as voluntary initiatives by corporations to improve environmental performance and to inspire a better way to move into the future. The development of these codes of conduct anticipated the use of much more formal management standards. In a way, all these codes of conduct mirrored the development of the TQM principles (Table 4.3).

Table 4.3 The principles of TQM in relation to the initial environmental codes of conduct.

Total Quality Management	CERES	Responsible care	ICC charter
Management responsibility	-1- Protection of the biosphere -2- Sustainable use of natural resources	-1- Commitment of the organization -13- Protection of groundwater	-1- Corporate priority -4- Employee education -10- Precautionary approach -15- Openness to concern
Training			
Process yields	-3- Reduction and disposal of waste -4- Wise use of energy	-2- Inventory of waste and releases	-14- Contributing to the common effort
Continuous improvement	-10- Assessment and audit	-3- Evaluation of potential impacts -5- Establishing a reduction plan, goals, and priorities -11- Facility evaluation	-2- Integrated management -3- Process of improvement -11- Contractors and suppliers -13- Transfer of technology
Structure and responsibility	-5- Risk reduction -7- Damage compensation -9- Environmental directors and managers	-6- Implementation of the reduction plan -9- Integration of reduction concerns in planning -12- Reviewing, selecting, and retaining contractors and toll manufacturers	-5- Prior assessment -8- Facilities and operation -9- Research -12- Emergency preparedness
Customer satisfaction	-6- Marketing safe products and services		-6- Products and services -7- Customer advice
Employee satisfaction			
Statistical analysis		-7- Measuring progress	
Customers	-8- Disclosure	-10- Outreach	-16- Compliance and reporting

Source: Authors' elaboration of different sources.

Starting with the formulation of quality management standards such as the ISO9001 and the mentioned environmental codes of conduct, the 1990s saw the development of a much more structured set of management options for corporations to deal with environmental issues. In the US and Europe, where stricter environmental regulations were in place and new advanced demands for environmental care were developed, several new formal standards for environmental management came about. Other regions of the world, including some already developed countries, showed less interest for these initiatives, although over the years they, too, adopted the tools promoted by the first group of countries.

In Europe, in 1992 the British Standard Institution published the world's first environmental management system standard, the British Standard 7750 (BS7750), as part of a response to growing concerns about the protection of the environment. The BS7750 is considered the father of the most widely used environmental management system in place today, the ISO14001 standard. A large part of the BS7750 was used and directly incorporated into this new developed standard (Technical Committee 207 of ISO). At that time, other European local, regional, and national schemes were considered (e.g., the Green Network in Denmark and the Stockholm environmental management systems for small and medium enterprises in the city), but, in 1996, all of these standards, including the BS7750, were phased out in Europe by accepting ISO14001 as the scheme that encompassed all aspects and that could be completely integrated into a global world norm. Mirroring the development of the BS7750, in 1993, the European Commission established another new standard for environmental management, the Eco-Management and Audit Scheme (EMAS). For some time, the EMAS and ISO14001 worked separately, but they later merged and now work together in Europe. ISO14001 was recognized by the EMAS regulation as an appropriate environmental management system within EMAS. The requirements of the EMAS standard oblige the owner, on the one hand, to have an environmental management system based on ISO14001, and, on the other hand, to comply with additional requirements discussed below.

In the US, the Environmental Protection Agency (EPA) helped in promoting ISO14001 as the standard to be followed and demonstrated its support for the development and use of those systems. The EPA implemented its own model of Environmental Management Systems (EMS) in 2002 for public sites, a system that has been followed for two decades. This system does not differ greatly from the ISO14001 (Gibson and Tierney, 2011). Today, ISO14001 is the world's most used standard supporting the development of appropriate environmental policies and ensuring their implementation in all types of organizations.

Environmental management systems (ISO14001 vs EMAS)

An EMS is a formal structured framework of policies, procedures, and practices to manage and reduce an organization's environmental impact. In the US, the EPA describes an EMS as "a set of processes and practices that enable an organization to reduce its environmental impacts and increase its operating efficiency" (US EPA, 2011). These systems were established to develop a continuous procedure for companies to set up objectives, targets and action plans for their environmental states, followed up by audits and corrective actions.

In many western countries, the license to operate of a company is largely tied to its environmental responsibilities. Due to this requirement, running an environmental management system is almost a need for companies. Initially, a distinction can be made between using an informal, self-established system that works just for a particular organization and using a formal standard following well-known, already-established norms such as ISO14001 or EMAS. In any case, adopting a formal or informal EMS helps organizations to:

- manage and improve their environmental performances,
- help in complying with environmental laws and regulations,
- improve eco-efficiency to generate cost savings,
- improve their reputations with stakeholders, and,
- adapt to further changes that will surely come in the near future.

All types of organizations can work with these standards.

The implementation of an EMS in an organization requires staff and employees to make time for both the management of the system and training. However, the benefits associated with using an EMS can be quite high, including improved environmental performance, improved compliance, reduced negative impacts, better use of resources, stakeholder engagement, increased credibility and, of course, the contribution to the protection of the social-ecological systems in which the organization operates.

A formal EMS follows a logic structure, as outlined below.

(a) The elements of the EMS largely follow the Deming cycle of management, which includes the PDCA conceptual framework (Deming, 1982). After committing to an environmental policy, the organization should:

   (1) **Plan, based on its policy, what it will do.** Planning includes identifying environmental aspects and establishing goals, objectives, and targets to improve the organization's environmental performance.
   (2) **Do what it planned to do.** Doing means implementing actions and includes training and operational controls.
   (3) **Check to ensure that it did what it planned.** Checking includes monitoring the organization's environmental performance to see whether the objectives and targets are being met and, if not, taking corrective actions.
   (4) **Act to make improvements.** Acting includes reviewing programs and taking action to make needed changes to the EMS. Table 4.4 shows the seven main stages of such a conceptual framework.

(b) The elements of the EMS are committed to a cycle of continuous improvement, a framework through which the organization can build a much better environmental performance using the revised plan.
(c) The elements of the EMS are designed to establish a formal approach that reflects the relationship between environmental issues (essentially, today, the ecological aspects of the social-ecological systems in which organizations operate) and the core mission of the business.
(d) The elements of the EMS help in the logic of a non-ending process. An EMS cannot be considered as simply a project with an ending point for the organization; it is a commitment to a long-term journey.

Typically, a formal EMS requires organizations to be confirmed by an external empowered third party. After this external confirmation, organizations can continue with the process of

Table 4.4 Seven stages of the Deming cycle to be used in environmental management systems.

PLAN	1. Establish policy.
	2. Carry out an analysis of the current position.
	3. Establish goals and objectives.
DO	4. Design and implement a management system to achieve the above plan.
CHECK	5. Audit the performance of the management system to check that it is achieving the stated goals and objectives.
ACT	6. Review and revise as necessary.
	7. Repeat 2–6 above.
	8. Prepare a public report (depending on the EMS selected).

Source: authors' elaboration.

registration and, finally, certification. In some cases, this certification is not a formal requirement (e.g., ISO14001 does allow organizations to self-declare that they have met all of the
requirements of the standard without being certified). However, by obtaining certification,
examining documents and records, and interviewing personnel, the perceived performance
by other stakeholders can reflect the real performance of the company. Certification provides
an independent demonstration that the management system of the organization is able to
develop and accomplish programs and objectives that have been effectively implemented.
Although empirical research still investigates the relationship between environmental performance and economic and competitive performance (see Chapter 2 for further discussion),
environmental performance does have a positive effect on technical and organizational innovation (Iraldo et al., 2009).

The two best-known environmental management systems, ISO14001 and EMAS, are based
on a series of basic principles.

- **Voluntary in nature.** They are open to all economic sectors, including local authorities,
  but in a voluntary way. There are no strict regulations forcing the implementation of
  these systems.
- **Credibility.** They include external verification by third parties.
- **Visibility.** They have visible and recognizable logos.
- **Applicability.** They can be applicable to an entire organization or specific sites. ISO14001
  is being used currently as the environmental management system of EMAS.
- **Transparency.** They strengthen the role of the environmental statement to improve the
  transparency of communication between parties. External communication in the EMAS
  cycle can act as a catalyzer for perceived company performance.
- **Control and verification.** They include independent third-party review and registration,
  and, in the case of EMAS, validation by independent accredited verifiers.

ISO14001:2015

ISO14001 is a global certifiable standard of an environmental management system that
forms part of the ISO1400 family of environmental management (see the link in Annex A.1.3
to this chapter and Figure 4.2.). ISO has developed a series of standards and guidelines in the environmental field that are known collectively as the ISO14000 series.

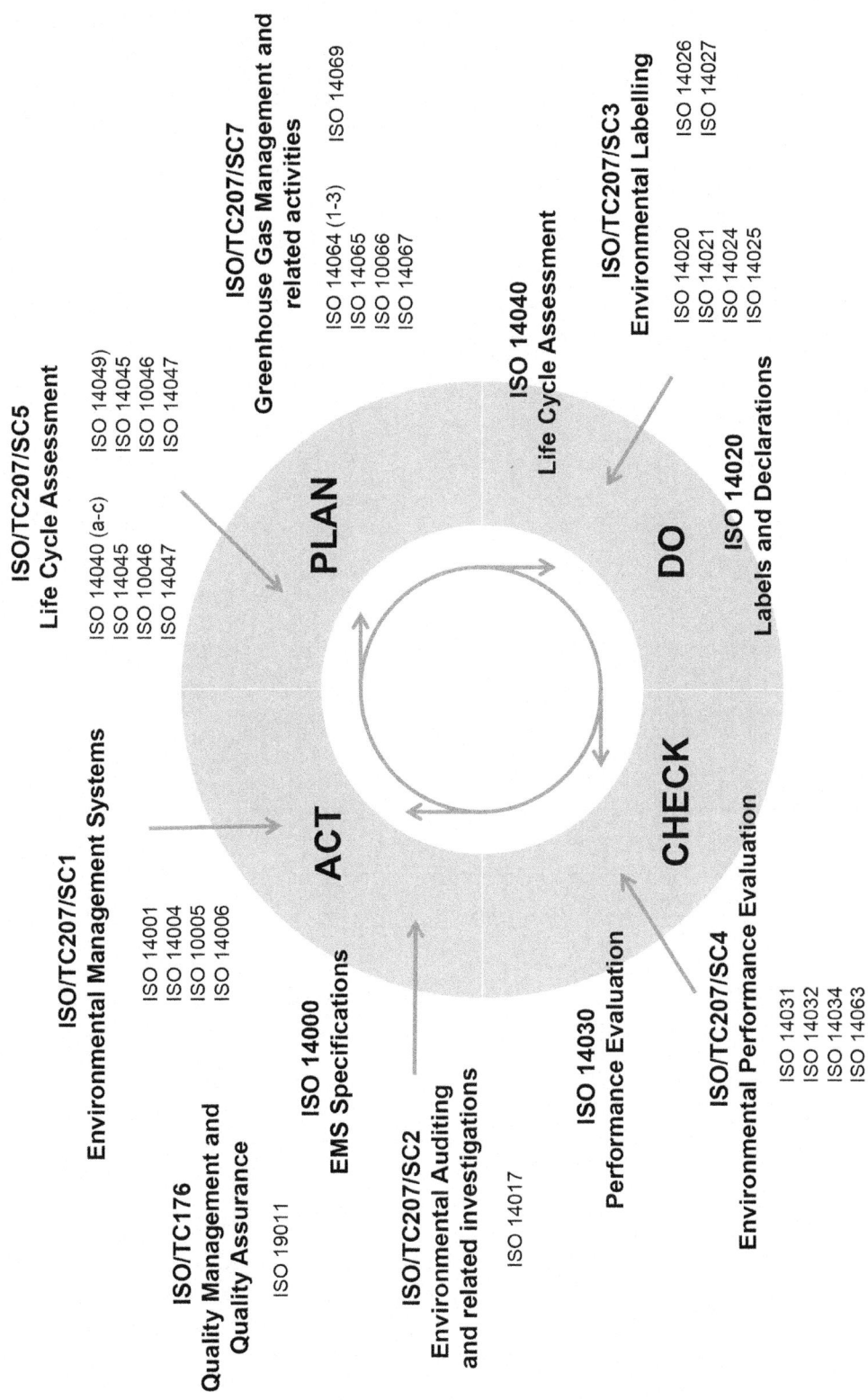

Figure 4.2 The ISO1400 family of environmental management.
Source: authors' elaboration.

ISO/TC207/SC5
Life Cycle Assessment

ISO 14040 (a-c)   ISO 14049)
ISO 14045   ISO 14045
ISO 10046   ISO 10046
ISO 14047   ISO 14047

ISO/TC207/SC7
Greenhouse Gas Management and related activities

ISO 14064 (1-3)   ISO 14069
ISO 14065
ISO 10066
ISO 14067

ISO/TC207/SC3
Environmental Labelling

ISO 14020   ISO 14026
ISO 14021   ISO 14027
ISO 14024
ISO 14025

ISO 14040
Life Cycle Assessment

ISO 14020
Labels and Declarations

PLAN

DO

ACT

CHECK

ISO/TC207/SC1
Environmental Management Systems

ISO 14001   ISO 14004
ISO 14004   ISO 14006
ISO 10005
ISO 14006

ISO/TC176
Quality Management and Quality Assurance

ISO 19011

ISO 14000
EMS Specifications

ISO/TC207/SC2
Environmental Auditing and related investigations

ISO 14017

ISO 14030
Performance Evaluation

ISO/TC207/SC4
Environmental Performance Evaluation

ISO 14031
ISO 14032
ISO 14034
ISO 14063

A.4. Context

A.4.1.
A.4.2.
A.4.3.
A.4.4.

A.5. Leadership

A.5.1.
A.5.2.
A.5.3.

A.6. Planning

A.6.1.　　A.6.2.
　A.6.1.1.　　A.6.2.1.
　A.6.1.2.　　A.6.2.2.
　A.6.1.3.
　A.6.1.4.

A.7. Support

A.7.1.　　A.7.5.
A.7.2.　　　A.7.5.1.
A.7.3.　　　A.7.5.2.
A.7.4.　　　A.7.5.3.
　A.7.4.1.
　A.7.4.2.
　A.7.4.3.

A.8. Operations

A.8.1.
A.8.2.

A.9. Performance evaluation　A.9.3.

A.9.1.
　A.9.1.1.
　A.9.1.2.
A.9.2.
　A.9.2.1.
　A.9.2.2.

A.10. Improvement

A.10.1.
A.10.2.
A.10.3.

Certification procedure
Communication
and ISO 14001 Declaration

PLAN

DO

CHECK

ACT

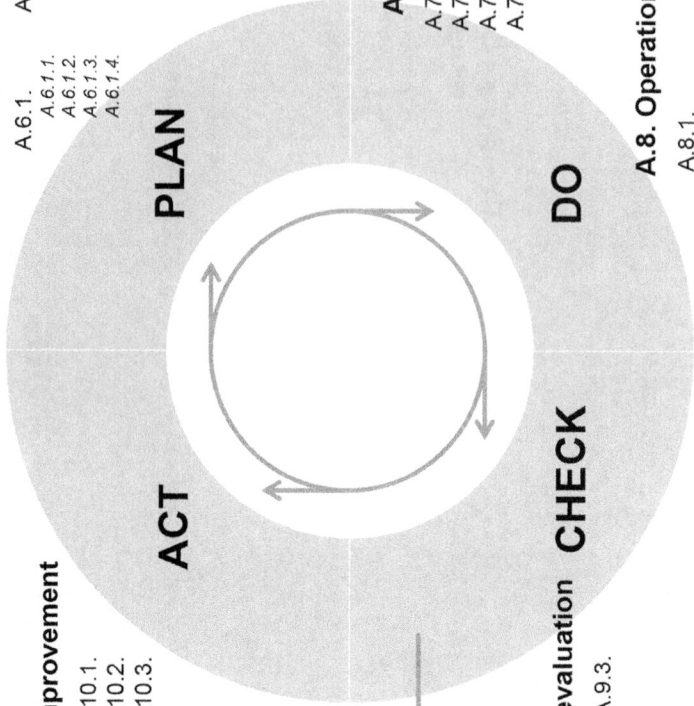

Figure 4.3 Schematic diagram of the ISO14001 norm.

Source: authors' elaboration.

The ISO14000 family of standards was developed by the ISO Technical Committee ISO/TC 207 and its various subcommittees on environmental management systems, environmental auditing and related environmental investigations, environmental labeling, environmental performance evaluation, life-cycle assessment, and greenhouse gas management (and related activities).

The ISO14000 family of norms supports the corporate goals of achieving compliance with legal requirements, establishing internal environmental quality policies, developing tools, and managing market expectations. These goals can be accomplished by implementing environmental quality management systems, environmental audits, environmental performance evaluations, product life-cycle assessments, and product labeling. All of these activities comprise a strategic environmental quality program that can be lead to distinct competitive advantage as well as the gateway to an organization's successful future. Of the ISO14000 family of norms, only one is certified, the ISO14001 EMS requirements, with guidance for use. The rest of the norms support organizations in effectively using this norm and structuring tools around the issues that relate to business and the environment in that organization.

The ISO14001:2015 specifies the requirements for an EMS that an organization can use to enhance its environmental performance by managing environmental responsibilities in a systematic manner. The first version of the ISO14001 norm was introduced in 1996 and, after a revision in 2004, the latest version was launched in 2015 with several new important aspects. The ISO is working to improve language commonalities by promoting a new high-level structure (HLS) for management system standards. This new common framework for ISO management system standards incorporates identical core text and common terms with core definitions, thus facilitating compatibility across standards. As issues related to the environment have escalated into firm planning processes, a new requirement to understand the organization's context has been incorporated to identify and leverage opportunities for the benefit of both the organization and the environment. In line with discussions in previous chapters of this book, the ISO14001:2015 standard adapted its structure to focus more on strategic issues: (a) an increased strategic approach through which environmental issues are identified to deal with risks and opportunities; (b) a new clause on leadership and responsibilities; (c) much more focus on issues related to environmental protection, environmental performance, and continuous improvement; (d) the introduction of a more detailed life-cycle perspective so that outsourced processes can be influenced and under control; and (e) better document information systems and communication. Figure 4.3. and Table 4.5. introduce the ISO14001:2015 norm (more information can be obtained in the official EU Commission Regulation 2017/1505; see the link in Annex A.1.3 to this chapter).

European Eco-Management and Audit Scheme (EMAS-III)

In the 1990s, the European Union launched its own EMS standard, EMAS. EMAS was developed as a voluntary but regulated tool (Regulation CE 1863/1993). Mirroring ISO14001, EMAS evolved through different revisions, and, in 2009, EMAS-III was launched (Regulation CE 1221/2009). Almost from the beginning, with the appearance of the ISO14001 in 1996, EMAS started to consider to have a close connection with it. Finally, since 2001, the ISO14001

Table 4.5 Structure of the ISO14001 norm and its additional requirements when implementing EMAS.

EMS requirements under ISO14001:2015	Additional EMS requirements for organizations implementing EMAS-III
A.1. General structure	
A.2. Context of structure and terminology	
A.3. Clarification of concepts	
A.4. Context of the organization	
A.4.1. Understanding the organization and its context	
A.4.2. Understanding the needs and expectations of interested parties	
A.4.3. Determining the scope of the EMS	
A.4.4. EMS	
A.5. Leadership	Continual improvement of environmental
A.5.1. Leadership and commitment	performance
A.5.2. Environmental policy	Management representative(s)
A.5.3. Organizational roles, responsibilities, and authorities	
A.6. Planning	
A.6.1. Actions to address risks and opportunities	
A.6.1.1. General	
A.6.1.2. Environmental aspects	Legal compliance
A.6.1.3. Compliance obligations	
A.6.1.4. Planning action	
A.6.2. Environmental objectives and planning to achieve them	
A.6.2.1. Environmental objectives	
A.6.2.2. Planning actions to achieve environmental objectives	
A.7. Support	
A.7.1. Resources	
A.7.2. Competence	Employee involvement
A.7.3. Awareness	
A.7.4. Communication	
A.7.4.1. General	
A.7.4.2. Internal communication	Communication
A.7.4.3. External communication	
A.7.5. Documented information	
A.7.5.1. General	
A.7.5.2. Creating and updating	
A.7.5.3. Control of documented information	
A.8. Operation	
A.8.1. Operational planning and control	
A.8.2. Emergency preparedness and response	
A.9. Performance evaluation	
A.9.1. Monitoring, measurement, analysis, and evaluation	
A.9.1.1. General	
A.9.1.2. Evaluation of compliance	
A.9.2. Internal audit	
A.9.2.1. General	
A.9.2.2. Internal audit program	
A.9.3. Management review	
A.10. Improvement	
A.10.1. General	
A.10.2. Nonconformity and corrective actions	
A.10.3. Continual improvement	

Source: authors' elaboration.

standard has been included as an integral part of EMAS, allowing many ISO-certified organizations to move into EMAS through a simplified process. The EMS requirements under EMAS include those observed by ISO14001:2015 as well as some additional requirements, listed below.

- From the very beginning of the exploration of the context, the organization needs to contact a National Competent Body in charge of the registration system.
- Organizations that want to set their standards through EMAS need to develop and document initial environmental reviews at the beginning of the process, as is demanded by the norm.
- Organizations registered with EMAS or wishing to register with EMAS need to demonstrate that they fulfill all applicable legal requirements: (a) they have identified and know the implications to the organization of all applicable legal requirements relating to the environment; (b) they ensure legal compliance with environmental legislation, including permits and permit limits, and provide the appropriate evidence; and (c) they have procedures in place that enable them to ensure ongoing legal compliance with environmental legislation.
- Organizations with EMAS shall commit to the continuous improvement of their environmental performance; this requirement is one of the most important differences between the two systems.
- Organizations must find innovative and creative ways to involve employees. Active employee involvement is a driving force of the process, a key resource in the improvement of environmental performance, and a critical anchor of all activities in the organization. This involvement becomes a necessity from the initial environmental review and should continue tightly throughout the process.
- There is an obligation to disclose information following a particular framework (an environmental report). Annually, an organization must produce an environmental statement in agreement with the rules of EMAS that is validated by an external body. This report documents environmental activities and performance.

To deal with the new aspects introduced by the ISO in the new ISO14001:2015, an amendment of the EMAS-III was introduced in August 2017 (EU Commission Regulation 2017/1505). Following that amendment, EMAS-registered organizations need to make some adaptations to comply with these changes. They will need to take into account additional elements as part of the environmental review process and when implementing the EMS. Some of these new concepts are aligned with the new "Business In Nature" principles, such as considering a life-cycle perspective when assessing the significance of environmental aspects or determining the risks and opportunities related to an EMS. Table 4.5. and Figure 4.4 introduce the structure of the EMAS-III norm by adding the distinct issues to the previously assessed ISO14001:2015 norm (more information can be obtained in the official EU Commission Regulation 2017/1505).

The European Union has a platform of information about EMAS and its implementation (see the link in Annex A.1.3 to this chapter). If a company wants to adopt to EMAS, its first step is to contact its National Competent Body and find technical support and information on the best consulting companies acting as auditors and verifiers. Once these external parties have been contacted and information obtained, a preliminary environmental review of the organization should be performed. The preliminary review is intended to establish a benchmark of the company's environmental performance, to understand the environmental impacts of the organization's processes and procedures as well as the legal environmental requirements, to check the organization's compliance with norms, and to provide information for the development of the environmental policy. This review is a way to assess the initial position, develop further improvement measures, and implement the EMS.

Prior to developing the core of the EMS, companies need to define the scope of the policy. The environmental policy prepared by the organization is a public document that describes commitments to the environment and specifies overall intentions and directions in terms of environmental performance. It also provides a framework for setting objectives and targets. The environmental policy is revised periodically and specifies compliance with legal and other requirements, the commitment to continuous improvement in environmental performance, and engagement in preventing pollution.

Once policy is set, it is time to implement the EMS – a set of environmental actions and management tools that depend on each other to achieve a clearly defined goal. The organization develops its environmental program, an action plan that translates its environmental policy into specific objectives. Then, the requirements under EMAS are those laid down in sections 4 to 10 of the ISO14001:2015 (see Table 4.5). The ISO establishes a continual cycle of planning, implementing, reviewing and improving the organization's environmental performance, ensuring the successful implementation of the organization's environmental policy and programs. During the process of running the EMS cycle, the checking phase includes a mandatory requirement to develop an annual internal audit that can be carried out by properly trained members of the organization's staff or with the help of outside experts. It is essential that the auditors are objective and properly trained.

In addition to running the EMS and different from the ISO14001, EMAS requires that a clear and coherent environmental report be produced annually, in electronic or print form, as a way to effectively communicate environmental performance to stakeholders. The environmental report contains an environmental statement and core indicators related to the organization's environmental performance.

Finally, to become EMAS registered, an organization needs to go through the verification, validation and registration process. The verification of the EMS is carried out by an independent environmental verifier who is accredited or licensed by an EMAS accreditation/licensing body of a member state. The verifier examines and verifies the organization's conformity to the EMAS in terms of the environmental review, environmental policy, compliance with environmental regulations, the environmental management system, and internal audit, and it validates the environmental report. All documents are presented to the national competent body that creates the European EMAS register for the organization. Then, the organization can use the EMAS logo to promote its registration and show its environmental commitment.

EMAS

Performance,
Credibility,
Transparency

Promotion of the
Environmental credentials

Environmental Statement
Report

Registration
Competent Body

Verification & Validation
External verifier

Compliance with all
relevant
environmental
legislation

Achieving continuous
improvements in
environmental
performance

Initial
Environmental
Review

Active
employee
involvement

A.4. Context

A.4.1.
A.4.2.
A.4.3.
A.4.4.

A.5. Leadership

A.5.1.
A.5.2.
A.5.3.

A.6. Planning

A.6.1.
   A.6.1.1.
   A.6.1.2.
   A.6.1.3.
   A.6.1.4.
A.6.2.
   A.6.2.1.
   A.6.2.2.

A.7. Support

A.7.1.   A.7.5.
A.7.2.   A.7.5.1.
A.7.3.   A.7.5.2.
A.7.4.   A.7.5.3.
  A.7.4.1.
  A.7.4.2.
  A.7.4.3.

A.8. Operations

A.8.1.
A.8.2.

A.10. Improvement

A.10.1.
A.10.2.
A.10.3.

A.9. Performance evaluation   A.9.3.

A.9.1.
  A.9.1.1.
  A.9.1.2
A.9.2.
  A.9.2.1.
  A.9.2.2.

PLAN

DO

CHECK

ACT

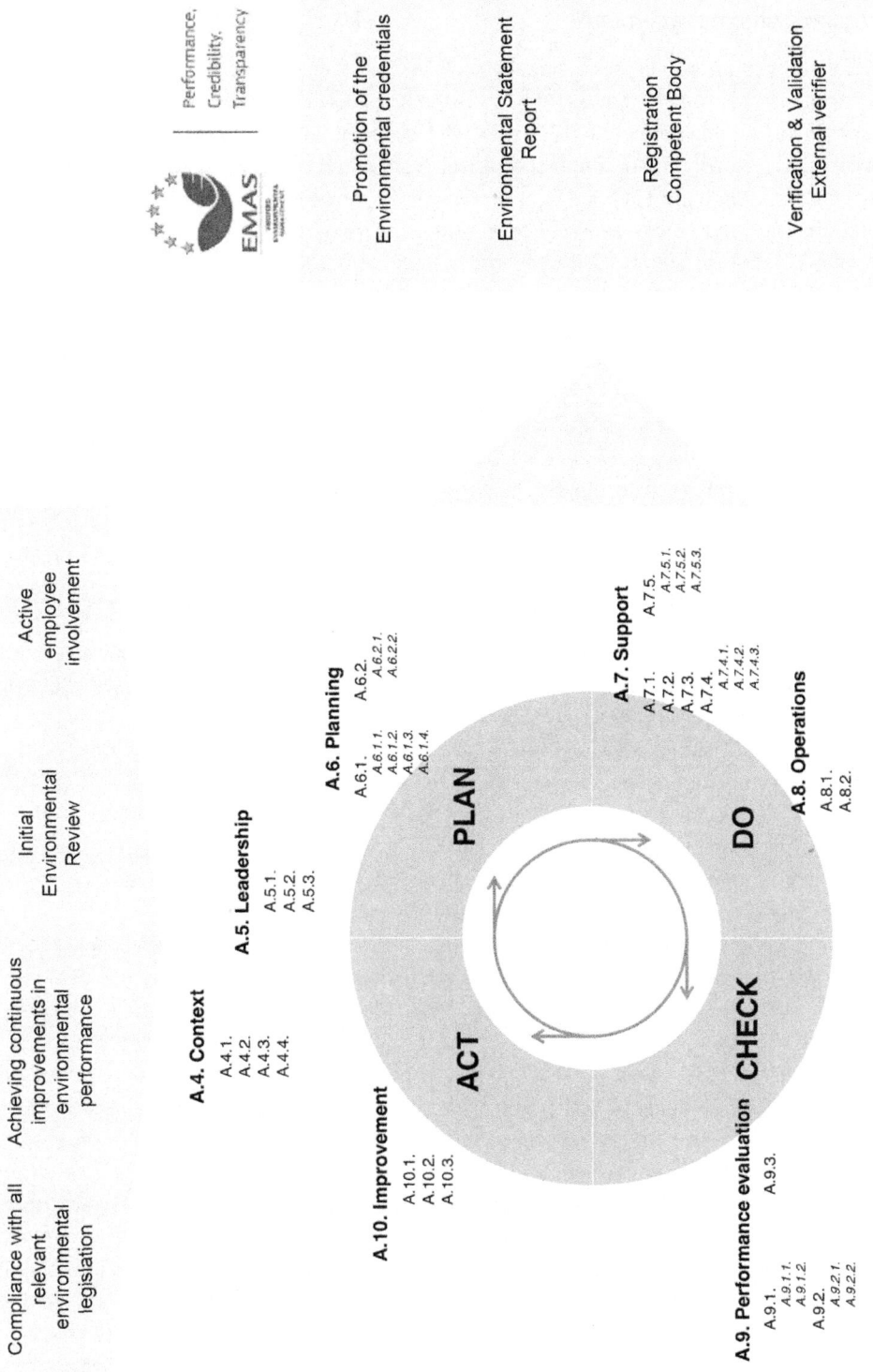

Figure 4.4 Schematic diagram of the EMAS management system.
Source: authors' elaboration.

The use of Risk Management Systems (RMS) for environmental management

Companies encounter many problems that involve environmental considerations and externalities along their value chains. Many of these considerations are risks that a company must manage. An environmental risk generically can be defined as the possibility of damage to ecosystems or to human health that arises from environmental pollution or conservational impacts. When used in the context of companies, however, this generic definition must include the fact that a company can lose significant value as a result of these problems. A complete set of tools and procedures that companies put forward to deal with these risks can be observed. Internally, companies can use different policies based on command and control procedures, they can prepare emergency response programs and plans of action, they can train people (e.g., staff and employees) to ensure knowledge about procedures, and they can even provide bonuses to facilitate better managerial behavior. Externally, they can transfer risk and insure against environmental liabilities. Sometimes, however, the exposure and vulnerability to these risks are so acute that companies are forced to implement some kind of specific informal or formal risk management system.

A risk can be defined as a scenario of an accident or other type of undesirable event evaluated based on its seriousness and probability of occurrence. When risks stem from the natural environment, scenarios like resource depletion, social-ecological process changes, pollution aspects, waste production and so on must be evaluated by organizations. Most environmental risks can be catalogued as ordinary risks. They can be anticipated because they are within the business frame of reference, and, even if some subjective aspects are involved, probabilities and values related to their outcomes can be computed. However, another type of risk may exist. This type of risk is referred to as major or unexpected risk and does not fit under the previous definition, as such risk exists outside the frame of reference. There are no rules for managing such risks, and they normally fall under crisis management practices (Laufer, 1993). Companies may face these risks due to future uncertainties related to present uncertainties coming from the new Anthropocene era. When a manager evaluates risks in a company, all of these factors should be revised.

Risk perception should be at the forefront of any risk-structured system. Companies need to know how their own employees perceive possible risks to provide good advice to deal with these before the introduction of different procedures or management systems. The perception of the notion of risk can vary between people. Although experts usually discuss risks using a basic risk assessment process, lay people normally have a common-sense notion of risk that is linked to numerous other social and psychological considerations. Reproducing the experiment of Paul Slovic (1987) can be a good way to understand this notion. Slovic presented a battery of human risks to different groups of people and experts to assess the risk perception and obtained very different results between groups. The first takeaway from these diverse results is that staff promoting and managing risks (public and/or private) need to understand the ways in which people (i.e., employees and/or different groups in society) think about and respond to risk. A second aspect of risk perception to consider is that the perceived risk of an activity or technology is higher when its associated dread risk is high; in such cases, strict regulations are employed to achieve risk reduction. Such regulations are especially important if people may die as a consequence of this type of risks. When conse-quences are delayed, the perception of the risk diminishes even if delayed response leads to higher negative consequences in the future. This notion is very relevant when dealing with risks associated with the natural environment (e.g., climate change, biodiversity loss, etc.). Finally, it is also important to consider that social concerns about environmental risks are cur-rently sharply increasing because of the belief that they affect human security and survival; this belief is based on the recognition that humans may be destroying their own environment, the consciousness generated by serious accidents and events, or increasing legislation and responsibilities. All of these factors combined lead organizations to consider how acceptable risks are defined and to analyze the social acceptability or rejection of these risks.

In order to make environmental risk management a more objective process, some orga-nizations have been able to implement informal risk management systems. Through these tools, these organizations are aimed to analyze the probabilities of possible events in relation to environmental considerations and externalities, measure which consequences can happen if these events occur, and derive responsibilities for the organization (e.g., among others, the Chevron corporation undertook a more systematic effort to deal with its environmental risks by introducing an informal system called Decision Making (DEMA) to make this process more efficient and objective; the Chevron case listed in Annex 3.1 at the end of this chapter explores this initiative). Despite the use of such informal systems and similar to the discussion on environmental management systems, today, there are also formal systemic ways to manage risk.

Risk Management (ISO31000:2009)

Formal risk management approaches can be used in a variety of management regimes, covering subject areas as such engineering, business, and, of course, human health and safety derived from environmental considerations. For years, the World Trade Organization (WTO) has embedded these activities into formal procedures, such as the risk analysis in the Sanitary and Phytosanitary Agreement for the protection of human, animal, and plant health in products traded internationally (Bjelić, 2012). Risk analysis was considered a sys-tematic way of gathering, evaluating, recording, and disseminating information, leading to a recommended position or action in response to an identified hazard (WTO, see the web link in Annex A.1.3 to this chapter), and it was developed within three sister organizations: (a) the Codex Alimentarius Commission; (b) the World Organization for Animal Health; and (c) the International Plant Protection Convention. This risk analysis framework can be used for other applications.

More recently, ISO has also developed guiding principles for risk management under ISO31000:2009 (Risk management - principles and guidelines). Although this guide can serve as a framework for organizations to manage risk, the guide is not intended to pro-mote the uniformity of risk management across organizations, and further plan and program design will need to take into account the varying needs of a specific organization. Follow-ing this guidelines document, ISO and the International Electrotechnical Commission (IEC) published a formal standard on risk management and risk assessment techniques, the

IEC/ISO31010:2009 standard, which is very relevant for the purpose of managing environmental risks (Figure 4.5). In this case, the standard assists organizations in implementing the risk management principles and guidelines provided by the guide itself and complemented by ISO Guide 73:2009 on risk management vocabulary. Although this standard is not intended for certification, regulatory or contractual use, it is designed to establish a series of steps and processes to structure and inform managerial decision-making. The processes and definitions are drawn from ISO, and, as described previously, this standard is also included in the committed communality of an HLS for management system standards.

The first aspect to be analyzed with the ISO31010:2009 is the analysis of the context. The geographical boundaries of the social-ecological systems touched by the organization along its value chain are studied in relation to the potential environmental effects to be assessed. The external context is considered in terms of key drivers and trends that affect the organization as well as social-ecological factors that can affect the assessment, and the internal context includes the organizational capabilities and the culture to be confronted. Once this context has been established, a risk assessment evaluation follows. Risk assessment is subdivided into risk identification, risk analysis and risk evaluation. Risk assessment computes the likelihood and magnitude of an event that can produce serious environmental effects and the loss of firm value; the assessment also identifies the consequences of not taking appropriate management action to avoid these effects in terms of social-ecological and management repercussions. The key output of risk assessment is the decision to either take or not take action based on the evaluation of possible existing organizational strategies and the level of risk.

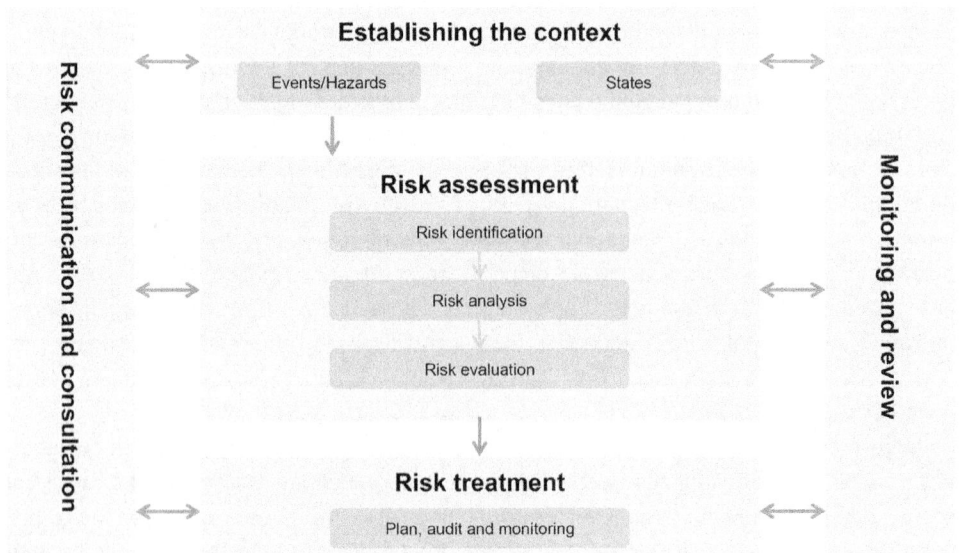

Figure 4.5 Schematic diagram of the IEC/ISO31010:2009 standard.
Source: authors' elaboration based on different sources.

Risk identification is used to identify the ecosystem components (structure and function) that can be affected by different aspects of the activities of the firm. Both the characterization of these aspects and the assessment of the pathway of effects into the ecosystem components are important. Two main groups of characteristics can be defined in this identification process. The first group includes those related to internal pressures that the organization makes on the social-ecological system in which it operates. The second group includes those aspects coming from the natural environment that are necessary to run the firm's operations throughout the analyzed scope of the value chain. The final output of the risk identification phase is a list of identified issues that can be converted into risks and a list of ecosystem components that can be altered. Risk analysis is then used to valuate and prioritize identified aspects through the associated pathway of effects that could put the organization's outcomes in danger. The final aim of this phase is to create a prioritized list of these aspects that need to be managed by the system. In addition to developing an understanding of the relationships between these aspects and the potential associated risks, causes, pressures and effects, the current management measures are also documented and analyzed in terms of their effectiveness for preventing or mitigating these risks. Finally, risk evaluation is the key decision step of the risk assessment process. Managers need to make a decision regarding necessary management actions with respect to the identified aspects/risks. Four typical strategies can be developed for each risk:

- **risk avoidance:** ceasing the activity associated with the risk,
- **risk reduction:** reducing probabilities and the contingent pressures,
- **risk retention:** bearing risk internally, and
- **risk transfer:** reducing the amount of damage for which the firm is responsible.

The key output of the risk evaluation phase is the selection of prioritized aspects/risks. If no new measures are needed or if the existing measures are adequate, the process does not move to the risk treatment phase. If new or enhanced measures need to be implemented, the risk management process identifies potential management options and moves to the risk treatment step to develop and implement new or enhanced management measures.

The risk treatment phase is the development and implementation of management strategies and measures to address the significant issues that have been prioritized. Moving into the risk treatment phase, the steps are similar to those that are commonly used in other formal environmental management systems.

The future of sustainable management systems

Although sustainable strategies provide the roadmap for orienting long-term decision-making, management systems are key to support the implementation of strategies into desired actions. Environmental and risk management systems applied to companies have been widely used in organizations for some decades, but some critiques of their use in corporate sustainability have arisen because their main bias towards more ecological issues result in inadequate focus on other social aspects (Steger, 2000; Esquer-Peralta et al., 2008), which must be considered as part of corporate social responsibility activities. Different management systems with different sectorial focuses can be found in practice in organizations, including ISO9001 for quality, ISO14001 or EMAS for the environment, OHSAS18001 for health and safety and SA8000 for social responsibility. As the main goal of each of these systems is essentially to improve the performance of the company, some demands for a comprehensive system considering all of these aspects jointly have been discussed (Noble, 2000). The tendency of companies to move to a single management system together with the idea of corporate sustainability has encouraged the development of the idea of sustainable management systems (Esquer-Peralta et al., 2008).

It is clear that companies need to find new ways to deal with the current social-ecological global paradigm. Integrating different management systems in a single framework may be an option. With such a framework, social and ecological issues could be treated together and continuous improvement could be measured and demonstrated. Recently, in a partnership with the Sustainable Development Goals (SDGs - #SDGAction1020), the United Nations is also promoting the development of sustainable management systems. One of the United Nations goals is to help businesses understand their impacts and implement strategies to transform them into positive impacts. The SDGs can facilitate the selection of sustainability targets by companies. In Chapter 8, we discuss how companies can try to do this in practice.

Clean technologies

The move from eco-efficiency to eco-effectiveness

The eco-efficiency revolution was proposed to stabilize and/or mitigate the global environmental impact of people. Many companies have opted for eco-efficiency as their guiding principle for improvement. Following the different applications of eco-efficiency in the business world, it was soon recognized that the concept should be applied across the entire life cycle of business activities at a level at least in line with the earth's estimated carrying capacity. Although the main paradigm of eco-efficiency - "creating more value with less impact or doing more with less" (Lehni et al., 2000) - was seen as the correct path forward, many authors soon started to claim that eco-efficiency itself is probably not sufficient to solve the large globalized environmental problems of today (Dyllick and Hockerts, 2002; Menoni and Morgavi, 2014). Two different arguments have been generally outlined.

The first argument is the so-called rebound effect. The rebound effect is similar to other concepts discussed in the literature: (a) the Jevon paradox, which states that an increase in efficiency is followed by increased usage (Jevons, 1865); (b) the environmental Kuznets curve, which again expresses a relationship between environmental impact and GDP (Stern, 1993); and (c) the technological factor, which implies that increased economic growth should be balanced by a factor of four by doubling wealth and halving resource use (von Weizsacker et al., 1998) or by a factor of ten, reducing human resource use by ninety percent (Schmidt-Bleek and Weaver, 2010). In any case, the rebound effect anticipates that the relative gain achieved by an increase in eco-efficiency does not always lead to an advantage in absolute terms. In the IPAT fundamental equation (see Chapter 2), increases in eco-efficiency in the

"T-term" do not result always in a reduction in the "I-term" if you find increases in the "P-term" or "A-term," due to a greater population of consumers and more consumption per capita (e.g., cars are much more efficient now than several generations ago, but the number of cars on the road and time spent driving have increased so that ultimately, in absolute term, there is not a positive gain). Eco-efficiency increases with increasing GDP per capita (relative measure), but materials use increases because total consumption also (absolute measure) increases. To explain this idea, Spangerberg (1995) identified relative and absolute decoupling in two situations: without an impact in absolute terms, environmental degradation does not end. The gap between the decreased use of resources that is expected from increased eco-efficiency and the actual utilization has been called the "gross rebound effect" (Vehmas et al., 2004 in Holm and Englund, 2009).

The second argument is the assumption that eco-efficiency is associated with a cradle-to-grave strategy (see Chapter 6 for a description of this strategy) and then, in business logic, it is necessary to find an equilibrium between the level of environmental protection and its costs because an improvement in one dimension can occur at the expense of the other dimension, and cost will then limit protection (Menoni and Morgavi, 2014). It is true that, initially, environmental protection and increased benefits/lower costs can occur in parallel, but, soon or later, a trade-off appears that requires a discussion on the modification of the business case. Specifically, the question to be discussed is how a company can continue to create sustainable value.

Based on these two arguments, several authors have contemplated the idea initially identified by Dyllick and Hockerts (2002) that "[c]orporate sustainability should not be only concerned with relative improvements. Due to the problem of non-substitutability, non-linearity, and irreversibility of some environmental vectors and components, it has also to consider absolute values." At that level, the term eco-effectiveness implies that, in addition to doing things right, organizations should be doing the right things. Coming from the technological side, eco-effectiveness has evolved to be used as a step further than eco-efficiency:

> [L]ong-term prosperity depends not on the efficiency of a fundamentally destructive system, but on the effectiveness of processes designed to be healthy and renewable in the first place. Eco-effectiveness celebrates the abundance and fecundity of natural systems, and structures itself around goals that target 100 percent sustaining solutions.
>
> (Dyllick and Hockerts, 2002: 146)

In a pure state, the concept of eco-effectiveness proposes the transformation of products and their associated material flows such that they form a supportive relationship with ecological systems and future economic growth. Eco-effectiveness works inside a cradle-to-cradle strategy (see Chapter 6 for a description of this strategy) with a broader focus on sustainability.

In the modern industrial system, in order to guarantee the maintenance of activity for future wellbeing and prosperity while ensuring the resilience of precious social-ecological systems, it is necessary to develop eco-effective processes that involve drastic and/or disruptive solutions. Using just eco-efficiency in the classical way through incremental developments has proven not to be sufficient. Coming back again to the useful IPAT equation, it may be time to replace the "I-term" with an "E-term" (i.e., an impact reduction that can guarantee the non-transgression of certain undesired limits able to maintain a resilient state of our precious natural systems). Then, either the "T-term" that reflects technological innovations or technological revolutions needs to be reduced, behaviors need to be developed that allow the reduction of the "A-term," or both. The following section introduces clean technologies as an open-door solution to contribute to eco-effectiveness in the material world.

Clean technologies

Although it is not always a strict rule, companies that introduce new technologies before their competitors can obtain a competitive advantage. It is clear that to address the environmental crisis, a technological revolution is needed to guarantee eco-effectiveness, or, at least, radical changes to make the social-ecological systems resilient. Putting these ideas together, it becomes evident than in order to work for the creation of a promising future, a move to a different set of revolutionary technologies is necessary now and at an accelerated pace. Clean technologies can offer such a move. Like other emergent technologies, clean technologies include both promises and risks in their implementation, but the continuous use of well-known mature technologies has proven not to be the solution, and new technologies are emerging for future use.

Around ten years ago, Pernick and Wilder (2007: 2) defined clean technologies as "any product, service, or process that delivers value using limited or zero non-renewable resources and/or creates significantly less waste than conventional offerings." In this definition, these technologies comprised systems that:

- harness renewable materials and energy sources or reduce the use of natural resources by using them more efficiently and productively,
- cut or eliminate pollution and toxic waste,
- deliver equal or superior performance compared with conventional offerings,
- provide investors, companies, and customers with the promise of increased returns, reduced costs, and lower prices, and
- create quality jobs in management production and deployment.

Today, many potential technologies serve as a road map into the future (e.g., genetics, nanotechnology, materials science, IT, renewable energy, water management, etc.), and many of them will form part of a new technological revolution intended to generate growth and prosperity with much lower environmental cost and degradation than in the past.

Interest in clean technologies (Clean-Tech) is increasing as every day the drivers of these technologies are clearer and stronger. On one side is a social phenomenon related to consumer behavior, the unacceptability of environmental risks, and people's concerns. On the other side is the rise of a middle-class that, because of resource scarcity, has no room for growth using conventional technologies. Finally, Clean-Tech is also growing because of economic incentives for research and development, lower costs, and more competitive prices to compete with traditional technologies.

Clean-Tech can be distinguished from traditional technologies by the fact that, in a way, it combines aspects of clean production and clean consumption, and, in order to be successful, these aspects cannot be separated from their business perspective. Value creation, for the user or the provider, is a "must have" for every new application of Clean-Tech, and it should include an economically viable business model. In general, a Clean-Tech innovation cannot be separated from its associated business model. Including these aspects, Clean-Tech can be defined as "a set of innovative technologies that aim at reducing environmental impacts by fostering a sustainable way of production, consumption and living or by optimizing the use of natural resources, applied in a profitable business model" (T. Wigniolle, personal communication). Figure 4.6 shows a classification of the most popular Clean-Tech sustainability-oriented innovations into six different groups (Cleverism, see the link in Annex A.1.3 to this chapter).

Cost, access to capital and a go-to-market approach are the three critical elements of Clean-Tech. In addition, these technologies and their more complex business models and finance possibilities generally need links with key partners in order to guarantee a successful implementation. Policy makers are seen as the fourth critical element. However, innovative sectors will not necessarily always depend on the regulatory system to survive. It is true that Clean-Tech must get approval from different legislative entities to reach the market, and, to gather momentum, they sometimes need incentives, but, as they become more mature, the relative importance of this change will become evident, and the possibilities for these technologies are enormous.

Although the introduction of Clean-Tech will vary significantly by industry and geography, there are cases in which it can completely alter industrial sectors. Very promising technologies in electrical utilities are linked to renewable energy, lighting, building technologies, the food industry and genomics, energy implementation, water treatment and mobility, to name a few examples. These technologies evidence a clear shift in Clean-Tech's application and usage in relation to:

- an increasing focus on the demand side and rather than just the supply side of the value chain,
- the appearance of Clean-Tech companies that are no longer centered only on their innovation and technology but now are better aligned themselves with business aspects through an efficient user-centric approach, and
- Clean-Tech companies that are now shortening their priorities to address first financial considerations.

As soon as these technologies move into mass markets, they will be ready to replace conventional technologies, deliver sustainable value, and start to provide more than a marginal contribution to sustainability.

The possibilities exist that clean technologies may evolve in the near future to work toward the development of society in an eco-effective scenario. Clean-Tech is knocking on the doors of consumers and other stakeholders and asking them to share the responsibility for its use. It is probably in this way that Clean-Tech will help with production, consumption, and life in the future.

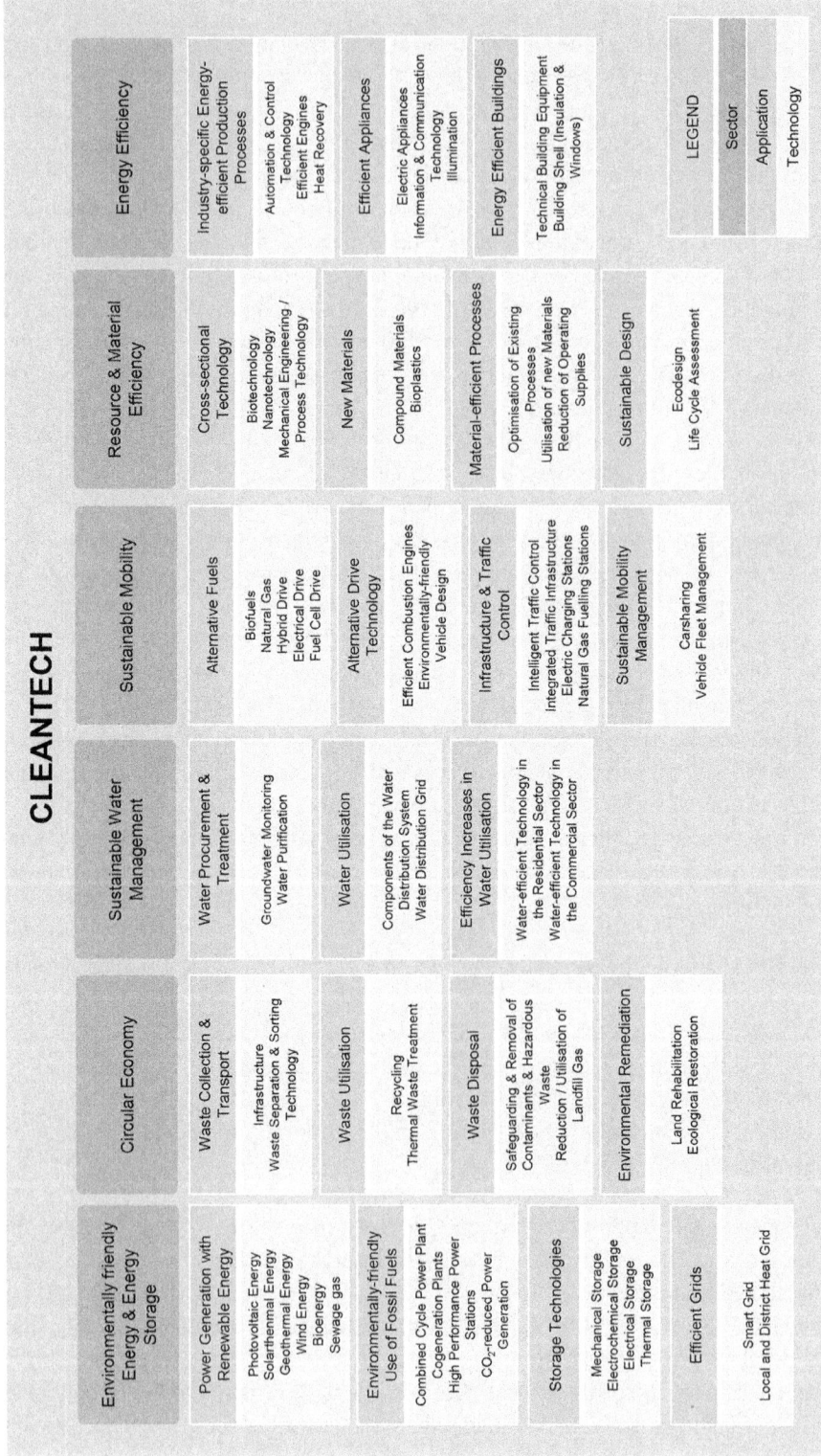

Figure 4.6 Classification of the most popular Clean-Tech sustainable oriented innovations.
Source: authors' elaboration based on DCTI, EuPD Research and KPMG (2013). Wikimedia Commons.

Summary

This chapter has analyzed the environmental sustainability of production units. We have focused on what has been considered inside the traditional boundaries of responsibility of an organization. The chapter described both technology as a powerful instrument to deal with the social-ecological crisis and management systems to ensure the continuous improvement and reduction of pressure by companies.

Regarding technology, we have concentrated mostly on technological advancements able to reduce companies' footprints on the environment. We initially reviewed pollution prevention technologies in the move to better and cleaner production and bringing innovation hand-in-hand with eco-efficiency to do things right. Then we described BATs and all the associated regulatory aspects. However, despite the successes they have achieved, we know that BATs are not enough. The "Business In Nature" (BInN) concept is aimed at achieving eco-effectiveness rather than just being eco-efficient, so this chapter argues for decisive actions on the use of Clean-Tech to effectively run production units in the future.

A second aspect of this chapter has been the analysis of management systems developed to solve environmental problems. We reviewed the evolution of EMSs and described the maximum use of these systems (ISO14001 and EMAS in Europe). We also focused on the use of risk management systems in companies that have operational concerns about their exposure and vulnerability to consequences coming from degraded environmental components and ecosystem services. Finally, we reviewed the introduction of suitable sustainable systems, mixed systems, which will be further discussed in Chapter 8 of this book.

The focus on making production units sustainable addresses the development of one of the basic principles of the (BInN) concept, the appearance of newer production processes and production systems. However, a life cycle analysis is needed to determine if these particular technologies can contribute more to a cleaner environment than the earlier alternatives can. These newer technologies can also revolutionize the way business is done, promote collaboration among stakeholders, and produce disruptive changes in industries.

Chapter 4 annexes

Annex 1

A.1.1: Questions for discussion

- What are the key steps of a waste management framework?
- What are the main differences between pollution prevention technologies, best available technologies, and clean technologies? How can these solutions be related to the IPAT equation?
- Describe the different phases used in the measurement of sustainable value using the Figge and Hahn (2006) methodology.
- What are the key differences between the ISO14001 and EMAS environmental management systems?
- What could be the potential interests of a sustainable management system in this century?
- Discuss the concepts of eco-efficiency and eco-effectiveness. How are these two concepts interrelated?
- Exemplify and discuss different typologies of clean technologies.

A.1.2: Assignments

Assignment 1: Eco-efficiency analysis and sustainable value calculations.

This exercise is aimed at allowing students to familiarize with the eco-efficiency and sustainable value calculation framework of Figge and Hahn.

- Select three companies of a particular industry (e.g., the automotive industry) for which you can easily download their environmental, economic and/or sustainability reports of the last three years. From these three companies, select which company you represent.
- Analyze the environmental parts of the reports and select some environmental variables (indicators) for which comparable data may be available (e.g., CO_2, SO_x, NO_x, VOC-emissions; water use, and energy use).

- Analyze the economic parts of the reports and try to find an economic figure (e.g., net value added) for comparisons.
- Perform the eco-efficiency measures for your selected company following the guide described in Box 4.2 (see the main part of Chapter 4). Calculate the average with the measures of the other two companies to make the comparison against your company.
- Finally, calculate how much return your company creates with its resource by comparing this data with the return created by the selected benchmark (the other two companies).

This exercise can be designed in a more elaborate way to allow calculations of the sustainable value and the return to cost ratio, but, in this case, further elaboration (including weighting factors) should be performed.

Note: You may use the ADVANCE document as guidance [Figge, F., Barkemeyer, R., Hahn, T. and Liesen, A. (2006). The ADVANCE guide to sustainable value calculations. ADVANCE project (www. advance-project.org)].

Assignment 2: Clean technologies

One of the emerging aspects of the relationship between business and the environment is clean technologies; the markets for clean technologies will grow significantly in the coming years and can contribute to alleviating pressures on energy, water and other natural resources.

In this assignment (preferably to be carried out by groups of students), each team will work on the issues needed to implement a new technological solution for a particular social-ecological problem. The groups will select one type of technology of interest (you can use Figure 4.6 in the main part of Chapter 4 to select the technology to work with). Each team will need to develop an analysis to be presented regarding the following aspects:

- explanation of the selected technology and its alleviation of environmental/social pressures,
- overview of governmental regulations that are relevant for this initiative,
- proposal for commercializing this selected technology,
- the present development phase of this technology,
- potential market analysis,
- company/ies (if any) behind its operational feasibility, and
- existence of government incentives for this technology.

A.1.3: Further reading and additional resources

- To gain more insights into the sustainable value calculation described in Box 4.2, a value-based perspective for the use of environmental resources, you can visit several web pages. The first one is related to the ADVANCE project (www.advance-project.org), where the ADVANCE guide for sustainable value calculations can be downloaded; the second one describes the sustainable value concept (www.sustainablevalue.com/).

- The European Integrated Pollution Prevention and Control Bureau (EIPPCB) is part of the Circular Economy and Industrial Leadership Unit of the European Commission's Joint Research Centre. The EIPPCB organizes and coordinates the exchange of information that leads to the drawing up and reviewing of BAT reference documents according to the dispositions of the guidance document on the exchange of information. Its web page (http://eippcb.jrc.ec.europa.eu/) includes plenty of information on European BATs and their reference documents.

- Here are the websites of famous organizations that developed environmental codes of conduct starting in the 1980s:

  - Responsible Care® program (https://responsiblecare.americanchemistry.com/)
  - The Coalition for Environmentally Responsible Economies (CERES) (www.ceres.org/)
  - International Chamber of Commerce - Business Charter for Sustainable Development (https://iccwbo.org/publication/icc-business-charter-for-sustainable-development-2015/)

- Environmental and Risk Management Systems has different web platforms that include much information and multiple resources to understand these concepts:

  - EMAS (http://ec.europa.eu/environment/emas/)
  - ISO1400:2015 (www.iso.org/iso-14001-environmental-management.
  - World Trade Organization (www.wto.org/english/tratop_e/sps_e/spsagr_e.htm)
  - ISO31010:2009 (www.iso.org/standard/51073.html

- Cleverism's web page (www.cleverism.com/cleantech-complete-guide/) presents information on clean technology, some key Clean-Tech categories and other information of interest, as well as Clean-Tech case studies.

Annex 2

A.2.1: References

Bjelić, P. (2012). World trade organization and the global risks. In A. Alemanno, F. den Butter, A. Nijsen and J. Torriti, (Eds.), *Better business regulation in a risk society*. New York: Springer.

Deming, W. E. (1982). *Out of the crisis*. Cambridge, MA: MIT CAES.

DeSimone, L. D., Popoff, F. and World Business Council for Sustainable Development (1998). *Eco-efficiency. The business link to sustainable development* (2nd edn). Cambridge, MA: MIT Press.

Dyllick, T. and Hockerts, K. (2002). Beyond the business case for corporate sustainability. *Business, Strategy and the Environment*, 11(2), 130–141.

Esquer-Peralta, J., Velazquez, L. and Munguia, N. (2008). Perceptions of core elements for sustainability management systems (SMS). *Management Decision*, 46(7), 1027–1038.

Figge, F. (2001). Environmental value added – Ein neues Maß zur Messung der Öko-Effizienz. *Zeitschrift für angewandte Umweltforschung*, 14(1-4), 184-197.

Figge, F. and Hahn, T. (2004). Sustainable value added – Measuring corporate contributions to sustainability beyond eco-efficiency. *Ecological Economics*, 48(2), 173-187.

Figge, F. and Hahn, T. (2005). The cost of sustainability capital and the creation of sustainable value by companies. *Journal of Industrial Ecology*, 9(4), 47-58.

Figge, F. and Hahn, T. (2013). Value drivers of corporate eco-efficiency: Management accounting information for the efficient use of environmental resources. *Management Accounting Research*, 24(4), 387-400.

Figge, F., Barkemeyer, R., Hahn, T. and Liesen, A. (2006). The ADVANCE guide to sustainable value calculations. Retrieved from ADVANCE project www.advance-project.org/.

Fox, M. J. (1994). *Quality assurance management*. EEUU Chapman & Hall.

Freeman, M. A., Haveman, R. H. and Kneese, A. V. (1973). *The economics of environmental policy*. New York, NY: John Wiley & Sons.

Gibson, K. and Tierney, J. M. (2011). The evolution of environmental management systems: Back to basics. *Environmental Quality Management*, 21(1), 23-37.

Hammer, M. (1990). Reengineering work: Don't automate, obliterate. *Harvard Business Review*, July-August, 104-112.

Hammer, M. and Champy, J. (2001). *Reingeneering the corporation: A manifesto for business revolution*. New York: Harper Business Essentials – HarperCollins Publishers.

Holm, S. and Englund, G. (2009). Increased ecoefficiency and gross rebound effect: Evidence from USA and six European countries 1960-2002. *Ecological Economics*, 68(3), 879-887.

Iraldo, F., Testa, F. and Frey, M. (2009). Is an environmental management system able to influence environmental and competitive performance? The case of the eco-management and audit scheme (EMAS) in the European union. *Journal of Cleaner Production*, 17, 1444-1452.

Jeffries, T. (1999). *Quality management system*. London: RIBA Publications.

Jevons, W. S. (1865). The coal question: *An inquiry concerning the progress of the nation and the probable exhaustion of our coal-mines* (3rd edn). London: Macmillan & Co.

Laufer, R. (1993). *L'entreprise face aux risques majeurs: A propos de l'incertitude des normes sociales*. Paris, France: L'Harmattan.

Lehni, M., Schmidheiny, S., Stigson, B. and Pepper, J. (2000). *Eco-efficiency: creating more value with less impact*. Genève, Switzerland: WBCSD.

McDonough, W. and Braungart, M. (1998, October). The next industrial revolution. *The Atlantic Monthly*, 282(4), 82-92.

McIntyre, R. J. and Thornton, J. R. (1978). On the environmental efficiency of economic systems. *Soviet Studies*, 30(2), 173-192.

Menoni, M. and Morgavi, H. (2014). Is eco-efficiency enough for sustainability? *International Journal of Performability Engineering*, 10(4), 337-346.

Noble, M. T. (2000). *Organizational mastery with integrated management systems: Controlling the dragon*. New York: John Wiley & Sons.

Orsato, R. J. (2006). Competitive environmental strategies: When does it pay to be green? *California Management Review*, 48(2), 127-143.

Pernick, R. and Wilder, C. (2007). *The clean tech revolution: the next big growth and investment opportunity*. New York: HarperCollins Publishers.

Porter, M. and van der Linde, C. (1995). Toward a new conception of the environment-competitiveness relationship. *Journal of Economic Perspectives*, 9(4), 97-118.

Schaltegger, S. and Burritt, S. (2000). *Contemporary environmental accounting: Issues, concepts and practice*. Sheffield, United Kingdom: Greenleaf Publishing.

Schmidheiny, S. (1992). *Change course: A global perspective on development and the environment*. Cambridge, MA: MIT Press.

Schmidt-Bleek, F. and Weaver, P. (Ed.). (2010). *Factor 10: Manifesto for a sustainable planet*. Sheffield, United Kingdom: Greenleaf Publishing.

Slovic, P. (1987). Perception of risk. *Science*, 236(4799), 280-285.

Spangenberg, J. H. (1995). *Towards sustainable Europe: The study. Sustainable Europe – Environmental space*. Luton, United Kingdom: Friends of the Earth Publications.

Steger, U. (2000). Environmental management systems: empirical evidence and further perspective. *European Management Journal*, 18(1), 23-37.

Stern, P. C. (1993). A second environmental science: Human-environment interactions. *Science*, 260(5116), 1897-1899.

United Nations Environmental Programme/Industry and the Environment (UNEP/IEO). (1990). Environmental Auditing. Workshop, 10-17 January 1989, Paris, France.

United States Environmental Protection Agency. (2011). Environmental Management Systems (EMS). Retrieved from http://www.epa.gov/EMS.

Vehmas, J., Luukanen, J. and Kavio-oja, J. (2004, 13-14 May). Technology development versus economic growth - an analysis of sustainable development. Presented at EU-US Scientific Seminar: New Technology Foresight, Forecasting & Assessment Methods.

von Weizsacker, E., Lovins, A. B. and Lovins, L. H. (1998). *Factor four: Doubling wealth, halving resource use*. London: Earthscan Publications.

Annex 3

A.3.1: Case study resources

In order to complement the learning experience, we suggest the following case studies which address the topics of environmental risk management and clean technologies.

Some interesting cases on environmental risk management:

- Reinhardt, F. L., Mandelli, M. M. and Burns, J. (1999). *Environmental risk management at Chevron Corporation*. Harvard Business School Case 799-062 (Revised April 1999).
- Andersen, T. and Andersen, C. B. (2014). *British Petroleum; from the Texas City to the Gulf of Mexico and beyond*. Reference no. 41 714-017-8. Available at The Case Center. Prize Winner.

In addition, a couple of interesting cases on managing operations and environmental risks that refer to the BP oil spill incident in the Gulf of Mexico (2010):

- Hoffman, A. J. (2012). *BP. Beyond petroleum*. Published by: WDI Publishing, William Davidson Institute (EDI), University of Michigan. Reference no. W92C29. The Case Center.
- Rotemberg, J. J. (2010). BP's Macondo: Spill and response. Published by Harvard Business Publishing. Reference no. 9-711-021. Version 2012, January 9.

Different cases on technological solutions and environmental problems offering examples of failure and success:

- Vietor, R. H. K. and F. L. Reinhardt. *Du Pont Freon Products Division (A)*. Harvard Business School Case 389-111, January 1989. (Revised March 1995.)
- Hoffman, A. (2010). *Molten Metal Technologies (A)*. Case study #1-429-049. Available at Global Lens Case Center.
- McDaniels, D. and Bowen, F. (2011). *Total's Carbon Capture and Storage project at Lacq (A): Risk and opportunity in Public Engagement*. Case study IVEY-8B10M105-E. Available at Richard Ivey School of Business, The Case Center.

5 Managing sustainable supply chains

Learning objectives

- Illustrate what is a sustainable supply chain.
- Link sustainable supply chain to the other options for corporate sustainability.
- Explain how to deal with environmental and social challenges that are localized outside corporate "traditional" boundaries, upstream in the supply chains.
- Explore how to design and manage a sustainable supply chain.
- Examine codes of conduct, auditing, and sustainable procurement.
- Link sustainable supply chain management to risk management and risk strategy.
- Analyze standards and certification and learn about transparency and traceability.

Chapter in brief

This chapter examines how green and sustainability issues are affecting the design and management of supply chains. It focuses on the upstream phase of the supply chain that links the extraction of raw materials and/or the utilization of ecosystem services to the focal firm through complex networks of suppliers often dispersed around the globe in developing and developed countries. It analyzes the implications of two major drivers: (1) increased resource scarcity and the risk for supply chain disruption; and (2) increased stakeholders' pressures for traceability and transparency of supply chains. Further, the chapter explores the type of actions multinational companies are implementing. We provide an in-depth look at codes of conduct, monitoring and control systems, new sustainability standards, certification systems, and collaboration strategies with suppliers and stakeholders.

Introduction

Toward the end of 2009, Puma, one of the leading sports brands in the world, joined forces with professional services firms PWC and TruCost in order to put a monetary value on its reliance on nature for the provisioning of key raw materials such as fresh water, productive land, or clean air. The outcome of this journey was the publication of the first environmental

profit and loss account - EP&L - in 2010 (see Box 5.1). The project adopted an extended supply chain-view, from raw materials to the point of sale, showing a total cost of €145 million
in terms of environmental impact. Looking at where the impact happened, 85% of the environmental footprint was generated outside what can be considered the traditional corporate
boundaries or the areas in which Puma has direct control or influence. More particularly,
Tier 4 suppliers were responsible for 57% of the overall environmental cost of producing and
selling Puma's products (Puma, 2010).

   Puma EP&L is not an isolated case. As previously illustrated (see Chapter 3), large sustainability impacts (not only water consumption, GHG emissions and ecosystem service deterioration, but also social aspects such as human rights abuse or corruption issues) occur along
the supply chains from the phase of raw materials cultivation and extraction to components
production, logistics and services, and extends beyond the traditional boundary of the firm.

   In this epoch of globalization, production processes are often dispersed around the globe,
and large manufacturing companies, suppliers, and customers are increasingly connected by
information, material, energy, and cash flows. Outsourcing and off-shoring have been strong
trends in the last decades (e.g., in the electronics, automotive, and apparel industries), with
implications on the level of tiers of supply chains and on the complexity of coordination and
control. The search for efficiency and cost reduction has been coupled with a growing rate of
innovations in product and services, and pressures on decreased time to market, as companies struggle to match the needs of new consumer segments and growing demands. In this
vortex of challenges, new and increasingly important risks are emerging for companies with
regard to environmental and social aspects. This is especially the case for large multinational
companies that own large brands that are asked to be accountable for their sustainability
impact by stakeholders.

   The 2013 Rana Plaza collapse in a suburb of Dhaka, Bangladesh, when more than 1,130
workers were killed in the deadliest disaster that ever happened in the garment industry,
was just the last of a series of incidents occurring along the supply chains of multinational
brands (Yardley, 2013). Other cases in the same period exposed the inner fragility of a loose
web of suppliers, built mainly on cost reduction purposes: the protest against "mass suicide"
at Foxconn, the Taiwanese electronic contract manufacturing giant (Barboza, 2010), or the
scandal of horsemeat contamination in frozen "lasagna" and spaghetti "Bolognese" that
spread around Europe in 2013 (Castle, 2013) heavily damaged the reputation of brands such
as Apple, Nintendo, Sony, Tesco and Nestlé, leading to reflect on the complexity of managing
globalized supply chain.

   The Rana Plaza collapse has contributed to shifting international attention to the unsustainable social and environmental conditions of large brands' suppliers. The consequence of
this tragedy was a policy shift for global corporations and a noticeable growth in stakeholders' awareness. Consumers, NGOs, governments, and increasingly, the financial community
are now demanding transparency along the entire supply chain and have started to ask for
full traceability of product ingredients and components.

   But it is not just a question of pressures from stakeholders, consumer preferences and corporate reputation. Ecosystem services degradation and natural resource scarcity may impact
the capacity of companies to produce and deliver products or services. For example, climate
change - rising temperatures, extreme weather events, droughts and floods - can alter the

growing conditions of crops, decrease yields, and affect the availability of raw materials, therefore disrupting the supply chains (Moore et al., 2017). These extreme events influence the volatility of commodity prices and condition the quality of final products and services. Companies have started to assess these risks, implementing tools that range from traditional risk management and insurance techniques to collaborating with suppliers across organizational boundaries, in order to build long-term resilient supply chains. These strategies have developed to the point that in several industries, competitors have joint forces with NGOs and agencies in designing and implementing programs to tackle environmental and social problems at the early stages of the supply chain, where natural resources are cultivated or extracted. This is the case of sustainability guidelines and standards for commodities such as cocoa, coffee, sugar, bananas, cotton and fisheries.

The growing interest from the community of practice toward supply chain and sustainability linkages (Bové and Swartz, 2016; Ernest and Young, 2016; WEF, 2013) has been coupled by a robust attention from business scholars (Brammer et al., 2011). This has led to the development of a novel and dynamic body of literature consolidating both theoretical and empirical terms (Carter and Easton, 2011). In a nutshell, supply chain management has become a material issue when it comes to corporate sustainability.

This chapter focuses on the upstream part of the supply chains, from focal companies to Tier n. suppliers, analyzing HOW companies respond to the challenges illustrated in the initial sections of this book. We do not include the examination of the retailer–consumer relations and issues such as recovery and recycling, take-back system and cradle-to-cradle design. That will be analyzed in the next chapters. Moreover, we deal mainly with the green aspects of supply chain design and management, but we acknowledge the linkages between the ecological and social issues and illustrate some of these interdependencies.

Box 5.1   Puma Environmental P&L and upstream impact (Tier 4 suppliers)

It was November in 2011 that Puma, the sports brand, published its first environmental profit and loss (EP&L) report providing a first economic measurement of the impact on the natural environment of its global supply chains. This was a first attempt by a multinational company to put a price on the value of ecosystem services used by business to produce products and services – e.g., fresh water, clean air, biodiversity and productive land. Jochen Zeitz, at that time the Executive Chairman of Puma and Sustainability Officer of the company, remarked: "I wanted to know how much we would need to pay for the services nature provides so that Puma can produce, market, and distribute footwear, apparel, and accessories made of leather, cotton, rubber, or plastic for the long run. I also wanted to know how much compensation we would have to provide if nature was asking to be paid for the impact done through Puma's manufacturing process and operations" (Puma, 2011: Foreword). Puma started the project in 2009. It was inspired by the release in 2008 of the "The economics of ecosystems and biodiversity" (TEEB), a global study on the economic benefits of biodiversity and the costs of natural capital loss. The EP&L was developed after a broad review of the main methodologies on ecosystem services and biodiversity evaluations. The company appointed PWC and Trucost, a specialized firm in assessing hidden environmental and social costs of business, to join forces and competencies - e.g., non-market evaluation techniques, ecosystem assessment, supply chain metrics - for designing and developing this new approach. The broad scope of the analysis extended from cradle to gate: from raw material cultivation, to production, to distribution and to sales, extending beyond the production plants directly controlled by the company. Puma, in fact, outsources the majority of its operations to a network of suppliers. The EP&L methodology grouped this web into four levels of suppliers: from Tier 1 suppliers, responsible for manufacturing activities, to Tier 4 suppliers, which included activities such as cotton farming, rubber plantation, or cattle rearing. The environmental impact areas included in the first report incorporated climate change, water use, loss of biodiversity and ecosystem services (land use); air pollution; and waste. The analysis covered all the business segments (footwear, apparel, accessories) and regional areas (Europe, Middle East and Africa - EMEA, Americas and Asia-Pacific).

The outcomes from the first EP&L showed that the costs of nature utilization were €145 million, but somehow surprisingly only 6% referred to the company direct operations, and only 9% was the responsibility of Tier 1 suppliers, the firms responsible for Puma's shoes and apparel manufacturing. On the other hand, 85% of the impact was generated outside areas were Puma had direct control. In particular, 57% (equivalent to €87 million) was generated by Tier 4 suppliers, at a supply chain stage where the company has low influence capacity, and innovative strategies are required to enforce sustainability practices and standards.

The EP&L methodology and report received large attention from the media, NGOs, authorities, and businesses, obtaining acknowledgment by the academic community and receiving important awards (e.g., The Guardian Sustainable Business award). The findings showed the importance of the environmental costs generated at the early stages of the supply chains. Managing and mitigating the ecological impacts, for a company that targets true sustainability goals, asks for innovative solutions at Tier 2 to Tier 4 levels, stages that have received less attention from companies when compared to manufacturing activities strictly owned or direct suppliers. To tackle these challenges, Puma began a collaboration with multiple stakeholders, adopting a new managerial paradigm grounded in a holistic view of the firm that acknowledges its interdependence with nature and the need to guarantee its long-term preservation. In the words of Jochen Zeitz: "Even those concerned only about bottom lines—and not the fate of nature—must now begin to realize that the sustainability of business itself depends on the long-term availability of natural capital" (Puma, 2011: Foreword).

Over the years, the EP&L methodology has been extended to the Kering Group, the global luxury company that owns Puma and several luxury houses in fashion, leather goods, jewelry, and watches like Gucci, Bottega Veneta and Saint Laurent. The Kering 2015 EP&L reported an environmental cost of €811 million, with revenues about €11,500

ILLUSTRATION OF PROCESSES AND IMPACTS THROUGH PUMA'S SUPPLY CHAIN

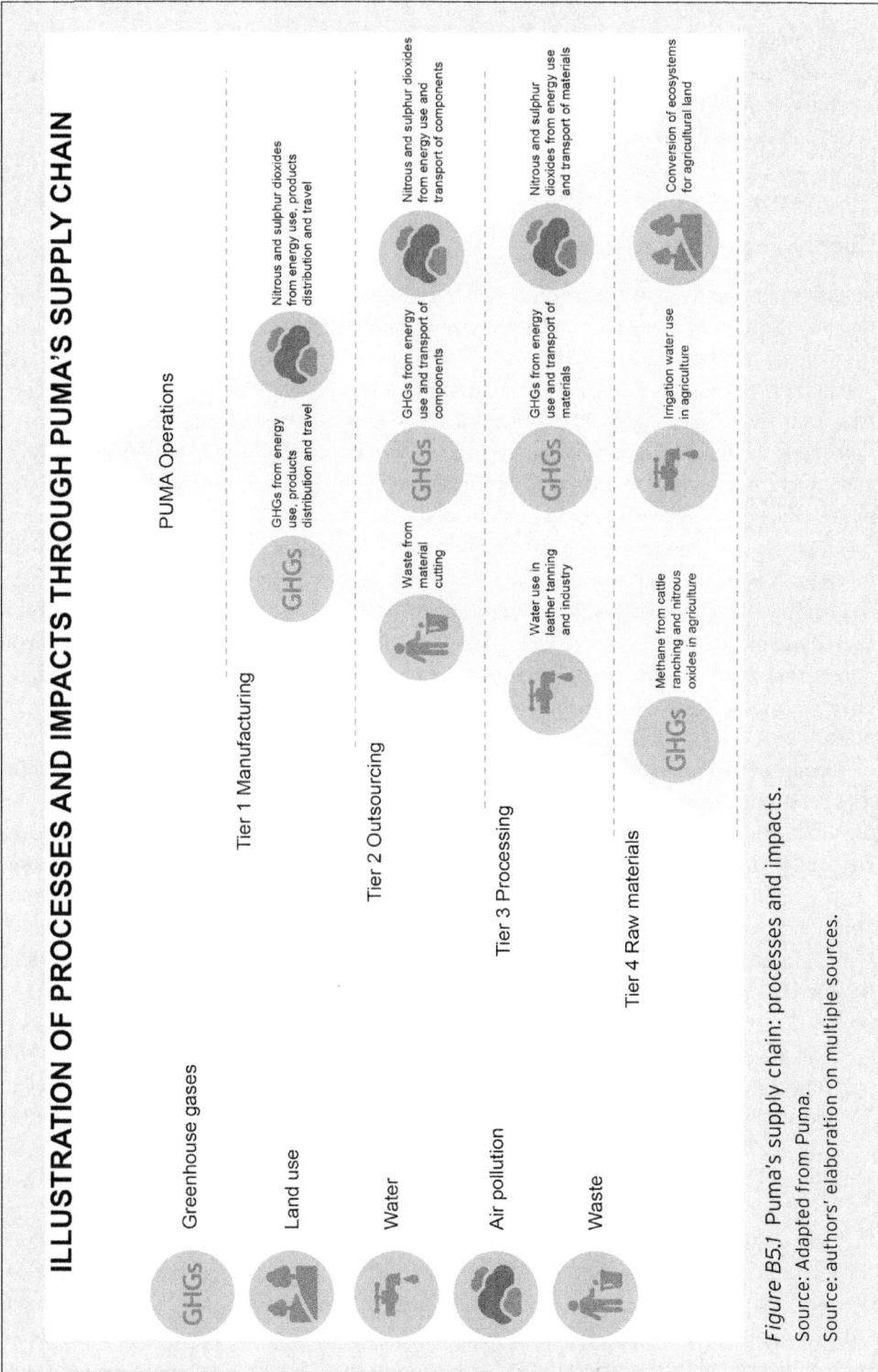

Figure B5.1 Puma's supply chain: processes and impacts.
Source: Adapted from Puma.
Source: authors' elaboration on multiple sources.

million (Kering, 2015). The company has further implemented the methodology that is now fully integrated into the sustainability strategy and supported decision-making with regard to raw material production and processing in order to reduce the impact on natural capital and the risks of supply chain management.

Supply chain management

Long-term trends such as globalization, increased competition and market uncertainty have impacted corporate behavior and strategy, and supply chain management that has evolved from an operational scope of activities to a critical success factor capable of contributing to competitive advantage. For example, the secret behind Inditex's success, the Spanish company that owns brands such as Zara and Bershka, lays in its capacity to quickly respond to demand with an agile, lean, fast and vertically integrated supply chain. In the case of Amazon, much of its achievements relate to logistics, shipping and inventory capabilities that allow fulfilling the online shopping experience of its customers.

As previously illustrated, sustainability provides additional challenges for supply chain managers who are increasingly required to address stakeholders' pressures for environmental and social issues while coping with cost reduction and efficiency targets or searching for flexibility and service quality. Before investigating these challenges and corporate responses, we will briefly answer the following questions: What exactly is a supply chain? How supply chain management can contribute to strategic goals and competitive advantage?

A supply chain consists of the network of individuals, organizations, resources, and activities that contribute to responding to the customer demand (Chopra and Meindl, 2013). This architecture includes the web of suppliers, the manufacturing and assembling centers, the warehouses and the transport systems, the information and technologies, the retailers and distribution outlets, the work-in process inventory and the finished products, the technologies and the data management systems. Another definition of supply chain provided by Handfield and Nichols focuses on the flow of information and goods and on the relationship between the different organizations:

> The supply chain encompasses all activities associated with the flow and transformation of goods from raw materials stage (extraction), through to the end user, as well as the associated information flows. Material and information flow both up and down the supply chain.
>
> (Handfield and Nichols, 1999: 2)

In the supply chain architecture, the focal company plays the role of designer of the network, orchestrator of the flows of goods and information, designer of the product and services offered. This can be the case of an original equipment manufacturing (OEM); large brands such as Apple, Toyota and Coca-Cola; retailers like Walmart and IKEA; and internet-based companies like Amazon and Facebook. The focal company also has the responsibility to align the interest of the multiple entities participating in the network that sometimes can be in contrast with the focal company since single organizations pursue their own agenda. From the perspective of the focal company (see Figure 5.1), the supply chain can be separated into two main parts:

- the **upstream phase** that extends from raw materials, through multiple tiers of suppliers, to the focal company, and
- the **downstream phase** that links the focal company and its final products to the final customers through a web of distributors and retailers.

Supply chains can be very long and geographically dispersed, with numerous ties and multiple entities involved around the globe, or short, with few ties geographically focused. For example, in the case of cotton, many different operations and entities are involved: farmers (cotton growers), ginners, yarn spinners, traders, transporters, mill and fabric workshops (dyeing, weaving, knitting, cutting and sewing), and finally retailers and brands. Raw materials and work-in-process items travel around the world, orchestration and control is complicated, and transparency is generally low. On the other hand, if we consider the case of cement, the structure of the supply chain is very short and much more simple: raw materials are taken from limestone quarries, combined with other materials usually coming from a short distance, and used for construction.

Another important consideration refers to the complexity of supply chains. Despite the fact that "chain" suggests the idea of linearity and simplicity, supply chains often hide complex architectures, with multiple interactions and transactions among entities at each level of the chain, and loops that connect suppliers–customers to different supply chains. Moreover, supplier–customer relations are not exclusive: for example, brands like Unilever, Procter and Gamble and Colgate-Palmolive buy raw materials and components (e.g., packaging and ingredients) from the same suppliers and sell to the same retailers.

As previously illustrated, supply chain management has become a critical function in the modern business and requires many strategic and tactic decisions with regard to the flow of goods, information, and funds. These decisions lead to the improvement of the performance of the focal company, and therefore need to be consistent with the corporate and competitive strategy, but also need to be consistent also with the performances of the supply chain as a whole (Mentzer et al., 2002). In other words, supply chain decisions cannot be taken in isolation and entail strong coordination both internal to the firm, with other functions and departments (e.g., strategy, marketing, finance, innovation), and also with other entities external to the focal company (e.g., suppliers and distributors).

Another important dimension in supply chain management is the time frame of the decisions. *Strategic decisions* are long-term investments and deal with the design of the supply chain network, including the level of outsourcing and the geographical distribution of the network architecture, number of suppliers, installed capacity and the type of shipping/transport. *Tactical decisions* are short term and refer to the management of everyday operations. They include managing incoming customer orders, lead-time quotations, inventory policies, and transport loading. Purchasing policies – which involve suppliers' selection, negotiation and collaboration – are also part of supply chain management processes.

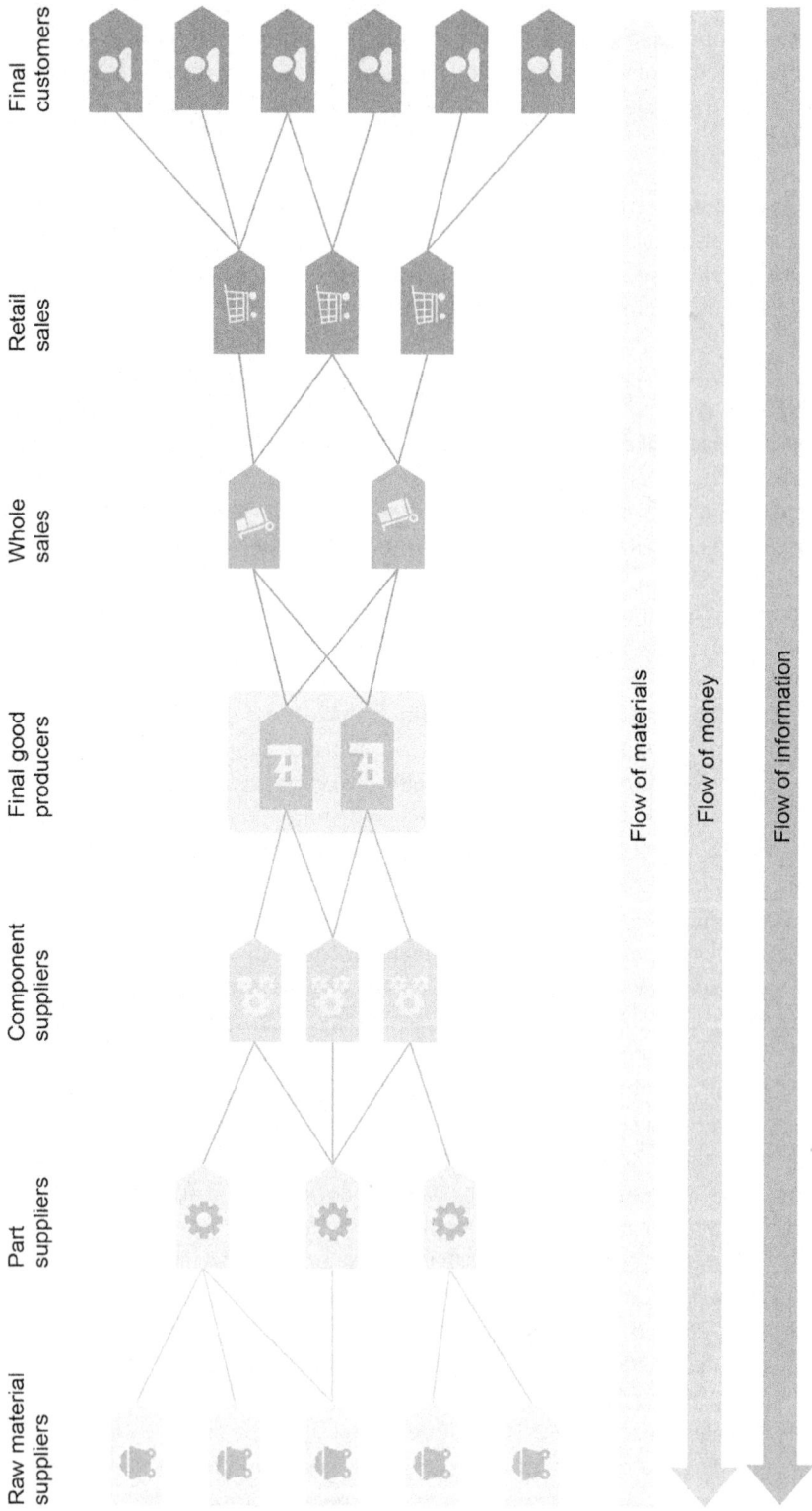

Raw material suppliers | Part suppliers | Component suppliers | Final good producers | Whole sales | Retail sales | Final customers

Flow of materials

Flow of money

Flow of information

Figure 5.1 Supply chain structure.
Source: authors' elaboration.

Another important aspect that we need to illustrate pertains to the relation with the competitive strategy of a company in order to develop a coherent supply chain. Chopra and Meindl (2013) highlight the importance of aligning the goals of the competitive strategy with the goals of the supply chain strategy. Two variables need to be considered: (1) the characteristics of the demand and its uncertainty; and (2) the structure of the supply chain and its capability to perform a specific set of tasks. On the one hand, the demand reflects the type of customers the company is addressing and the level of predictability or uncertainty. On the other hand, the supply chain structure and capabilities to perform reflect the type of responsiveness and the level of efficiency. Responsiveness, which is the capacity to be agile and flexible, and providing a variety of products with high level of service usually come at a cost. Efficiency, therefore, cannot be matched with responsiveness, leading to different strategic approaches to supply chain management that attempt to optimize the fit with the competitive strategies: the request for cost leadership pairs with efficiency-driven and lean supply chains, while product differentiation is coupled with fast, flexible and service-driven ones.

For decades, the concepts we have briefly discussed in this illustrative section have marked the debate on supply chain management, leading to a strong orientation toward the economic dimensions as a target. At the same time, the supply chain is not only a key determinant of the competitive performance of a firm, but it also deeply influences its environmental and social performance. Supply chains are the roots that materially ground companies to social–ecological systems, linking the corporate strategy to the utilization of natural resources and ecosystem services through the web of suppliers and distributors. Decisions such as the number of tiers in the supplier network, the level of globalization and the location of the manufacturing plants, the system of shipping and transport, the orientation toward efficiency or responsiveness, and the type of collaboration with the different supply chain members impact the sustainability profile of the company. Accordingly, turning to supply chain management is a logical step toward corporate sustainability. We will now look at the components of a sustainable supply chain, focusing on the upstream phase.

Sustainable supply chain

What is a sustainable supply chain?

Over the last 25 years, sustainable supply chain management has become a mainstream topic in both academic and practitioner literature, as witnessed by several publications and special issues in management and organization journals (Krause et al., 2009; Carter and Easton, 2011; NBS, 2011; Sarkis et al., 2011). This literature shows that a multitude of labels have been used, including green supply chains, socially responsible supply chains, closed-loop supply chain, resilient supply chain, just to name a few (see Table 5.1.). It therefore seems critical to try to better clarify what is a sustainable supply chain: How can we define it? How is it different from a traditional supply chain? What are the main dimensions?

Seuring and Müller (2008: 1700) define sustainable supply chain management as:

the management of material, information and capital flows as well as cooperation among companies along the supply chain while taking goals from all three dimensions of sustainable development, i.e., economic, environmental and social, into account which are derived from customer and stakeholder requirements.

Pagell and Wu (2009: 38) provide another definition elaborating the concept of "truly" sustainable supply chain as:

> a supply chain would at worst do no net harm to natural or social systems while still producing a profit over an extended period of time; a truly sustainable supply chain could, customers willing, continue to do business forever.

United Nations Global Compact (UNGC) and Business for Social Responsibility (BSR) in a key publication dated 2010 introduce the concept of "supply chain sustainability" as "the management of environmental, social and economic impacts, and the encouragement of good governance practices, throughout the lifecycles of goods and services." And they add: "The objective of supply chain sustainability is to create, protect and grow long-term environmental, social and economic value for all stakeholders involved in bringing products and services to market" (UNGC and BSR, 2010: 5). Other definitions can be found in Table 5.1.

These definitions share some distinctive features:

- they focus on the three dimensions of sustainability – the environmental, the social and the economic impact generated,
- they connect supply chain management with the stakeholders' view,
- they adopt a long-term view incorporating a performance perspective that extends over time.

Another important, and somehow controversial, aspect of sustainable supply chain management is the prevalent focus on an instrumental perspective that translates into the search for specific business cases and win-win opportunities. Authors such as Montabon et al. (2016) have recently raised this debate providing a broad review of the main publications in the field. According to their analysis, sustainable supply chain management builds on the older vision of sustainability (as identified as business sustainability 2.0 in Chapter 2), adopting an approach oriented to maximizing the value of the company while meeting customer needs. This often translates in postponing the environmental and social dimensions to the economic one. In other words, when a business is facing trade-offs between the ecological, social and economic dimensions, and when win-win perspectives are not emerging and cannot be easily captured by the firm, the predominance of profit orientation still drives supply chain decisions to the detriment of the preservation of social-ecological system resilience.

Moreover, another important aspect identified by these authors is a prevalent attention to the focal firm and to the minimization of the impact generated by its direct suppliers (e.g., focusing on the relations with Tier 1 suppliers), while underestimating the social and environmental impacts produced by the extended supplier network. On the other hand, as we have seen with the case of Puma EP&L, and with several other examples, for many companies,

Table 5.1 Some definitions of sustainable supply chain.

Green supply chain	Green supply chain focuses mainly on the environmental impacts of supply chain activities (e.g., air and water pollution, consumption of natural resources) and the relation with the economic dimension. The social and ethical aspects are not addressed.
Responsible supply chain	Similar to the sustainable supply chain, the responsible supply chain is a term broadly used in literature (e.g., OECD Guidelines for Multinational Enterprises, 2011). It refers to the concept of responsible business and corporate social responsibility and focuses on the social and environmental dimensions of the relationship between a corporation and its suppliers. Strong attention is paid to aspects such as labor conditions and human rights protection.
Resilient supply chain	A resilient supply chain refers to the capacity of the supply chain to cope with, resist and recover from events such as natural (fires, earthquakes, floods, droughts) or man-made disasters (equipment failures, terrorist attack, suppliers default). The concept links to the idea of risk of disruption, which will be further discussed in a later section (see pages 195–203).
Closed loop supply chain	Closed-loop supply chains focus mainly on the down-stream phase and on taking back schemes for products from customers. It is linked to the notion of life cycle and to the idea of generating value by reusing or recycling the entire product and/or some of its parts.

Source: authors' elaboration using multiple sources.

much of the impact is generated upstream and downstream from the operations that the focal firm directly controls. Several solutions that at first glance might seem good for the focal firm could prove unsustainable when we look at a broad picture, including the web of suppliers.

In this book, our approach based on the Business In Nature (BInN) concept acknowledges the social and ecological embeddedness of firms. We therefore share the critical view by Montabon et al. (2016) and propose a holistic approach to sustainable supply chain management that extends upstream to the first stages of the supply chain where raw materials are extracted and cultivated. Sustainable supply chain management is not just about minimizing the impact, doing less bad, and operating with integrity; it is about managing the extended supply chain network within the natural and social thresholds.

Distinctive features of sustainable supply chains

Sustainable supply chain management incorporates the environmental and social impacts generated by the company along the entire web of suppliers, but what are the specific ingredients of these two dimensions?

One example can help us. One of the industries where supply chain management has changed more rapidly over the last two decades is the garment and footwear industry (OECD, 2017). The emergence of fast fashion has required the development of supply systems that need to combine quick responses to uncertain demand and match the product design with minimal lead times and low inventories. These supply chains have become geographically dispersed, with a varied range of entities operating at all stages, including farmers and fiber producers, buying agents, traders, textile manufacturers, etc. Big brands and retailers operating in this industry have developed different sourcing models that translate in different business relationships (e.g., transactional or long-term and stable relations) and diverse capability to monitor and influence the activities of their suppliers and intermediaries.

Risks of adverse impact on the social and environmental dimensions are well documented and emerge at multiple stages of the supply chains across different contexts. This includes, for example, the ecological impacts of cotton cultivation and leather production, use of chemicals in the textile finishing, forced and child labor issues at multiple stages, restriction to freedom of association for workers, discrimination and low wages, fraud and corruption (OECD, 2017). Other industries face similar problems, with risks that range from human rights abuse in the case of conflict minerals (e.g., tantalum, tin, tungsten or gold) used by companies in electronics and communications, aerospace, or automotive, to deforestation, ecosystem deterioration, and labor issues in the case of palm oil, natural rubber or forestry.

We identify three main types of risks that result from globalized supply chains, often deeply interconnected and interdependent.

- **Environmental issues.** Companies need to address the different kinds of environmental impacts generated by their network of suppliers. As previously illustrated with Puma EP&L (see Box 5.1), the damage to the ecosystems generated by the extended supply chain can be extremely severe. Environmental impacts might have different levels of intensity according to the characteristics and the resilience of the local/global ecosystems affected by the industrial processes. Therefore, specific assessment techniques are required to better understand what is (see Chapter 8 for further explanation) and what needs to be prioritized in terms of actions and interventions. Areas of impact include air emissions that affect air quality (e.g., pollutants like NO_x, SO_x, VOCs) and contribute to climate change (GHGs); waste generation that enhances the risks of soil and water contamination (e.g., toxic waste) and signals inefficiencies in the transformation processes; water consumption and water emissions; and biodiversity loss, land use and consumption of ecosystem services (e.g., soil utilization, use of marine natural resource) that can reduce the health of the ecosystems with severe implications for land or marine productivity.
- **Labor issues and human rights.** Labor issues and human rights are extremely important when analyzing sustainable supply chains management (see Table 5.2 for some facts and figures). Here, we refer mainly to the ILO (International Labor Organization) landmark Declaration on Fundamental Principles and Rights at Work that was adopted in June 1998, ratifying a list of previous ILO Conventions on labor aspects. This Declaration highlights a set of core principles and establishes a common "social floor" in the world of work (ILO, 2002: 7). In the last two decades, the Declaration has become a point of reference for multinational companies with global supply chains, where stakeholders and final consumers have started to focus on fair treatment of workers and respect of human rights.

Table 5.2 Labor, human rights, corruption and bribery (some facts and figures).

Forced labor	According to the ILO, 21 million people are victims of forced labor, of which 11.4 million are women and girls and 9.5 million are men and boys. A large proportion of forced labor victims are exploited by private firms. Forced labor in the private economy generates about US$150 billion per year of illegal profits.
Child labor	According to ILO statistics, the number of children in child labor stands at 168 million despite a decline of 246 million since 2000. Asia and the Pacific are the regions with the highest numbers (78 million) and agriculture is by far the most important sector where child labor can be found.
Freedom of association	According to the ITUC – International Trade Union Confederation – Report on Global Rights Index (2016), workers' rights have been weakened in many regions of the world. This situation is particularly severe in the Middle East and North Africa where no freedom is guaranteed. The worst countries for workers' rights are Belarus, Cambodia, China, Colombia, Guatemala, India, Iran, Qatar, Turkey and the UAE. Out of the 141 countries analyzed, workers don't have the right to strike in 68%. Several trade union activists are killed, imprisoned, blacklisted, dismissed and injured every year.
Corruption and bribery	According to a survey by the OECD (2014) on enforcement actions for foreign bribery activities (427 cases examined), two-thirds of the corruption occurred in four sectors: extractive (19%); construction (15%); transportation and storage (15%); and information and communication (10%).

Source: authors' elaboration on multiple sources.

- **Forced labor** is defined as: "all work or service which is exacted from any person under the menace of any penalty and for which the said person has not offered himself voluntarily" (ILO, 2002: 25-26). In other words, forced labor means that workers can be threatened with severe deprivations such as withholding food or land or wages, physical violence or sexual abuse, restricting peoples' movements or locking them up (ILO, 2002: 23). The issue of *child labor* is more complex to define since certain forms of work can be acceptable for children at different stages of their development. The principle of the effective abolition of child labor means: "ensuring that every girl and boy has the opportunity to develop physically and mentally to her or his full potential" (ILO, 2002: 43). The aim, in this case, is to stop all types of work that can endanger children's education and development. Forced and child labor represent main risks to supply chain management when companies are subcontracting in sensitive countries with national laws that are less stringent than ILO principles (e.g., some Asian or African countries) and when business models are based on multiple product lines and loose networks of suppliers. As a result, a reduction of the transparency and of the possibility to monitor and control the supply chain calls for innovative policies for supplier procurement.

Another important dimension of a sustainable supply chain is the protection of the *human rights* of workers and other stakeholders involved directly or indirectly by the network of suppliers. The major principles of human rights were defined in 1948, when the United Nations adopted the Universal Declaration of Human Rights. Particularly important is the case of freedom of associations and collective bargaining that were adopted by the International Labour Conference (ILC) in 1948 and 1949 with two break-through Conventions (N. 97 and N. 98). *Freedom of association* is a basic human right linked to freedom of expression and forms a part of the foundation of democratic representation principles. It means that all workers and all employees have the right to "freely form and join groups for the support and advancement of their occupational interests" (ILO, 2002: 9). *Collective bargaining*, instead, refers to the right of the workers (their organizations, representatives, or trade unions) to discuss and negotiate their relations, in particular terms and conditions of work. Respect of both principles is an essential feature for today's business and requires a broad supply chain oversight by corporations.

The fourth pillar of this Declaration is the elimination of any *discrimination at work*. This principle encompasses aspects such as gender, race or skin color, religion, political opinion, social extraction, disabilities, age, and status. Every worker, in fact, needs to have the same opportunities – equality principle – and receive the same treatment to develop skills, knowledge, and competencies relevant for his/her contribution to the economic activity of the organization, career, and personal wish.

The list of labor issues that companies need to tackle can be further extended. Other critical features include the respect of *fair wages* or minimum living wages at all stages of the supply chain; the question of *working time*, which consider the need to protect workers' health by limiting working hours; and the broad area of *occupational health and safety* risks, which comprises a number of specific aspects related to the protection of workers, ranging from the adoption of protective equipment, to the implementation of specific procedures for handling toxics and solvents, factory exits, fire risks and programs for employees' training.

- **Corruption and bribery.** In a globalized supply chain, suppliers that engage in corrupt practices or fraud might significantly affect product quality, increase costs and liability, and damage reputation. This misbehavior can be particularly relevant in specific countries and jurisdictions where anti-bribery legislations are not in place or are not enforced. They can also occur when government, media, local firms and civil society fail in promoting transparency and responsible business practices. Similarly, some industrial sectors and activities, such as mining and extraction, construction and transportation and storage (see Table 5.2 and OECD, 2014) are more exposed than others and require high attention. Anti-corruption procedures are therefore useful to create conditions for more sustainable businesses.

To conclude, sustainable supply chain management means addressing the environmental and social challenges emerging along the network of suppliers. Labor issues, human rights, GHG emission, ecosystem services consumption, corruption and bribery when not tackled properly can result in severe risks for the company with the capacity to affect its competitiveness, reputation and the bottom line. How can business manage these risks? How can a company transform from minimizing the harm generated to the natural environment and society through its supply chain to "doing good" to the social-ecological system? We will discuss these topics further in the next sections.

Baseline approaches to sustainable supply chain management

When we think about sustainable supply chain management, probably the first thing that comes to mind is a "code of conduct" and, more specifically, the levels of engagement. Supplier code of conduct, or code of ethics, are broadly diffused tools among large multinational companies and organizations (e.g., international agencies like UN or NGOs such as Oxfam and Amnesty) and support the process of procurement. They are created for the purpose of ensuring that a company's suppliers adopt and integrate a set of procedures in order to comply with environmental and social principles and requirements of the focal company.

According to a recent benchmark study published in 2017 by the École des Hautes Études commerciales de Paris (HEC) and EcoVadis (a specialized service company providing sustainability ratings for global supply chains), Fortune 500 companies seem heavily interested in ensuring that sustainability is embedded into their procurement operations and across the supply chains. About 97% of the managers surveyed think that sustainable procurement is critically important (23%) or important (74%) to their business activities. Fifteen years ago it was only 40%. At the same time, 88% of respondents have in place a supplier code of conduct (or similar instruments, like a contract clause).

Supplier codes of conduct, anyway, are only a partial response to sustainable supply chain management. In order to be effective they must be incorporated into a formalized supply chain policy and linked to the broad sustainability strategy. Building on the OECD guidelines for responsible supply chain management (2010) and on the UNGC and BSR supply chain sustainability practical guide for business (2014), we identify four main steps for developing

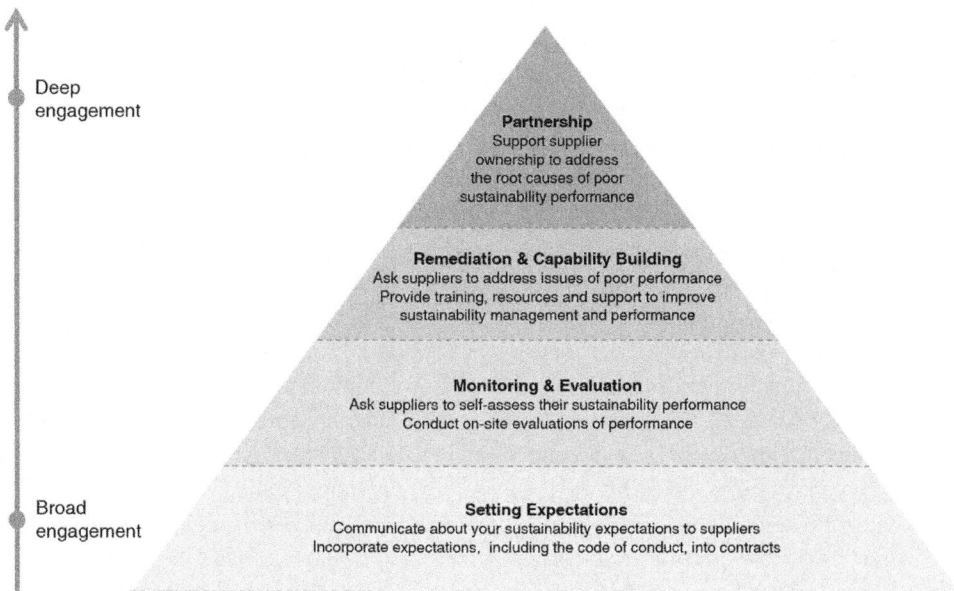

Figure 5.2 Tools for engaging with suppliers on sustainability.
Source: authors' elaboration from UNCG and BSR, 2010: 33.

and implementing a sustainable procurement process. These four steps correspond to a growing level of suppliers' engagement and a higher ability to influence their environmental and social decisions (see Figure 5.2).

- **Setting expectations.** The first level of engagement corresponds to the establishment and communication of the focal company expectations in terms of environmental and social sustainability through the adoption of a code of conduct for sustainable procurement. Codes of conduct are essential documents that allow to align the two parties involved in the exchange. They "create a shared foundation for sustainability from which supply chain management professionals, suppliers and other actors can make informed decisions" (UNGC and BSD, 2014: 23). The design of a code of conduct can be inspired by some of the existing guidelines and international standards: for example, the UNGC, the OECD Guidelines for Multinational Enterprises, the ILO Declaration on Fundamental Principles and Rights at Work. Industrial organizations and business coalitions also provide support to drafting effective codes of conduct. To be effective, a code of conduct must obtain the endorsement of the Board of Directors and the support of the high-level executives, and it should be developed with the involvement and support of key company managers from different functions (procurement and operations, sustainability, communication, institutional affairs). Moreover, it should engage external stakeholders such as suppliers, peers and civil society through a process of formal consultation. Finally, companies should incorporate the codes of conduct into their purchasing orders, asking the suppliers to commit to them in the contract and in the social and environmental requirements and specifications.
- **Monitoring and audit.** In order to assess the effective implementation of a sustainable business conduct, the focal company needs to put in place a system of monitoring and auditing the portfolio of suppliers. A variety of approaches can be implemented to verify the level of compliance, and companies must carefully select what methods to use and what suppliers to apply this to since costs, product quality and reputation can be affected. *Supplier self-assessment*, for example, is evidently cheaper than an on-site audit and can be considered a useful approach to start with and to clearly set expectations. On the other hand, self-assessment procedures can produce inaccurate data when no other verification system is in place. Lack of competencies, lack of trust or fear of being penalized and excluded from the business can motivate loose data collection or misbehavior from suppliers.

  On the opposite side, the supplier's evaluation can be performed through *on-site audits* at facility or headquarter levels. Audits usually respond to a compliance need and, depending on frequency and the length, they can be useful tools to check the implementation of the code of conduct and to monitor the sustainability performance with regard to key indicators (e.g., environmental, labor, human rights). In sum, they are useful methods to identify problems. However, on-site audits may be insufficient to effectively motivate suppliers toward better sustainability practices and to drive a real change in their behavior. Two types of audit can be distinguished: *internal*, performed by the company staff, and *external* (or third party), carried out by independent organizations like auditing firms, NGOs, trade unions and stakeholders. The decision to rely on external auditors or internal ones depends on a number of things such as the level of required competencies, expertise and knowledge, geographical availability, reputation and cost issues (UNGC and BSR, 2014).

- **Remediation and capacity building.** When the monitoring and auditing process reveals non-compliance with the code of conduct, what is the right strategy to implement? Should the focal company delist the suppliers and exclude it from its portfolio? Terminating the relation with the supplier is probably the last available option when inadequacies and misbehavior are repeated and considered not remediable. Instead, when the expectation is clearly defined between the buyer and the supplier and they are contractually shaped, the focal company should use remediation and capacity building to engage the supplier in actions for improving its conduct. *Remediation* consists in collaborating with the supplier in order to design a corrective action plan to address the problems of compliance. A joint roadmap for improving the supplier capacity to match the social and environmental expectations is also useful. *Capacity building* aims at increasing the knowledge and capabilities of the supplier in order to improve the environmental and social performance. There exist several methods and techniques for capacity building, ranging from training and workshops to call centers and use of distance learning resources.
- **Partnership.** The last step of this process is establishing a true partnership with the suppliers with regard to sustainability. Actions at this level range from: (1) establishing common sustainability targets (e.g., emission reduction targets for GHG, water consumption, or increase share of renewable energy over energy use) that integrate suppliers into the focal company sustainability strategy, to (2) developing collaborative innovations in order to tap into the supplier capabilities to identify and implement high-impact and scalable solutions to improve the overall environmental and social performance (see Box 5.2 for a couple of examples).

Partnership for sustainability is still not a common practice as highlighted by several reports and surveys (CDP, 2017; HEC and EcoVadis, 2017; UNGC and BSR, 2014), but it becomes more and more critical as large brands and retailers are increasingly considered accountable for the impacts of the entire supply chain. Moreover, since large environmental and social impacts are often several links away from the focal company, sustainability calls for innovative approaches and new forms of collaborations. These issues will be further discussed in the last paragraph.

Box 5.2   Examples of partnership with suppliers on sustainability

BMW

BMW works with over 13,000 suppliers around 70 countries. The carmaker company has a sustainability policy for its supplier network that includes "a commitment to environmental and social responsibility and to the ten principles of the United Nations

Global Compact, as well as compliance with all internationally recognized human rights and labor and social standards" (BMW, 2017: 1). For the top 100 suppliers, a yearly based assessment on carbon emission key indicators is performed (e.g., emission targets, reduction initiatives, and changes in absolute emissions) (CDP, 2017).

LEGO

In June 2015, the LEGO Group announced it would invest about US$152 million in sustainable materials to manufacture LEGO® toys and packaging. The Danish company, which produces more than 60 billion elements each year, has developed an ambitious sustainability strategy aimed at reducing its carbon footprint through investing in renewable energy and Forest Stewardship Council (FSC)-certified packaging. Finding alternative materials is paramount to its success: this goal requires the company to establish strong partnerships with suppliers since only 10% of LEGO's carbon impact of the supply chain originates at the factories producing LEGO bricks (LEGO, 2015). The company is therefore joining forces with suppliers and World Wildlife Fund (WWF) to research, identify and implement low-carbon materials (e.g., new bio-based solutions) into its products.

Source: authors' elaboration on multiple sources.

In conclusion, supply chain and procurement objectives such as efficiency, quality, speed, and flexibility are complemented by the quest to include the environmental and social dimension as well. Still the prevalent focus is on more baseline actions such as the introduction of codes of conduct and monitoring of supplier compliance, but among large multinationals and global brands, there is a broad recognition that "within their vast, complex and sometimes opaque supply chains there are a number of sustainability challenges, resource risks, and efficiency opportunities" (CDP, 2017: 6). Despite this awareness, more intense action is essential, and a large majority of companies must move upstream in the network of suppliers, beyond the organizational boundaries or the Tier 1 relationships. Sharing sustainability strategies with sub-contractors, intermediaries and Tier n. suppliers is the missing link to create sustainable and resilient social-ecological systems as indicated in one of the main principles of the BInN concept (collaboration). In paragraphs that follow, we will illustrate in detail how companies can transform their supply chain in the earliest phases of cultivation and extraction of raw materials. We will look first at supply chain risks and resilience, and then into multi-stakeholder initiative and certification standards, traceability and transparency.

Supply chain risks and sustainability

It was in late October 2012 that hurricane Sandy heavily hit the Caribbean and the Northeast coast of the US leaving dozens dead, thousands homeless, and millions without electricity.

The storm generated fuel shortages that blocked ports and airports resulting in huge consequences on supply chains and logistics. The overall cost topped US$70 billion (WEF, 2013: 9). In the last few years, prolonged droughts and worsening water scarcity are impacting crops and harvest and agriculture productivity in many regions. According to CERES, a business partnership focusing on sustainability based in Boston, several multinationals in the agro-food industry are unprepared to face water shortages and will suffer increased costs and a decline in profit if they do not address these challenges with specific strategies and partnerships with suppliers (Ceres, 2010).

Supply chain vulnerability has been addressed by many recent publications, acquiring momentum among managers and executives (Simchi-Levi et al., 2014; WEF, 2012, 2013). Operational risks such as supplier lead-time reliability, transportation delays, demand and price volatility have been successfully addressed with specific strategies aiming at maintaining safety stocks, splitting orders among suppliers, and developing hedging techniques. Companies, knowing the likelihood of occurrence and the magnitude of the impact, have addressed these risks building on historical data and management techniques in order to increase efficiency and minimize the costs for the focal company.

Today, new risks have emerged linked to transformations occurring in the social-ecological systems: natural disasters and extreme weather, global energy shortages, international conflicts and terrorism, viral epidemics, sudden demand shocks and price volatility, cyber attacks and ICT disruption (WEF, 2013, 2018). In this new scenario, risks have become much more interdependent, wide-reaching and "contiguous," therefore amplifying the potentially disruptive effects on firms. Data on these events are often not available, and it is not easy to foresee the consequences in terms of the damage and financial impacts of these events.

The goal of this book is to focus on the new BInN paradigm and to illustrate corporate responses to the interdependencies between business and social-ecological systems. Therefore, we will not discuss further the effects of geopolitical, technological and economic transformations on supply chains and corporate sustainability. Still, it is paramount to briefly recall the linkages among these issues. For example, several papers by the UN, World Bank and OECD have investigated and discussed the effect of climate change and water scarcity on increased poverty, geopolitical instability, conflicts and migrations.

Climate change is having a significant impact on meteorological and hydrological events, increasing the frequency and intensity of storms, hurricanes, floods, heat waves and droughts. As in the case of Hurricane *Sandy*, these extreme events can damage infrastructures, interrupt transport and manufacturing activities, and finally disrupt the supply chains with significant impact on companies' economic and financial performances. Mining can be severely affected by storms and heavy rains, but also droughts that limit the availability of water can impact the costs of extraction, which is a water-intensive operation. Oil and gas operations through offshore platforms and onshore activities can also be damaged by hurricanes, cyclones, and extreme weather events. Probably the most seriously affected industry is agro-food. According to a recent report by the Food and Agriculture Organization (FAO), this sector will bear the major brunt of the impact of climate change (FAO, 2015; Moore et al., 2017), with immediate shortage of crops, livestock, fisheries and forestry. Rice, maize, wheat, coffee, tea, soybean and cotton are all commodities affected by temperature rise and water scarcity. Supply chain risks for agro-food companies can increase significantly in the next decades depending on:

- how much a specific crop is vulnerable to temperature variation, pests and occurrence of storms, flood and droughts,
- what type of climate adaptation policies countries and local governments are implementing with regard to regions where their suppliers are localized, and
- how much the company supply chain is concentrated in specific countries and geographical areas, and on a small number of suppliers.

Resource scarcity and ecosystem services degradation represent another important risk factor for supply chain management and companies. We have seen in Chapter 1 that entering the Anthropocene, ecological systems are under heavy pressure due to human activities, with the risk that several ecosystem services cannot be any longer taken for granted (MA, 2005). In other words, when ecosystems are stressed persistently and cumulative effects reach a certain threshold, they may suddenly shift to another state with unpredictable effects on the capacity to provision services (Winn and Pogutz, 2013). The consequences of ecosystem degradation can dramatically influence the price of raw materials and their availability, with severe risks of supply chain disruption. In Box 5.3, we briefly illustrate the case of Unilever and the North Atlantic codfish collapse.

Box 5.3   Ecosystem degradation and supply chain (the case of Unilever and codfish)

Unilever is now known as the "sustainable-living" brand thanks to a landmark strategy unveiled in 2010 by it visionary CEO Paul Polman. The ambitious long-term plan to decouple growth from environmental impact, integrating sustainability goals into all products and operations, led to excellent financial results, huge media visibility, and reputation coupled with significant environmental and social benefits. The Anglo-Dutch company, with global revenues about €52.7 billion in 2016 and 2 billion consumers worldwide, is a giant in consumer goods, but it is also highly dependent on the availability of natural capital for the provisioning of raw materials like crops, palm oil and water.

In the 1990s, Unilever was one of the world's largest buyers of fish for its several leading brands in the market: iglo, Findus and Birds Eye. In particular, cod was the main fish species used in the company's premium frozen food products (fish sticks). Due to over-fishing through the 1900s, the cod stock declined abruptly in the western North Atlantic to the point of an almost permanent collapse. As a result, cod prices increased dramatically over the years, with relevant effects on the margins of Unilever's cod-related range of products and serious risks for the supply chain (Winn & Pogutz, 2013). The company, one of the main players in the market, decided to engage in the issue taking the lead in encouraging more sustainable fishing. They decided to start a dialogue with the WWF in order to establish a new market-based mechanism to address the challenge of sustainable fishing. WWF at that time had succeeded in introducing sustainability practices in the forestry sector with the FSC. Unilever and WWF in 1996 started to work at the Marine Stewardship Council (MSC), an independent organization launched in 1997 with the aim to certify that the fish comes from sustainable fishery practices and to orient consumer decisions toward seafood from responsible fisheries thanks to an eco-label on the packaging. The company also committed to purchase all the fish from sustainable sources by 2005 (Unilever, 2003). Unilever ended up divesting of the European frozen food business in 2006 for about €1.7 billion. After 20 years of activity, MSC is the standard for sustainable fishing. In 2016, 296 fisheries in 35 countries were certified with the MSC Fisheries Standard.

Source: authors' elaboration on multiple sources.

Natural disasters, climate change, resource scarcity and ecosystem degradation are increasing the risks for supply chain management and require companies to be prepared for these events. As previously illustrated, it is not enough to focus on Tier 1 suppliers. Many events happen upstream in the supply chain where companies usually suffer from low visibility and have a limited capacity to assess the risk exposure. It is therefore paramount to start collecting and sharing data in order to improve risk management capability to cope with these new events, and engage suppliers and stakeholders in order to increase the transparency of the supply chain operations.

The approach that has emerged recently in supply chain management in order to address these new challenges is *supply chain resilience*, or the *resilient supply chain*. We have defined resilience in the initial chapter of this book applied to the notion of social-ecological systems (Holling, 1973; Folke, 2006). A resilient supply chain refers mainly to the capacity to reorganize and deliver its core function continuously, despite the impact of external and or internal shocks to the system (WEF, 2013). In order to increase resiliency, companies will have to further differentiate the portfolios of suppliers and consider the sustainability of the whole supply chain. Moreover, partnerships with suppliers might create the conditions for sustainable innovations and responsive supply systems, with benefits for competitiveness and long-term performance.

Supply chain management: standards, certification, transparency and traceability

The last section of this chapter aims at illustrating the new approaches that leading companies and organizations are implementing to increase supply chain sustainability. We have seen that environmental and social impacts are largely a result of activities outside the boundaries of corporations, or outside their direct control. CDP supply chain report 2016/2017, a document written with 89 leading multinational corporations representing US$2.7 trillion of procurement spending, highlights that 4:1 is the "average ratio of indirect supply chain emissions compared to direct operational emissions" (CDP, 2017: 6). Climate change and ecosystem protection call for intensive transformations of the networks of suppliers. The adoption of codes of conduct and compliance monitoring systems, the threat of delisting suppliers, the request to implement sustainability management systems such as ISO14001 discussed in Chapter 4, are necessary but not enough.

How can companies involve the entire supply chains in the journey toward sustainability? How can Tier n. suppliers (e.g., farmers and smallholders) become partners and collaborate for innovative solutions?

Engaging with "Tier n" suppliers might require involvement in contexts that are very remote, socially and ecologically complex, and somehow distant from the mind-set of managers and executives. Skills and competencies to address these sustainability problems might not be available within the company, or might be difficult to access, requiring large investments and time. Alliances and partnerships with specialized NGOs, industry peers and stakeholders can provide important inputs beyond just sharing ideas and advices. Independent organizations such as Fairtrade, Rainforest Alliance and UTZ have become key players for certifying suppliers (from smallholders to large cooperatives) on sustainability standards, with very specific competencies on issues such as combating deforestation, climate change, poverty, unsustainable farming and improving the well-being of people living in rural communities. These players help companies in building sustainable supply chains contributing to designing effective responses and acting as local implementing partners. Market-based mechanisms based on multi-stakeholder platforms such as the Roundtable for Sustainable Palm Oil (RSPO), the MSC, the FSC or the Conflict-Free Sourcing Initiative (CFSI) are also becoming increasingly important among companies, developing an emerging body of data and knowledge and driving real impactful results at supply chain level. In the following sections, we will discuss sustainability standards and certification, transparency and traceability.

Sustainability standards and certification

What does it mean that a company is sourcing its raw materials from sustainable sources? There are no general criteria to define what is a sustainable commodity: it depends on several factors related to the type of raw materials considered and to the social-ecological system conditions of its origin. It is therefore necessary to conform to a sustainability standard system that establishes a set of minimum requirements for the ecological, social and economic dimensions and assure the accomplishment with these requirements. Only raw materials or components that achieve these minimum requirements can then be sourced by the company.

A sustainability standard can be defined as a set of "voluntary predefined rules, procedures, and methods to systematically assess, measure, audit and/or communicate the social and environmental behavior and/or performance of firms" (Gilbert et al., 2011: 24). As previously illustrated, they are innovative market-based mechanisms to achieve sustainability impact, engage responsible suppliers along the entire supply chain, and promote sustainable production among consumers. Here are some other features that are very common in sustainability standards (UNGC and BSR, 2014; IISD, 2014).

- They are **voluntary** in the sense that firms and organizations can decide to participate on a voluntary basis.
- They are usually developed and managed by **multi-stakeholder** groups that include firms along the supply chain (from farmers to large brand and retailers), industrial associations, civil society (NGOs and unions), often international and national agencies and standardization organizations. This participatory governance has the capacity to incorporate into the decision-making processes stakeholders who have no voice with regard to international supply chains.
- They are not static or they do not provide a collection of principles and requirements written in stone. On the other hand, they are **dynamic** and subject **to evolution** and **improvement** thanks to the participation of different stakeholders.
- They usually provide a **system** of elements and tools that integrate the standard itself. Examples of these components are the certification process to assure the conformity with the standard, the label that goes on the product packaging to orient consumer decisions, the traceability to guarantee the origin of the product, the capacity building to support and educate farmers and producers (Komives and Jackson, 2014).

Standards are very common in agricultural commodities like coffee, tea, palm oil, soybeans, rice and cocoa, but they also extend to other sectors such as forestry, fishing, mining (e.g., aluminum, gold, steel), hospitality and textile. Ecolabel Index, the largest global directory for sustainability market-based mechanisms, recently reported 465 ecolabels referring to 25 industry sectors and 199 countries (as of June 2017). Table 5.3 offers a list of international sustainability standards.

Sustainability standards, in order to acquire credibility among stakeholders and consumers, need an assurance system in order to check if growers or producers (farms, fisheries and factories) meet the criteria. But how will the assurance system work? Certification is the traditional procedure of assurance: an independent third-party auditor checks for conformity with the standard of a product, process, or organization. The process ends with a written certification (e.g., the use of a label on the product package), which guarantees the buyer that the supplier complies with the requirements. The third-party auditor, in order to perform the certification and assure credibility, has to be a part of a certification body and should be independent from any direct interest in the economic relationship between the parties (supplier and buyer).

Another approach that has acquired relevance with regard to some sustainability standards does not include the provision of a certificate or of a label and relies on some degrees of self-assessment. This approach is called verification and is used by some entry-level standards where the aim is to rapidly enter and scale up the market, mainstream sustainability practices into the sector, and push capacity building with farmers and producers (e.g., BCI or 4C Association). The advantages of this approach are related to the fact that the observations and remarks provided by the independent auditors can be used to strengthen practices

Table 5.3 Examples of standards and multi-stakeholder platforms for commodities.

Forest Stewardship Council – FSC	FSC is the pioneer in voluntary sustainability standards. Founded in 1993, it is a multi-stakeholder initiative operating today in 102 countries. FCS promotes producer-consumer traceability of forestry products and certification according to environmental, social and economic criteria. It also enables consumers to make responsible purchasing decisions by marketing the FCS label. (ww.ic.fsc.org)
Roundtable on Sustainable Palm Oil – RSPO	Palm oil is a controversial commodity: very versatile and with multiple applications in food, detergents and energy; cheap and high yielding. At the same time, palm oil production has negative environmental impacts on deforestation, GHG emission and biodiversity loss. RSPO was founded in 2004 to promote the production of sustainable palm oil. This multi-stakeholder initiative provides a standard based on sustainability principles and provides a certification system and supply chain traceability. As of 2016, about 19% of palm oil produced is RSPO certified (2.57 million hectares of land), with about 139,120 smallholders engaged in the program. Several global brands like Unilever, P&G, Ferrero, Nestlé, Carrefour, Mars, L'Oreal and Starbucks are committed to 100% RSPO-certified palm oil and use the RSPO logo in their products. (www.rspo.org)
Better Cotton Initiative – BCI	BCI was launched in 2005 by WWF as a multi-stakeholder platform to promote better sustainability standards in cotton farming and to make Better Cotton standard a mainstream commodity by creating a large market demand. BCI incorporates players from the entire supply chain, including large brands such as H&M, IKEA, Levi Strauss. BCI has now developed a system to trace cotton from the farm to the gin. In 2016, BCI covered about 12% of the world cotton production in 23 countries and involved about 1,584,900 farmers. (www.bettercotton.org)
Global Coffee Platform – GCP	GCP aims at promoting the diffusion of sustainability practices in a non-competitive approach with a long-term agenda. It is a multi-stakeholder platform with more than 300 members from coffee farmers (small and large), traders, industry players (roaster and retailers) and civil society (NGOs, trade unions, etc.). GCP has established a Baseline Common Code: a set of globally referenced baseline principles and practices for coffee sustainable production and processing. GCP was launched to combine and build on the achievements of two other well-known coffee standards: the 4C Association and the Sustainable Coffee Program. The GCP platform aims at complementing other voluntary sustainability standards that use a "pull" approach (e.g., 4C or coffee certification) with a "push" approach" to mainstream a minimum level of sustainability practices among coffee producers. (www.globalcoffeeplatform.org)
Conflict-Free Sourcing Initiative – CFSI (today, the Responsible Minerals Initiative)	This initiative was started in 2008 by the Electronic Industry Citizenship Coalition and the Global e-Sustainability Initiative with the purpose of providing independent, third-party audit to validate "conflict-free" operations. CFSI is now the most diffused and respected resource for companies addressing conflict mineral issues in their supply chains and operates in line with current global standards and protocols. It is a multi-stakeholder initiative, with over 350 members among companies, from seven industrial associations and other organizations. (www.conflictfreesourcing.org)

Bonsucro	Founded in 2008, Bonsucro is a multi-stakeholder initiative aimed at developing sustainable producer communities, assured supply chains and marketing through a label to the final consumer. It has developed a standard system for sustainable sugarcane practices that encompasses areas such as pesticide, GHG and water use reduction, improved labor practices and reduced farm accidents, increased yields. Supply chain transparency and traceability are guaranteed through third-party certification. Bonsucro has now 470 members, reaches about 159.200 farmers and mill workers around the world and engages 25% of the world sugar cane land. (www.bonsucro.com)
Marine Stewardship Council – MSC	Started in 1996 by its founding members WWF and Unilever with the purpose of providing a standard system for sustainable fishing, the MSC became an independent and multi-stakeholder organization in 1999. MSC is based on three core principles that cover the preservation of sustainable fishing stocks, minimization of the environmental impact and effective management. An independent certification process aims at guaranteeing the full traceability of supply chain to the fishery and allows seafood to display the MSC logo in the package. More than 24,000 products are on sale with the MSC logo in 100 countries; this represents an amount close to the 10% of the total annual wild seafood harvested. About 312 fisheries in over 30 countries are certified with the MSC standard. (www.msc.org)

Source: authors elaboration from different sources.

and improve the sustainability performance over time instead of policing the organization audited. A second advantage is that the costs of the system are generally lower than with certification. The main disadvantages are the higher risks of non-compliance with reduced credibility in the marketplace and among consumers.

To conclude, sustainability standards, in order to be successful and generate environmental and social impacts, must cope with a subtle balance between deep conformity with the principles and the requirements to protect trust and credibility, and containing the costs to acquire and maintain attractiveness for the adopters and the market.

Transparency and traceability

Consumers, NGOs and policy makers are demanding more information about the origin and the ingredients of the products they are consuming. They want to know the exact geographical origin of the coffee they are drinking, whether their shampoo contains harmful ingredients or palm oil, and if their smartphone is using only workforce from factories where human rights are protected. Conversely, some leading companies are starting to make virtue of the transparency of their supply chains to attract consumers and build legitimacy (see Box 5.4 for two short stories). Some questions to start with: How can companies acquire complete visibility on practices that might have taken place thousands of kilometers away from the focal company? What is a transparent supply chain? What is traceability?

Box 5.4   Transparency and partnership with suppliers

H&M

H&M has recently been ranked in the top three of 100 companies scrutinized by the latest Fashion Transparency Index (resulting from the partnership between Fashion Revolution and Ethical Consumer), a ranking released every year according to how much information companies disclose about their suppliers, supply chain policies and practices (Fashion Revolution, 2017). H&M claims that it was the first company in the sector to make the full supplier list public in 2013, including 56% of Tier 2 suppliers. Moreover, it counts on this commitment to differentiate the product and it thinks that better informed consumers can put more pressure on companies to act sustainably (Fashion Revolution, 2017).

Barilla

Barilla, one of the leading Italian groups in the food industry and a worldwide leading pasta company, has developed a supply chain system based on stable collaborations with their suppliers and network of farmers. Building on specific long-term supplier contracts with the growers combined with innovative technologies, the Italian company has developed a sustainability program named Sustainable and Integrated Supply Chain (Pogutz & Winn, 2016). This program actively engages farmers in quality and sustainability and guarantees the efficient monitoring of the entire supply chains of Barilla's key raw materials such as durum wheat, tomatoes, and sugar. The outcome is a transparent and stable supplier network that has protected the company against reputational scandals such as the 2013 horsemeat crisis that forced many major bands to withdraw millions of meat-based products.

Source: authors' elaboration on multiple sources.

Transparency is the quality of being easy to see through, the opposite of being opaque. With regard to corporate sustainability, transparency refers to the extent to which the information about the company, its products, its supply chain network (manufacturing, warehousing and transportation up to Tier n. suppliers) and sourcing locations is made available to its stakeholders. Transparency is becoming increasingly important for business as a consequence of pressing requests by consumers, governmental agencies, and civil society at large. In the digital epoch, with large access to information available on the web and the proliferation of mobile devices and social media, transparency is essential for establishing and maintaining trust and securing reputation. Therefore, large brands need to know exactly who made their products and their components and must share this information publicly. New technologies provide one of the keys to transparent supply chains, the other being collaboration and partnership with stakeholders and peers along the supply chain.

For many years, data and information have been shared in sequential ways along the supply chains, with evident loss of quality and distortion in each node of the chain, like in the "telephone game." Today, companies still struggle in accessing data about both their network of suppliers and raw materials and components that travel around the world. In any case, with Internet and the Cloud, there are several new technologies and data platforms that can offer innovative solutions to increase the visibility over the supply chain (New, 2010). For example, assessment tools that help companies understand environmental and social risks are available online and are easy to access (e.g., Sourcemap); specialized service providers (e.g., Verisk Maplecroft; EcoVadis; Committee on Sustainability Assessment – COSA) can support companies in auditing and engaging their suppliers. Specialized technological devices such as radio-frequency identification (RFID) tags are becoming smaller, cheaper and more flexible, and are thus helping companies in tracing the products from the origin to the end. The same devices can be used to collect and store historical data that can be easily accessed by consumers who can dig deep into the supply chain and the origin of the specific components of the product.

Collaboration along the supply chain is the other key to transparency. We have illustrated the emergence of several multi-stakeholder initiatives that support the diffusion of sustainability standards. These platforms provide access to information among peers at the pre-competitive stage with regard to opaque areas of sourcing or inefficient stages of the supply chain. Sharing experiences and best practices help overcoming the challenges linked to the visibility over environmental and social problems in remote sourcing locations, in particular at Tier 3 or 4 levels.

Complementary to transparency is traceability, defined as "the ability to identify and trace the history, distribution, location and application of products, parts and materials, to ensure the reliability of sustainability claims, in the areas of human rights, labor (including health and safety), the environment and anti-corruption" (UN Global Compact and BSR, 2014: 4). In order to assure traceability, a system is needed to record and follow the products, its ingredients, components and parts from the origin of the sourcing till the ultimate distribution as end products. In sum, traceability assures that the quality and the sustainability claims associated with raw materials, transformation and manufacturing, assembling, distribution and transportation are respected all along the supply chain (UNGC and BSR, 2014: 4).

Track and trace have become vital capabilities in many industries, such as food and pharma, where consumers require very specific information about product ingredients, and safety issues are paramount. It is also becoming extremely important in other sectors influenced by the enforcement of laws and regulations that address specific sustainability issues like conflict minerals or palm oil. Companies can rely on the sustainability standards that contribute to implementing traceability thanks to assurance systems, or can develop their own traceability programs and schemes for specific business-critical commodities. Credibility, costs and efficiency are elements that companies need to balance carefully before deciding on a well-known scheme or designing a new one (Box 5.5 illustrates the case of the development of the Responsible Down Standard).

Box 5.5   The North Face and the Responsible Down Standard

In some cases, private companies decide to start a sustainability standard to increase the traceability of the supply chain of a critical raw material. A few years ago, The North Face, an iconic outerwear brand and a part of the VF Corporation (with about US$12.0 billion in revenues, the VF Corporation is a worldwide leading group in apparel and footwear sectors) became aware of the potential risks related to lack of transparency in the supply chain of the down. Down is largely used by the company in outerwear products such as sleeping bags, footwear and accessories because it guarantees superior properties compared to other materials. The North Face was sourcing down from the food supply chain when problems emerged regarding the mistreatment of animals. The complexity of the supply chain, with several obscure steps where materials changed owners, required an effort involving experts and specialized organizations. In order to guarantee full traceability, The North Face decided to establish a new sustainability standard system engaging two key partners, Control Union Certifications and Textile Exchange (The North Face, n.d.). In 2014, the Responsible Down Standard (RDS) was launched to safeguard the welfare of the geese and ducks that provide down and feathers and to track the entire supply chain from farms to the finished products. As at 2016, around 3,080 farms were certified with RDS in Asia and Europe, and 400 million birds fell under the program (Responsible Down, n.d.).

Source: authors' elaboration on multiple sources.

We identify three types of supply chain traceability models (UNGC and BSR, 2014; Komives and Jackson, 2014). The difference refers both to the facts that certified and non-certified materials are allowed to mix, and the sustainability claims associated with the final product. Schemes with lower controls on certified and non-certified materials are usually cheaper and can be scaled up more easily. Anyway, the issue of cost and efficiency is not the most important feature to be aware of when a company decides to select and adopts a traceability scheme. The sustainability claims, the type of materials used and consumer expectations are the very important features that need to be addressed. For example, in the case of some type of food products, the chosen approach should guarantee the complete physical separation of supply chains in order to increase the product safety.

- **Product segregation.** This approach implies that certified materials and products are physically separated from non-certified ones at all stages of the supply chain. With this model, the consumer knows that the products purchased consist of 100% of certified materials. Product segregation can be further divided into two different models: bulk commodity, where certified materials and products from different producers can be mixed and sold together, identity preservation, where certified materials and products cannot be mixed preserving the identity of the specific farm, plantation or production site where the material originated. As previously mentioned, in the case of food such as organic food, consumers want to know the exact origin of the product, the name of the farm or the identity of the producers.

- **Mass balance.** In this case, the certified and non-certified materials may be mixed, but the certified volumes are tracked and the amount entering the supply chain must be equivalent to the quantities that are sold at the end of the supply chain as certified. This approach is common for some commodities such as cocoa, cotton or sugar, where it is very complex to maintain a segregation of the physical flows.
- **Book and claim.** The last approach does not imply traceability along the supply chain. In this case, sustainability certificates are associated to the certified materials at the origin of the supply chain. The company that obtained the certificates can sell them via a trading system. Companies interested in making sustainability claims can buy certificates on the market, for example, on online trading platforms. Certificates that are negotiated must be equivalent to the amount of material certified, but certification is completely decoupled from the product. This approach is common for some raw materials like palm oil, where the cost for segregation and tracing is very high due to the complexity and fragmentation of the supply chain. The price of traded certificates provides a premium price to the producers that have decided to address sustainability challenges acting responsibly.

Transparency and traceability provide responses to the increased consumer demand about product sustainability and what is inside what they purchase and to stakeholders' pressures on the reduction of corporate environmental and social impacts. It is no longer acceptable to focalize only on Tier 1 suppliers, and companies need to acquire visibility over the entire supply chain in order to favor the adoption of sustainability practices in the earliest stages. Looking at the near future, it is possible to see novel technologies enabling companies to improve their ability to trace every ingredient of their product portfolio, knowing the exact provenance and the social and environmental footprint associated to the generation of each single product. Big data and cheaper devices will help reduce costs of certification and increase the credibility of sustainability standards, and therefore facilitating the diffusion of responsible principles and practices along the supply chain and favoring their market diffusion among more responsible consumers.

Summary

This chapter has explored in depth the upstream phase of the "Business In Nature" (BInN) framework, namely the notion of sustainable supply chain design and management. Companies are embedded in social-ecological systems, and for many sectors, a large part of the environmental and social impact is generated outside the boundaries of the firm, upstream in the supply chain. The typical supply chain of a consumer company accounts for 80%–90% of the overall environmental impact produced by the firm. We have illustrated the case of Puma EP&L to exemplify how important it is to engage Tier 4 suppliers through collaboration and partnership if a firm wants to reduce its impacts. We have then provided a résumé of supply chain management terms and concepts to facilitate reading.

We started the main part of this chapter defining the concept of sustainable supply chain and exploring the distinctive features: environmental issues, labor issues and human rights;

bribery and corruption. We have then introduced the baseline approach to sustainable sup-ply chain management discussing codes of conduct and sustainable procurement. Supply chain risks and sustainability have been addressed in the following paragraph, where we have connected natural disasters, climate change, and loss of ecosystem services with the perspective of risk management and the notion of supply chain disruption. The last para-graph has offered a broad analysis of strategies and tools to move upstream and trans-form the entire supply chain. We first introduced sustainability standards and certification processes, providing several examples of multi-stakeholders' initiatives that have developed sustainability standards in the field of commodities, and, finally, we have examined supply chain transparency and traceability.

In sum, this chapter has covered a great deal of novel topics; we hope we have been able to deliver enough theoretical frameworks and stories to illustrate HOW companies can con-tribute to social-ecological resilience with regard to designing and managing their extended network of suppliers.

Managing sustainable supply chains offers a view that is consistent with the four main principles of the BInN concept: (1) it extends the boundaries to the origin of raw materi-als; (2) it helps to move from a focus on impacts of operations to a focus on effects; (3) it adopts a the life cycle thinking perspective; and, finally, (4) it highlights the importance of multi-stakeholder collaboration.

Chapter 5 annexes

Annex 1

A.1.1: Questions for discussion

- How can a company ensure that its supply chain is sustainable?
- Why do companies need to address the impact of their extended supply chains?
- What are codes of conduct and how can they contribute to improve the supply chain sustainability? Please design the main elements of a supplier code of conduct for a company such as an international luxury hotel chain or a fast-food chain (e.g., Hyatt, Sheraton, McDonald's or Taco Bell).
- Consider the case of a company in the textile industry that is approaching the challenges of supply chain sustainability for the first time. Please suggest the main steps to develop and implement a sustainable procurement process.
- Ecosystem services are affected by globalized supply chains, but the sudden deterioration of ecosystem functioning represents a major risk for supply chain management. Explain the mechanism through which ecosystem service loss can affect supply chain leading to disruption. Please consider the case of the agro-food and mining industry.
- What are sustainability standard systems? How and why do they work? How can they help business sustainability?
- MSC, Bonsucro, Down Responsible Standard: What do these initiatives have in common?
- With reference to supply chain traceability models, discuss the pros and cons of product segregation and book and claim. What would you recommend in case of organic food or coffee? Why? And what would you suggest in the case of a commodity such as palm oil or sugar?

A.1.2: Assignments

Assignment 1: Code of conduct

Many companies have developed codes of conduct to deal with sustainability along the supply chains. A supplier's code of conduct is a document realized for the purpose of ensuring that a company's suppliers adopt and implement actions and procedures for protecting their employees (labor issues, human rights, safe working conditions) and the natural environment. This assignment aims at analyzing how codes of conduct are designed and how they work.

Individual students can select a couple of codes of conduct from similar companies (same industry) and confront how they are structured, the fundamental requirements included (e.g., UN principles on human rights, ILO Declarations, ethics and anti-corruption measures, environmental protection policies), the review and documentation policy. Guidelines for the analysis and comparison can be established with regard to the comprehensiveness, readability and clarity, accessibility and top-down commitment.

Consider that industries (e.g., Electronic Industry Citizenship Coalition) have developed several initiatives to develop homogeneous supplier conduct standards. Students can check these frameworks and use them as a model to benchmark the company code of conduct.

A final report can be about 3,000 words in length.

Assignment 2: Supply chain traceability: risks and mitigation

This assignment addresses the problem of the traceability and transparency of the supply chain. Teams must play the role of a consultant and must identify and suggest a valuable strategy (technological or organizational innovation, certification, etc.) that can help a specific company to mitigate supply chain risks (business interruption, price volatility, reputation and image).

The aim is to provide a solid *business case* to support the decision-making (key objectives, outline the business needs, describe the proposed solution, analyze the risks, provide a robust estimate of the cost and benefit of the proposed solution, etc.).

Organize your class in teams of four or five students. These are the main steps for the project:

Identify the supply chain

Each team focuses on a specific supply chain - for example, food upstream (raw materials coffee, cocoa, seafood), food downstream (retail), luxury/fashion products, sports apparel, electronics, and personal care.

Select the company

Each team selects a company in each supply chain.

Supply chain mapping

Map the supply chain structure, the different stages of value creation, and the main stakeholder involved. Analyze the main problems related to traceability and transparency.

Develop your innovative solutions

Search and develop your innovative idea (technological or organizational), explain your approach, provide a plan for its implementation, and prepare a solid business case to justify the decision.

Identify the social and environmental benefits

Identify and briefly analyze the benefits that result from the implementation of your idea.

The expected outcome of this assignment is a group report of 4,000 words, and a group presentation that is no longer than 15 slides.

Additional questions that can help students with the analysis:

- What technologies can help your company and supply chain in increasing the traceability and transparency?
- Is your consumer interested in knowing the origin of your products?
- Is certification important for your product?
- What are the best practices with regard to traceability and transparency in your industry?

A.1.3: Further readings and additional resources

- The World Wide Fund for Nature (WWF) has an amazing website rich of information and resources with a special focus on transforming business. WWF took an active leadership in collaborating with companies and seeking to reduce the negative impact of commodities and supply chains. For further resources, please visit: www.worldwildlife.org/initiatives/transforming-business.
- UN Global Compact (UNGC) participants rank sustainability in supply chain practices as the most important challenge in corporate sustainability. The website provides interesting reports on sustainable procurement, traceability and sustainability standards, and partnership with smallholders. Please visit: www.unglobalcompact.org/what-is-gc/our-work/supply-chain.
- The CDP (former Carbon Disclosure Project) has developed a specific section on the impact of climate change, water, and forest risks on supply chains. Every year they publish a report based on a large survey with business organizations mapping and analyzing how companies tackle these risks, take advantage of opportunities, and ensure business continuity. Please visit: www.cdp.net/en/supply-chain.
- In 2014, the International Institute for Sustainable Development (IISD) released the report "The State of Sustainability Initiatives (SSI). Standards and the Green Economy". This document provides a "bird's-eye view of market and performance trends across 16 of the most prevalent standards initiatives operating across ten different

commodity sectors." Please visit this link to access the report: www.iisd.org/library/state-sustainability-initiatives-review-2014-standards-and-green-economy.

- International Labor Organization (ILO) website has a broad and detailed section on labor issues and human right in supply chains. Please visit: http://libguides.ilo.org/global-supply-chains-en.

Annex 2

A.2.1: References

Barboza, D. (2010, June 4). After suicides, Scrutiny of China's grim factories. *The New York Times*. Retrieved from www.nytimes.com.

BMW. (2017, May). BMW Group supplier sustainability policy. Retrieved from www.bmwgroup.com/content/dam/bmw-group-websites/bmwgroup_com/responsibility/downloads/en/2017/BMW%20GROUP%20Supplier%20Sustainability%20Policy.pdf.

Bové, A. T. and Swartz, S. (2016, November). Starting at the source: Sustainability in supply chains. *McKyinsey on Sustainability and Resource Productivity*. Retrieved from www.unilever.com/Images/2003-fishing-for-the-future-ii-unilever-s-fish-sustainability-initiative_tcm244-409706_1_en.pdf.

Brammer, S., Hoejmose, S. and Millington, A. (2011), Managing sustainable global supply chains. A systematic review of the body of knowledge. *Network for Business Sustainability*. Retrieved from http://nbs.net/wp-content/uploads/NBS-Systematic-Review-Supply-Chains.pdf.

Carter, C. R. and Easton, P. L. (2011). Sustainable supply chain management: Evolution and future directions. *International Journal of Physical Distribution and Logistics Management*, 41(1), 46–62.

Castle, S. (2013, February 19). Nestlé Removes 2 Products in Horse Meat Scandal. *The New York Times*. Retrieved from www.nytimes.com.

CDP. (2017). Supply Chain Report 2016/2017. Missing link: Harnessing the power of purchasing for a sustainable future. Retrieved from www.cdp.net/fr/research/global-reports/global-supply-chain-report-2017.

Ceres. (2010, 6 April). Murky waters? Corporate reporting on water risk. A benchmarking study of 100 companies. Retrieved from http://waterfootprint.org/media/downloads/Barton_2010.pdf.

Chopra, S. and Meindl, S. C. (2013). Supply chain management: Strategy, planning, and operation (5th edition). Essex: Pearson Education.

FAO. (2015, May). The impact of natural hazards and disasters on agriculture and food security and nutrition: A call for action to build resilient livelihoods. Retrieved from www.fao.org/3/a-i4434e.pdf.

Fashion Revolution. (2017). Fashion Transparency Index. Retrieved from https://issuu.com/fashionrevolution/docs/fr_fashiontransparencyindex2017?e=25766662/47726047.

Folke, C. (2006). Resilience: The emergence of a perspective for social-ecological systems analyses. *Global Environmental Change*, 16(3): 253–267.

Gilbert, D. U., Rasche, A. and Waddock, S. (2011). Accountability in a global economy: The emergence of international accountability standards. *Business Ethics Quarterly*, 21(1), 23–44.

Handfield, R. B. and Nichols, E.L. (1999). *Introduction to supply chain management*. New Jersey, NJ: Prentice-Hall.

HEC and EcoVadis. (2017). Scaling up sustainable procurement. A new phase of expansion must begin. White paper based on the 2017 HEC/EcoVadis Sustainable Procurement Barometer 7th Edition. Retrieved from www.ecovadis.com/new-release-2017-sustainable-procurement-barometer/.

Holling, C. S. (1973). Resilience and stability of ecological systems. *Annual Review of Ecology and Systematics*, 4, 1–23.

IISD. (2014). The State of Sustainability Initiatives (SSI). Standards and the green economy. Retrieved from www.iisd.org/library/state-sustainability-initiatives-review-2014-standards-and-green-economy.

ILO. (2002). The International Labor Organization's fundamental conventions. Retrieved from www.ilo.org/wcmsp5/groups/public/@ed_norm/@declaration/documents/publication/wcms_095895.pdf.

ITUC. (2016). TUC Global Rights Index. The worlds' worst countries for workers. Retrieved from www.ituc-csi.org/IMG/pdf/survey_ra_2016_eng.pdf.

Kering. (2015). Kering Environmental Profit & Loss (EP&L), 2015 group results. Retrieved from www.kering.com/sites/default/files/kering_group_2015_environmentalpl_0.pdf.

Komives, K. and Jackson, A. (2014). Introduction to voluntary sustainability standard systems. In C. Schmitz-Hoffmann, M. Schmidt, B. Hansmann, D. Palekhov (Eds.), *Voluntary standard systems: A contribution to sustainable development* (pp. 3–19). Berlin Heidelberg, Germany: Springer-Verlag.

Krause, D. R., Vachon, S. and Klassen, R. D. (2009). Special topic forum on Sustainable Supply Chain Management: introduction and reflections on the role of purchasing management. *Journal of Supply Chain Management*, 45(4): 18–25. https://doi.org/10.1111/j.1745-493X.2009.03173.x.

Lambert Douglas, M. (Ed.). (2008). *Supply chain management. Process, partnership, performance*. Sarasota, Fl: Supply Chain Management Institute.

LEGO. (2015, 16 June). LEGO Group to invest 1 Billion DKK boosting search for sustainable materials. Retrieved from www.lego.com/en-us/aboutus/news-room/2015/june/sustainable-materials-centre/.

Mentzer, J. T., DeWitt, W., Keebler, J. S., Min, S., Nix, N. W., Smith, C. D. and Zacharia, Z. G. (2002). Defining supply chain management. *Journal of Business Logistics*, 22(2), 1–25.

Millennium Ecosystem Assessment, MA. (2005). Living beyond our means. Natural assets and human well-being. Retrieved from www.millenniumassessment.org/en/Reports.aspx#.

Montabon, F., Pagell, M. and Wu, Z. (2016). Making sustainability sustainable. *Journal of Supply Chain Management*, 52(2), 11–27.

Moore, F. C., Baldos, U., Hertel, T. and Diaz, D. (2017). New science of climate change impacts on agriculture implies higher social cost of carbo. *Nature Communications*, 8(1), 1607.

Network for Business Sustainability (NBS). (2011). *Executive Report: Sustainable global supply chains*.

New, S. (2010). The transparent supply chain. *Harvard Business Review*, 88, 1–5.

OECD. (2011). OECD Guidelines for Multinational Enterprises. 2011 Edition. Retrieved from www.oecd.org/daf/inv/mne/48004323.pdf.

OECD. (2014). *OECD Foreign Bribery Report 2014. An analysis of the crime of bribery of foreign public officials*. Paris, France: OECD Publishing.

OECD. (2017). OECD Due Diligence Guidance for responsible supply chains in the garment and footwear sector. Retrieved from https://mneguidelines.oecd.org/OECD-Due-Diligence-Guidance-Garment-Footwear.pdf.

Pagell, M. and Shevchenko, A. (2014). Why research in sustainable supply chain management should have no future. *Journal of Supply Chain Management*, 50(1), 44–55.

Pagell, M. and Wu, Z. (2009). Building a more complete theory of sustainable supply chain management using case studies of 10 exemplars. *Journal of Supply Chain Management*, 45(2), 37–56.

Pogutz, S. and Winn, M. (2016). Cultivating ecological knowledge for corporate sustainability: Barilla's innovative approach to sustainable farming. *Business Strategy and the Environment*, 25(6), 435–448.

Porritt, J. and Goodman, J. 2005. Fishing for good. Report commissioned by Unilever. Forum for the Future. See: www.forumforthefuture.org.uk.

Puma. (2010). PUMA's Environmental Profit and Loss account for the year ended 31 December 2010. Retrieved from http://about.puma.com/damfiles/default/sustainability/environment/e-p-l/EPL080212final-3cdfc1bdca0821c6ec1cf4b89935bb5f.pdf.

Sarkis, J., Zhu, Q. and Lai, K. (2011). An organizational theoretic review of green supply chain management literature. *International Journal production Economics*, 130: 1–15.

Seuring, S. and Muller, M. (2008). From a literature review to a conceptual framework for sustainable supply chain management. *Journal of Cleaner Production*, 16: 1699–1710.

Simchi-Levi, D., Schmidt, W. and Wei, Y. (2014). From superstorms to factory fires: Managing unpredictable supply chain disruptions. *Harvard Business Review*, 92(1–2), 96–101.

UNGC and BSR. (2010). Supply chain sustainability. A practical guide for continuous improvement. Retrieved from www.unglobalcompact.org/library/205.

UNGC and BSR. (2014). A guide to traceability. A practical approach to advance sustainability in global supply chains. Retrieved from www.unglobalcompact.org/docs/issues_doc/supply_chain/Traceability/Guide_to_Traceability.pdf.

Unilever. (2003). Fishing for the future II. Unilever's Fish Sustainability Initiative (FSI). Retrieved from www.unilever.com/Images/2003-fishing-for-the-future-ii-unilever-s-fish-sustainability-initiative_tcm244-409706_en.pdf.

Winn, M. and Pogutz, S. (2013). Business, ecosystems, and biodiversity: New horizons for management research. *Organization & Environment*, 26(2), 203–229.

World Economic Forum, WEF with Accenture. (2012). New models for addressing supply chain and trans-
    port risk. Retrieved from www3.weforum.org/docs/WEF_SCT_RRN_NewModelsAddressingSupply
    ChainTransportRisk_IndustryAgenda_2012.pdf.
World Economic Forum, WEF with Accenture. (2013, January). Building resilience in supply chains. Retrieved
    from www3.weforum.org/docs/WEF_RRN_MO_BuildingResilienceSupplyChains_Report_2013.pdf.
World Economic Forum, WEF. (2018). Global Risk Report 2018. Retrieved from www3.weforum.org/docs/
    WEF_GRR18_Report.pdf.
Yardley, J. (2013, May 22). Report on deadly factory collapse in Bangladesh finds widespread blame. *The
    New York Times*. Retrieved from www.nytimes.com.

Annex 3

A.3.1: Case study resources

In order to complement the learning experience, we suggest the following case studies that
address the topic of designing and managing a sustainable supply chain.

- Russo, M. and Crooke, M. (2016). *Guayakí: Securing supplies, strengthening the mission*.
  Oikos Case Writing Competition 2016, Corporate Sustainability Track. Available at Oikos
  International Homepage website. Prize winner.
- Purkayastha, D. and Rao, A. S. (2015). *Apple and conflict minerals: Ethical sourc-
  ing for sustainability*. Published by: IBS Center for Management Research. Reference
  no. 615-066-1. Available at The Case Center. Prize Winner.
- Pogutz, S. (2015). *When supply chain management drives Environmental sustainability:
  The case of Barilla*. Published by Università Bocconi, Bocconi Graduate School.

A.3.2: Case study from this book

We suggest the case of IKEA and Better Cotton Initiative (BCI). Below, we provide a brief version of the case and the list of questions for guiding the discussion. For the extended version
and the teaching notes, contact the authors:

- Pogutz, S. (2014). *IKEA and Better Cotton Initiative*. Published by Università Bocconi,
  Bocconi Graduate School. Oikos Case Writing Competition 2016, Corporate Sustainability Track. Available at Oikos International Homepage website. Prize winner.

Case study: IKEA and BCI

Founded in 1943 in Sweden, IKEA is the world's largest furniture retailer with more
than €36 billion revenues in 2016, 355 stores around the world and 783 million store
visits (IKEA, 2017). IKEA vision is "To create a better everyday life for the many people," and, from the early stages of its story, it has been associated to the combination
of functionality, quality, and low prices in home furniture. In addition, its commitment
to sustainability is embedded in the original business model developed by its founder
Ingvar Kamprad: "doing more with less" has always been its mantra.

Sustainability at IKEA

As a matter of fact, IKEA's conscious environmental strategy was launched in the early 1980s, and its first environmental policy was approved by the board in 1990, just one year after the appointment of its first environmental manager. Throughout the years, the company has progressively integrated sustainability into its organizational culture, permeating every activity and making it a cornerstone of strategy. Examples of sustainability-oriented initiatives include:

- the introduction of the Sustainability Product Scorecard (SPS). From the initial design, and throughout the entire life cycle, every IKEA's product is assessed with regard to product's safety, quality, and environmental profile. In 2010, IKEA launched the SPS, an assessment system based on 11 criteria to measure the sustainability of the product range. In 2016, 55% of IKEA's sales came from product classified more sustainable on the basis of the SPS, and the goal is to reach 90% in 2020 (IKEA, 2017),
- the launch in year 2000 of the IWAY – the "IKEA Way" – a code of conduct and a program based on the UN's Universal Declarations (e.g., Human Rights, Fundamental Principles and Rights at Work, Environmental Declarations) that aims at diffusing sustainable and responsible practices among suppliers (products, materials, and services) that work with IKEA.

The cotton challenge

Despite its strong history of sustainability, in 2005 IKEA found itself facing a massive and new challenge: cotton. With cotton representing the second most used raw material by the company (wood being the first), and one of the most extensively cultivated crops of the world, addressing the issue of its sustainability was critical. Just to put things into perspective, the water used to produce cotton for IKEA amounted to 2,890 billion liters (equivalent to the drinking water consumption of a country such as Sweden for 176 years), together with 170 million kg of chemicals (about 10% of the world's pesticides). The opportunity for improvement was clear: transform its cultivation to make it more economically, environmentally and socially sustainable. How could this be achieved?

Making cotton production more sustainable was a global challenge. Nearly 35 million hectares are actively cultivated (around 2.5% of the world cultivated land), 99% of the world's cotton farmers are from developing countries, and about 300 million people work in the wider cotton industry (Rai, 2010). Moreover, cotton for some of the world's poorest countries represents a vital link to the global economy. At the same time, its cultivation and manufacturing is responsible for several social and environmental problems. In response to this, the BCI was established in 2005 with the goal of making "global cotton production better for people who produce it, better for the environment it grows in and better for the sector's future" (www.bet tercotton.org). The BCI is a multi-stakeholder platform that works along the cotton supply chain, including farmers, producers, manufacturers, retailers, major brands (e.g., Adidas, Levi Strauss, H&M), and NGOs such as WWF (see Box 5.6. in the main part of this chapter).

Making cotton production more sustainable

As a first step in its path toward sustainable cotton, IKEA decided to join the BCI and become one of the founding members of the platform, enabling it to confront and collaborate with competitors in order to find a common solution. Obviously, just joining the BCI was not enough, IKEA's path toward mainstreaming Better Cotton was still long and full of obstacles, but the company was ready to address the challenge and created a specialized team to develop capacity and build linkages with the supply chain.

The first big problem was the knowledge gap in the production of cotton. IKEA had no control over its suppliers, and the lack of integration in the supply chain made it nearly impossible to obtain information on the upstream phases of the production process (see Figure A5.1). In addition, pressure from interest groups was rising as awareness of the environmental and social impacts of textile production emerged. The problem was framed, and the stage was set; it was time to start doing something concrete.

IKEA's decision was to collaborate with WWF to launch a pilot program in Pakistan, in order to test whether producing Better Cotton at an affordable price was feasible. The chosen districts were those of Bahawalpur and Yazman in South Punjab, one of the most important cotton-growing regions of Pakistan. The results were impressive, and the pilot project became vital to IKEA's strategy as it proved it was possible to cultivate and process cotton with significantly lower impact on the environment (a reduction in the use of fertilizers, pesticides and water) and at a lower cost, with an increase in the gross margins for the farmers. Moreover, it was possible to trace cotton throughout the pipeline. As a result, the project was extended to other regions in Pakistan as well as to India.

But the goal hadn't been reached yet, the next step for IKEA was to scale it up: it had to build a strategy to mainstream Better Cotton and make a commodity. The company decided to do this setting ambitious goals and targets and identifying key actions in order to study the supply chain and secure capacity. These actions included contributing to identify the quality standard for BCI cultivation and manufacturing (BCI standard system), assuring traceability of the raw materials, scaling up the production and create BC capacities. After the initial reluctance of cotton suppliers to disclose information because afraid of compromising their competitive advantage, IKEA managed to bridge the knowledge gap it had started with. Other key issues that emerged in this process included ensuring that sustainable cotton moved through the supply chain and entered the final products and communicating to customers in order to grasp competitive advantage.

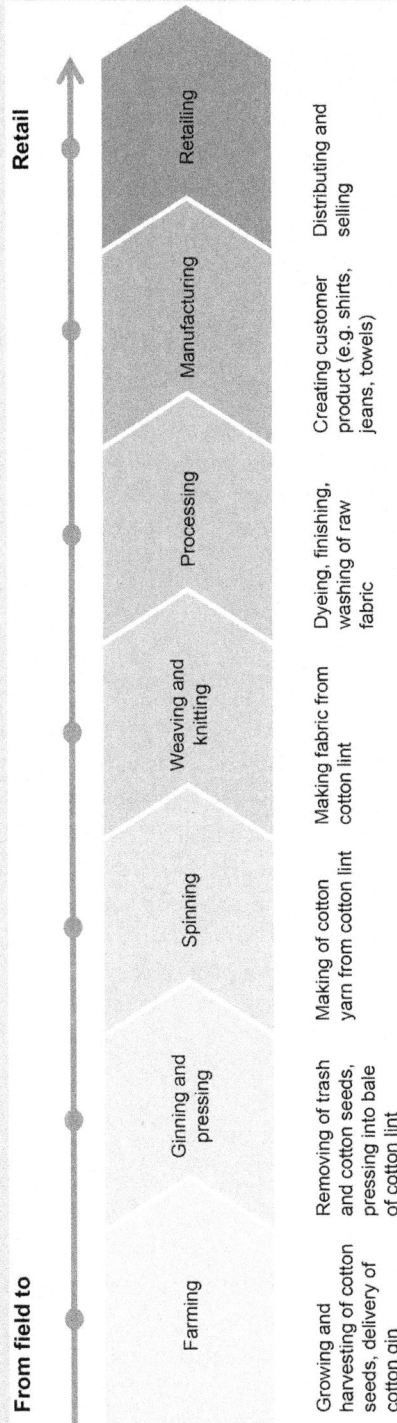

Figure B5.2 Cotton supply chain simplified.
Source: authors' elaboration.

The results were measurable: in 2012, thanks to IKEA's continuous investment in the issue, 30% of its cotton was sourced from "preferred" sources and approximately 10,000 farmers were engaged in the production of sustainable cotton in India and Pakistan. In 2014, 75% of cotton was obtained from more sustainable sources, and at the end of 2015, the target of 100% was finally reached. Today, IKEA uses in its products about 130,000 tons of cotton, and that amounts to around 1% of the world cotton production. As of 2016, 17.8% comes from recycled cotton, while the remaining 82.2% is BCI or other programs for more sustainable cotton (such as the "e3 Cotton Program" in the USA) (IKEA, 2016).

Future outlook

Despite these results, room for improvement still exists and new sustainability challenges continue to emerge. For example, as stated in the 2020 Sustainability plan, IKEA aims to source 100% of wood, paper and cardboard from sustainable sources and produce as much renewable energy as it consumes in its operations by the year 2020 (IKEA, 2016).

Link to supply chain sustainability

IKEA's case offers an opportunity to analyze how sustainability issues can quickly emerge and impact businesses, thereby affecting firms and generating new risks. Moreover, the case illustrates the complexity involved in dealing with supply chain sustainability while demonstrating the existence of underlying opportunities in such challenges. By adopting a holistic approach toward the cotton supply chain, IKEA was able to identify where the criticalities lied, looking beyond the walls of the company. In addition, the fundamental role of the focal company emerges from this success case: IKEA was one key enabler of the supply chain progress toward sustainability, contributing to move Better Cotton from a pilot project to commoditization.

Source: authors' elaboration.

Questions for discussion

(1)   IKEA has solid experience with sustainability. What is new with cotton compared with other sustainability challenge?
(2)   What is BCI? How is Better Cotton different from organic cotton?
(3)   What is the role of IKEA in BCI?
(4)   IKEA's pilot project in Pakistan verified that Better Cotton is feasible (business case). How can IKEA commoditize Better Cotton? Design the main areas of IKEA's strategy to make Better Cotton a commodity and to reach its target of a 100% share of sustainable cotton in IKEA products in 2015.

(5) How can IKEA use the BCI case to increase motivation and commitment to sustainability within the company and industry as a whole?

(6) IKEA needs to transfer the value of Better Cotton to end consumers. How can Better Cotton be used to differentiate the IKEA products? What are the main risks and opportunities? How can Better Cotton be communicated to consumers?

References

IKEA. (2017). Yearly Summary 2017. Retrieved from www.ikea.com/ms/it_IT/pdf/yearly_summary/YS17_Final_highres.pdf.

IKEA. (2016). Sustainability Report FY 2016. Retrieved from www.ikea.com/ms/en_US/img/ad_content/IKEA_Group_Sustainability_Report_FY16.pdf.

Rai, K. J. (2010). The IKEA experience in moving towards a Better Cotton supply chain. Making sustainability work, Case Study, Dutch Sustainable Trade Initiative. Retrieved from www.idhsustainabletrade.com/uploaded/2016/08/Ikea-booklet-def.pdf.

6 Making sustainable products and services

Learning objectives

- Highlight the importance of sustainable consumption.
- Analyze the evolution of the green consumer.
- Introduce the reader to the concept of life cycle thinking.
- Illustrate the Life Cycle Assessments (LCA) methodology and its links with the Environmental Product Declaration (EPD).
- Explain what the product environmental footprint is.
- Discuss the concept of sustainable product and services.
- Analyze the concept of Design for the Environment (DfE) and explain the relationship between product and service innovation.
- Clarify what an ecolabel is and its function in favoring sustainable consumption.

Chapter in brief

This chapter focuses on the development of sustainable products and services, addressing the environmental and social challenges (localized mostly downstream) that result from the consumption and utilization phase. Many products and services (e.g., electronic appliances and cars, but also shampoos, detergents and apparel) generate most environmental externalities when they are utilized or in the last phases of their individual life cycle, when waste is generated, management is required to recovery and recycle, and final materials need to be disposed. Life cycle thinking appears then as a powerful strategy: it facilitates the reduction of impacts on ecosystems and the management of wastes. In addition, a life cycle thinking approach opens up opportunities for companies to innovate and address emerging requests from a growing demand for and interest in green and ethical products and services. The chapter initially investigates the role of sustainable consumption and the "flashing arrow" of consumerism. A large part of the chapter is devoted to illustrating concepts such as Design for the Environment (DfE) and green product innovation, linking environmental tools such as Life Cycle Assessment (LCA) to innovation processes and strategic decisions. Finally, we discuss ecolabels and product certification systems, exploring the role of sustainable consumption initiatives that connect the public sector and civil society to consumers.

Introduction

To produce a typical cotton t-shirt, we consume about 2,700 liters of water, which is enough for one person to drink for about two and a half years (Drew and Yehounme, 2017). Fast fashion compared to traditional processes has shortened the cycle of the utilization of clothing, and in 2014 the average consumer bought 60% more products than in 2000, keeping them half as long (Remy et al., 2016). T-shirts made of synthetic fibers such as polyester have a smaller environmental footprint than cotton, but they generate more than twice the carbon emissions (5.5kg vs 2.1kg). The apparel industry is expected to grow rapidly in the coming decades, driven by countries such as China and India, where hundreds of millions of consumers are entering the global middle class. Nevertheless, it is not just about clothes and fast fashion; these trends refer to many other types of goods and services that we consume every day. By 2030 the size of the global middle class is expected to reach 5.4 billion people worldwide, up from 3 billion in 2015, with growing urbanization, which further increases the needs for transportation and the distribution of goods. To sum up, if consumption continues at the current rate, in 2050 we will need three times as many natural resources as we used in 2000 (Drew and Yehounme, 2017).

It is not surprising, then, that consumption has begun to attract a lot of attention in the field of sustainability; it is clear that current trends can no longer be maintained. The efficiency gain obtained through improvement in production processes and technological advances that we have illustrated in the previous chapters has not been enough, and the goal of Sustainable Development will require radical changes in our lifestyle – including the way we consume and use products (WBCSD, 2008). On the other hand, our relentless aspiration for social status and new products seems to lock us to an escalating spiral of consumerism (Jackson, 2008), putting unsustainable stress on the planet. How can we escape from this trap? What can we do?

Of course, there is no easy way out. Market logics are deeply embedded in social-economical systems, but at the same time, when an individual decides whether or not to purchase a product or a service, he or she has the potential to orient the patterns of consumption towards a higher or lower range of sustainability. Therefore, the question becomes: How can we get consumers to change their habits and lifestyle? How can policy makers, NGOs and businesses engage them in sustainable consumption? Consumer behavior is complicated, influenced by multiple internal and external factors, such as individual values, beliefs and attitudes, personal concern, contextual variables, habits and routines. At the same time, it looks promising that due to the increase of societal awareness and knowledge about environmental and social issues, there is a global movement of consumers who are now demanding new products that can address sustainability challenges. These consumers are concerned about climate change, ecosystem degradation and human rights; however, they have still not translated the demand for sustainability into consumption patterns due to several barriers, such as skepticism, product availability and convenience. Companies are therefore being called to break these barriers, increase transparency and communicate in order to raise awareness, engage consumers, and, of course, design and innovate goods and services to reduce their impact and support efficient end-of-life management.

There is another major reason for the importance of changing consumption behavior. In general, although probably counter-intuitively for many people, most of the impacts associated with several typologies of goods and services occurs during the consumption phase, not during the manufacturing activity. For example, 87 and 86% of the energy demand/carbon consumption and water requirements respectively in the use of a powder dishwasher comes from its usage phase (PCF Pilotprojeckt, 2008). Similarly, about 80% of the CO_2 emissions of a conventional combustion-based vehicle are produced during its utilization (WWF and SustAinability, 2007). We could continue with many other examples from different product categories, such as white goods, shampoo and personal care products and electronics. Therefore, the design of new products and services able to respond to the above considerations has become one of the main priorities for corporate sustainability, and is one of the critical points driving innovation today (Nidumolu et al., 2009). To address future challenges companies, need to change their view. They need to understand the new demand profiles and customers' concerns, they must figure out the ways in which new consumers are willing to accept new sustainable products and services, they need to identify innovation priorities and use the right competencies and tools, they need to increase their collaboration with other stakeholders and team up to generate new ideas, and they need to address their responsibilities. Finally, as important as the development of new products and services that can deliver maximum societal value at minimum environmental cost, companies must remove non-sustainable products and services from their portfolios.

Product and service innovation to increase sustainability require the development and application of new competencies and tools:

- A strong orientation to "life cycle thinking," moving beyond the traditional focus on manufacturing processes at the production site, to include all social, ecological and economic impacts of a product over its entire life cycle. LCA and key performance indicators are expected to become a fundamental component of decision making, supporting innovation and strategic choices.
- A better understanding of product recovery, reuse and recycling is needed. Resource scarcity and ecosystem degradation require companies to focus more to downstream activities (see also the concept of circular economy in Chapter 7). Business must start to use Design for the Environment (DfE) principles at the beginning of the development of new products and services.
- A focus on communicating environmental and social information relating to products and services, and on learning how to apply specific methodologies such as the Environmental Product Declarations (EPD), which can be used as a standardized way to disclose and report information to stakeholders.
- Finally, the adoption of ecolabels that can serve to inform and engage consumers in sustainable purchasing. Ecolabels set the requirements for specific features and properties of products/services, and through a process of independent verification, they assess whether specific environmental or social quality criteria are met.

We will go through these topics in the following sections.

Sustainable consumption

During the 1992 Rio Conference, the term sustainable consumption was first introduced to recognize the importance of addressing consumer behavior and market demands in the sustainability agenda. The conference aimed to create a bridge between sustainable production and sustainable consumption patterns, launching a call for future responsibilities. In 1994, during a symposium hosted in Oslo, the United Nations Commission on Sustainable Development (UNCSD) provided a definition of sustainable consumption and production:

> The use of goods and services that respond to basic needs and bring a better quality of life, while minimizing the use of natural resources, toxic materials and emissions of waste and pollutants over the life cycle, so as not to jeopardize the needs of future generations.

Since then, sustainable production and consumption have been consolidated as a single concept that should always be named together.

Prior to these global conferences, pioneering work on the subjects of ethical and responsible consumption was carried out, starting in the late 1960s and early 1970s. These studies contributed with a first profile of the demographic features of green consumers: female, relatively young, above-average socio-economic status and with high level of education (Berkowitz and Luttermann, 1968; Anderson and Cunningham, 1972). Despite this early work, it was only in the late 1980s that the concept of green-marketing emerged. In 1988, John Elkington, Julia Hailes and Joel Makower published a successful book titled *The Green Consumer* (Elkington et al., 1988). At that time sustainable consumption was not yet a topic; however, on the front cover of the book it was written: "you can buy products that don't cost the Earth." A few years later, the same group of authors introduced *The Green Consumer Supermarket Guide* (Makower et al., 1991), another popular book that opened with the statement: "Let's start with the basics: Every time you open your wallet, you can cast a vote 'for' or 'against' the environment." In this vein, Jacquelyn A. Ottman published another key book, *Green marketing. Challenges & opportunities for the new marketing age* (1992), foreseeing environmentalism as a new business trend for the following decades, and identifying business opportunities for ethical purchasing.

These books served to develop initial arguments for sustainable consumption, raising interest among marketing managers and communication experts towards green consumers. This interest was mirrored by companies, which started to introduce and develop green brands such as The Body Shop (the worldwide ethical beauty retailer) and Ecover (a Belgium-based manufacturer of cleaning products), and to address questions such as packaging waste, chemicals and product efficiency. An upsurge in academic research on green business and green marketing was also registered in this period (Peattie and Crane, 2005). However, conscious green consumers in the 1990s were not a large group, representing only around 10% of the total consumer population, and over the course of the decade the hopeful forecast of the emergence of a "green tide" of consumers was disregarded. Green marketing somehow underperformed with regard to the capacity to engage consumers in purchasing green products. The consequences were twofold:

- On the one hand, for several years business and marketing departments have been disillusioned and skeptical with regard to the possibility of addressing green and ethical market segments.
- On the other hand, scholars have spent resources and time trying to understand why personal environmental concerns fail to translate into green purchasing, building a rich body of knowledge on barriers to green consumption.

The rapid deterioration of the earth's conditions and the impressive upsurge of global consumptions with more and more people expected to join the middle class has again turned the attention towards the necessity of transforming our lifestyles. New interest has grown, and business has been put at the center of the action to find new solutions with regard to innovative low-impact products and services. Business, moreover, can also use its power and competences to drive consumers to responsible purchasing, successfully translating willingness into real sustainable behaviors (WBCSD, 2008) and inspiring individuals with novel sustainable lifestyles. The relevance of this topic has been underlined by the UN SDGs, Goal 12 of which is focused on Responsible Consumption and Production (see also previous chapters).

How can we define sustainable consumption? There are many implications behind the logic of sustainable consumption: consumption that considers social and ecological constraints, consumption that favors more efficient production of more sustainable products, consumption aligned with the pursuit of better lifestyles, and consumption directed either to consume less or to consume in a different and novel manner. In any case, a consumption pattern that bypasses traditional market logics and that allows well-being and prosperity to be decoupled from the intensive use of the earth's ecosystems is needed.

Consumers are one of the most important forces able to modify the course of the environmental crisis in which we are immersed. If the consumer buys a product, that product contributes to sustaining the industry. If the consumer does not buy a product, the industry that produces that product is in trouble. Therefore, in a way, a well-educated consumer able to make choices based on different product's attributes, which includes the protection of the environment, could be considered the optimal option. Unfortunately, as previously highlighted, there is a gap between attitudes and behavior, between saying one is willing to pay for sustainable products and actually deciding to in real life. Several barriers have been identified and broadly discussed in academic and practitioner literature; among them, we have outlined below what we consider most important (Ottman et al., 2006; Gabler et al., 2013).

- **Lack of understanding of the problem.** The first factor is knowledge of behavior-specific environmental and social issues. In general, knowledge is an important variable that influences consumer attitudes and actions. With regard to sustainability, the information related to the effects of human actions on the planet is considered an important element when we are looking for environmentally conscious consumptions. Knowledge of climate change, ocean litter and biodiversity loss can increase individual concern and induce people to behave responsibly. However, multiple studies have shown that other internal and external constraints influence these relationships and that knowledge alone may not be enough to change behavior.

- **Selfishness.** A second dimension refers to the fact that individuals do not act if others – one's peers – do not acting as well. This problem, also referred to as the "tragedy of the commons" (see Chapters 2 and 3), is an important barrier when purchasing and using sustainable products requires a change in one's individual habits or lifestyle.
- **Convenience and perceived higher purchasing costs.** Another important element that contributes to the gap is the perception of higher prices associated with this typology of products and services when compared with traditional brands. Moreover, the lack of availability of these products in some of the usual distribution channels contributes to increasing the idea of incurring extra costs, for example the time spent in finding specialized stores.
- **Skepticism and lack of credibility.** Another major barrier, and the object of many studies in green marketing and sustainability, is the lack of credibility about the real commitment of companies, and the fact that sustainable products provide effective responses to sustainability challenges. In the past, many companies and brands have attempted to position themselves in the green space through marketing claims, often unsubstantiated and without any real transformation in the characteristics of their market offer. These marketing-style actions, conducted mainly for pure reputational purposes, have been identified with the term "greenwashing." The result, in many cases, has produced confusion and increased consumers' suspicion of environmentally oriented products, negatively impacting variables such as trust in and legitimacy of brands.

The barriers identified are responsible for causing even conscious consumers to consider with caution the purchasing of sustainable products and services. It is therefore extremely important for business to work on an array of levers to address these obstacles and engage consumers. These actions include the definition of the best communication and marketing techniques, credibility through transparency and information sharing, and the design of innovative products and services that respect the environment and social demands.

In Chapter 2, we introduced Yvon Chouinard, former CEO and founder of the company Patagonia (see Annex 2.1 at the end of this chapter for video links). In a very interesting video interview filmed a few years ago, he illustrates his vision for business and focuses on company relationships with consumers in today's markets (7th Generation, 2010). He says that the "consumer has choices now (. . .) and if you know that one company is more responsible in its business and the price is the same, everything is equal, you buy that" (7th Generation, 2010, 9:10–9:22). Consumers have the ability to drive the change we need, but sustainability must be fully embedded in products, which need to provide the same functions as competing products at the same price. He continues:

> Our customers buy Patagonia's products because they like the values behind the company (. . .) that's 10 percent of the people (. . .) the other 90 percent they buy because

they like the color or the style. If you wait for the consumer to tell you to green your company, you're way too late.

(7th Generation, 2010, 9:57-10:10).

This is another stimulating argument. Chouinard is straightforward and underlines again that even if consumers can call on companies to act responsibly, brand values and green product features must be incorporated in a more comprehensive market proposition, associated with quality, style and other attributes. Second, he explains to the interviewer that entrepreneurs and managers would do better to not wait too long to address sustainable production if they feel that this move is necessary – which it is. In other words, the question is not to wait or not wait for the consumer to awake; the urgency of our social-ecological problems requires a prompt response.

To conclude, it seems that in today's markets we face a kind of dichotomy in consumption patterns. On one side is a segment of environmentally oriented and eco-conscious consumers who are introducing green and social considerations in their buying preferences; on the other are consumers who are environmentally illiterate with a low level of awareness and the new consumers from the growing global middle class who are interested in joining our environmentally costly lifestyles. Although it is true that we are seeing an increase in the number of people who are more conscious in their purchasing behavior, we like Yvon Chouinard when he says that we cannot wait too long to find solutions to the environmental crisis – we cannot wait until a majority of consumers ask for change. We think that firms, both SMEs and multinationals, need to take the lead and act first.

From the green consumer to the LOHAS consumer and more

As previously illustrated, the implications of green consumerism in a competitive market place have been investigated for a long time (Peattie and Crane, 2005; Dangelico and Vocalelli, 2017). Several studies have analyzed the evolution of the green consumer segmentation. Back into the 1990s, one of the most prominent such works, by Roper Starch Worldwide, investigated whether American consumers were engaged in environmentally friendly behaviors, such as recycling, reading product labels, or paying more for ecological friendly products. The study identified five distinct groups of consumers: the true-blue greens (11%), who are true environmental activists and leaders; the greenback greens (5%), who express a strong commitment towards the environment and are willing to pay a premium for green products, but are not involved in pro-environment activities such as recycling due to restricted time; sprouts (33%) who do not have a very high concern about environmental problems, but do involve in some kind of environmentally responsible activities; grousers (15%) who are relatively uninvolved in pro-environmental activities and who think that they do not contribute to environmental problems; and basic browns (34%) who are convinced that they cannot make a difference and do not feel a need to rationalize their behavior.

During this period of time, many other studies tried to capture the characteristics of ecologically conscious consumers. Different demographic variables and behavioral indicators were used to describe this segment, the basic characteristics of which were as follows (Straughan and Roberts, 1999; Ottman et al., 2006; Gabler et al., 2013).

- **Age and sex.** Pre-middle aged people and females were identified as key demographics, although they did not have a clear significance in every survey.
- **Income and education** were associated with environmental concerns and behavior.
- **Place of residence**, which seems to favor urban living conditions as more promising for these attitudes.
- **Political orientation**, liberal people being the most strongly associated with this movement.
- **Environmental concerns** and environmental knowledge.
- **Psychographic characteristics** such as altruism and openness.
- **Perceived consumer effectiveness.** The premise that consumers' attitudes and responses to environmental appeals are a function of their belief that individuals can positively influence the outcome of such problems.

Since these early studies, many researchers have tried to come up with numbers to analyze the evolution of the green consumer. In the United Kingdom, Gilg et al. (2005) provided a market segmentation, according to which 23% of people could be called committed environmentalists and 33% mainstream environmentalists. Most studies carried out later recorded more or less similar data: Approximately 10% of people were really committed (super greens) and another 20%–30% could act responsibly (greens), but around 50% were concerned but not willing to act for multiple reasons (cannot afford to pay, do not want to compromise quality or convenience, lack of knowledge, price issues). We can conclude from these studies that even if significant shifts in levels of concern and general attitudes toward environmental and social issues were truly expressed and awareness raised, this would not be reflected in significant behavior and purchasing changes. Consumers were more likely to adopt environmentally responsible behaviors if both cost-efficient and convenience attributes were also present.

At the turn of the century, a new concept gained popularity, the so-called Lifestyles of Health and Sustainability (LOHAS) consumer, which is pushing a new wave of interest in sustainable consumption. LOHAS describes a type of consumer who is actively searching for a healthier and more sustainable lifestyle. They are considered to be the leading edge of the population, they share a solid belief and values system (personal, family and community health, environmental sustainability and social justice), and they make their purchase decisions with these criteria in mind. LOHAS consumers are also used as predictors of upcoming trends, as they are early adopters of many attitudinal and behavioral dynamics. LOHAS consumers play the role of the true-blue greens of Roper Starch study, or the super-greens identified by Gilg et al. (2005). In the United States in recent years, approximately 20%–25% of the population can be associated with this profile (although estimates vary according to each survey).

The consolidation of this trend towards green and sustainable purchasing attitude was finally confirmed at the end of 2015 by a large global survey conducted by Nielsen (2016). Their data showed a positive tendency, with an increase in consumers willing to pay more for products and services from responsible companies (66%, versus 50% from two years before). A large percentage (50%) is influenced by sustainability related to the product, the brand, or the company (e.g., the product is made from fresh, natural and/or organic ingredients; the company is environmentally friendly or is commitment to social values). Another novelty refers to a generational issue. Millennials and Generation Z are very different from the previous generations; they are extremely interested in sustainability and in healthy and green lifestyles, and they have the will, knowledge and capacity to promote transformative changes. Brand trust, transparency, health and wellness benefits, organic ingredients, the eco-factor and commitment to social justice are important drivers referenced today in purchasing decisions. As a result, companies that focus on environmental and social issues and build a sustainable reputation among today's youngest consumers have the chance to increase their market share and the possibility to build brand loyalty among future consumers.

Environmental campaigns

Environmental campaigns use marketing communication tools and awareness-raising actions to call attention regarding environmental issues, and encourage consumers to choose and use products more efficiently and sustainably, with the final goal being to influence their perspectives and choices. Environmental campaigns are used these days by very different groups of actors such as non-governmental organizations, companies and industrial associations, to criticize or support the use of products or practices related to the various aspects of ecological sustainability. Environmental campaigns can have a powerful influence on company behavior, individual decisions and consumer purchasing patterns.

In this section we briefly focus on two groups of environmental campaigns: those promoted by environmental activist organizations and those related to NGOs that aims at increasing transparency and information disclosure through collaborative actions among various stakeholders (see Annex 1.2 to this chapter). We will not focus here on other campaigns promoted by companies: e.g. GE "Ecomagination"; Timberland "Earthkeepers"; Patagonia "Don't buy this jacket" (see Chapter 7); and Nespresso, "The choice we make."

The environmental movement is a complex and varied group of activists, sympathizers and non-governmental organizations. Environmental campaigns can have a strong influence on business practices and can promote changes in company behavior. Although there is an enormous range of viewpoints among environmental and social activists and campaigners, there are also very powerful international groups such as Greenpeace or Friends of the Earth that can directly influence company responses (see Table 6.1). These campaigns can have also positive impacts on the way firms deal with environmental problems, forcing companies to find solutions for environmental crises, as it was done with the "Roundtable on Sustainable Palm Oil" (RSPO; see Chapter 5 for a full description of this event).

In other cases, environmental campaigns are promoted for awareness generation or educational purposes. For example, a remarkably successful one was the "Follow the frog" campaign, promoted by the Rainforest Alliance, who tried to engage everyday citizens in saving the rainforests. The ad was tapping on authenticity and humor instead of emotional appeal, to show that everybody can make a difference to protect nature making small changes on a daily basis. It is enough to purchase Rainforest Alliance certified products, that can be easily recognized by the green frog seal on the packaging of an array of agriculture and forestry goods - tea, coffee, chocolate, paper, etc.

Table 6.1 Greenpeace environmental campaigns.

Kit Kat and the palm oil dilemma	Nestlé, the Swiss food giant producer of Kit Kat bars and other confectionery, like many other companies in the world, used palm oil from companies exploiting Southeast Asian rainforests, threatening the social-ecological systems in these areas. In its "Have a Break. Give orangutans a break" campaign (2010), Greenpeace launched a 60-second clip showing an office worker opening a Kit Kat chocolate bar and finding an orangutan's finger. The viral video, that was a parody of a Nestlé commercial, aimed at highlighting how the Swiss company was buying palm oil (Armstrong, 2010). Moreover Greenpeace wanted to put pressure on Nestlé to find solutions for not destroying the precious rainforests, the homeland of protected orangutans and many other species. As a result of the campaign, Nestlé joined with other organizations to take action to ensure that none of its palm oil came from destroyed rainforest. The company committed using only sustainably certified palm oil by 2015 and engaged in strengthening the standard for sustainable palm oil cultivation (the Roundtable for Sustainable Palm Oil).
LEGO, Shell and the Arctic environment	In its "Everything is NOT awesome" campaign, Greenpeace wanted to send a wake-up call to oil, gas and other energy companies to cease lobbying the youngest generations through the use of toys. At that time, LEGO and Shell had a partnership that dated to the 1960s to sell LEGO toys at Shell petrol stations (Vaughan, 2014). Greenpeace launched a YouTube viral video that received 7 million views, showing a pristine Arctic built of LEGO, being covered with oil. The campaign successfully ended the partnership between LEGO and Shell, but could not interrupt the oil giant plan to drill the Arctic (Gunther, 2015).

Source: authors' elaboration on multiple sources.

Life cycle thinking

Since we began to worry about the impact that man had on the planet, we have started to develop a broader approach looking at every aspect involved in our production and consumption patterns. Since the Rio Conference in 1992, we were called to "develop criteria and methodologies for the assessment of environmental impacts and resource requirements throughout the full life cycle of products and processes" (Rio Conference-Agenda 21: Chapter 4), promoting what today is called life cycle thinking. Life cycle thinking is an essential requirement in the move to sustainable production and consumption, and to reach sustainable decision making in times in which we are facing high demands for better information; everyone in the whole chain of a product's life cycle has a responsibility and a role to play, taking into account all the relevant external effects (de Leeuw, 2005).

Life cycle thinking aims to reduce products' and/or services' use and waste of resources, as well as to improve socio-economic performance throughout a resource's life cycle. The impacts of all life cycle stages need to be considered comprehensively when taking informed decisions on production and consumption patterns, policies and management strategies. During the last few years, LCA, the life cycle approach, life cycle management and life cycle costing have become regularly used terms in environmental management jargon. As one of the principles of the "Business In Nature" concept is to broaden the boundaries of intervention by the companies, life cycle thinking is one of its most powerful aspects. An organizational approach based on life cycle thinking uses different procedural and analytical tools. LCA has become the most important of these management tools; it can be used for many different purposes, but above all it has become a powerful support system to better inform managerial decisions and to support credible and fact-based communication with different stakeholders.

Life Cycle Assessment (LCA)

LCA is a classical tool in environmental management. The concept was developed back in the 1980s. It provides a set of articulated techniques with the objective to identify, classify and quantify the pollutant loads, environmental impacts and material and energy resources associated with a product, process, or activity from its conception until its elimination or later use. It is based on the collection and analysis of the inputs and outputs of the system (natural resources, emissions, waste and by-products) to obtain quantitative data relating to their potential environmental impacts, in order to develop strategies for their minimization or reduction.

An LCA framework can be applied to different activities. Its more classical application is in the ecodesign of products and services; for this purpose, the LCA is one of the major tools used in the "Design for the Environment" activities to be revised in the next section. However, LCA frameworks can also support comparative analyses between similar products and the elaboration of criteria to be fulfilled for ecological labels, can serve as information for the implementation of environmental management systems, and can be used to obtain specific information about the environmental performance of a product or service.

Even if many organizations dealt with principles and criteria to develop LCAs and companies can adapt the tool to their own requirements, today LCA has been standardized by the International Organization for Standardization (ISO) through two norms, the ISO14040:2006 (Environmental management analysis of the life cycle: Principles and framework of reference) and the ISO14044:2006 (Environmental management life cycle analysis: requirements and guidelines). While the first covers the principles and framework for LCA, the second gives indications about the requirements and offers specific guidelines for the realization of an LCA: (a) the definition of the goal and scope of the LCA; (b) the analysis of the life cycle inventory; (c) the phase of life cycle impact assessment; (d) the life cycle interpretation phase; and (e) the final report and critical review. Based on the ISO14040, an LCA is defined as:

> A tool to study the environmental aspects and potential impacts throughout a product's life (i.e. cradle-to-grave) from raw material acquisition through production, use and disposal. The general categories of environmental impacts needing consideration include resource use, human health and ecological consequences.
>
> (ISO14040 norm)

Although LCA was initially developed as an environmental management tool designed mostly to deal with ecological aspects in organizations, today this framework can be used for other applications. It can be combined with the analysis of:

- **Social Issues (S-LCA):** "the assessment of the social and sociological aspects of products, their actual and potential positive as well as negative impacts along the life cycle" (UNEP/SETAC - Society of Environmental Toxicology and Chemistry - Life Cycle Initiative at UNEP, 2009,

- **Organizational Issues (O-LCA):** "a life cycle approach that aims to support the iden-tification and quantification of environmental aspects within and beyond the gates of the organization" (see Martinez-Blanco et al., 2015 for a further explanation on this matter), and
- **Sustainable Development** through Life Cycle Sustainability Assessment (LCSA).

All of these frameworks can be combined in various ways and with other techniques such as life cycle costing (LCC).

LCA makes holistic evaluations that requires the assessment of raw-material production, manufacture, distribution, use and disposal including all intervening transportation steps necessary to make, distribute and dismantle the product. One of the first decisions when making an LCA is to define the scope of application of this tool; based on this decision, the terminology used to define the scope of an LCA can vary, as follows.

- **Gate-to-gate:** A door-to-door analysis. The LCA is limited to the company's production process, mostly factories and distribution centers.
- **Cradle-to-gate:** The LCA is made from obtaining the needed resources to the produc-tion process included.
- **Gate-to-grave:** In this case the LCA is made around the productive process and the waste phase of the product.
- **Cradle-to-grave:** The LCA goes from obtaining the needed resources until the waste phase of the product, including its recycling and/or final management disposal.
- **Cradle-to-cradle:** Full life cycle thinking developed in a cyclical way with the reintroduc-tion of the product out of use (residue) in the same productive cycle or another.

For the purpose of this book and its "Business In Nature" approach, even if we gain in complex-ity, the best use of the LCA framework is to analyze cradle-to-cradle applications (McDonough and Brangart, 2003). Figure 6.1 illustrates the different options available.

An LCA is a tool that through the years has gained relevance and legitimacy. Besides ISO, SETAC and other guidelines published by different agencies and independent organi-zations, there are plenty of resources on which companies and organizations can base their assessments. There is a specific scientific journal, *The International Journal of Life Cycle Assessment*, which has been publishing papers on this subject since 1996. Other detailed publications, such as the work of Walter Klöpffer and Birgit Grahl (2014), can provide guid-ance. In the paragraphs below we review the key stages of a LCA (see Figure 6.2.).

Definition of the goal and scope

An LCA starts with the explicit statement of the goal and scope of the study: Why is the study being conducted, what is the objective, and for whom is it being developed? Finally, how are the results being communicated?

The scoping process identifies the system to be studied, the system boundaries (the group of operations connected to each other by the flow of materials and energy), and the data that will be analyzed (including also the definition of temporal limits of the assessment).

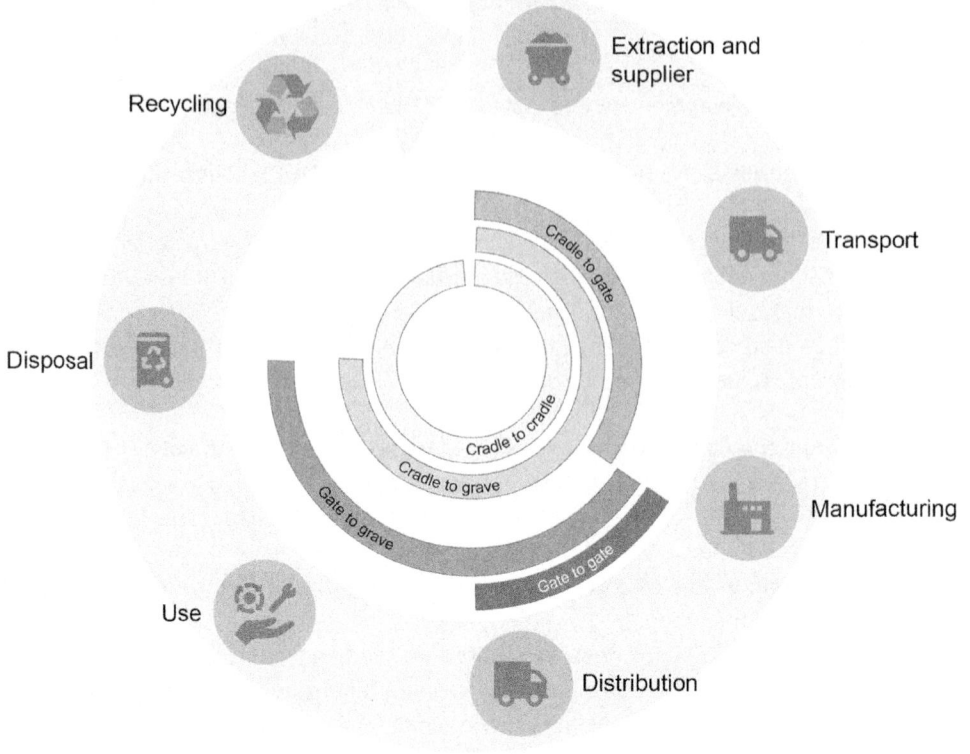

Figure 6.1 Different LCA's options.
Source: authors' elaboration.

Based on this scoping process, the measured parameters are collected and stored in the inventory (second phase). The definition of the system boundaries influences the typology of assessment to be done (gate-to-gate, cradle-to-gate, gate-to-grave, cradle-to-grave, or cradle-to-cradle). The limits should always satisfy the comparability principle – in general, the use of these techniques always includes (explicitly or implicitly) some kind of comparison, and when making comparisons between different options the scope should be similar for the units to be compared.

Another important aspect to consider is the establishment of a "functional unit" to which all system data will be referenced. The selected unit can be either physical (e.g., a dishwasher of appropriate characteristics) or functional (e.g., a functional unit is the amount of time needed to move from one site to another by a different transport system). In any case, the selection of the functional unit conditions the rest of the assessment.

Finally, it is important to consider that an LCA is a linear analysis and is based in the functional aspect of a product or service, so what we are going to analyze is not just the product but the entire system needed to make that product functional in the market. It is necessary, then, to establish a procedure to guarantee data management and data quality (a quality indicator) with a clear delimitation of the time in which the assessment was done.

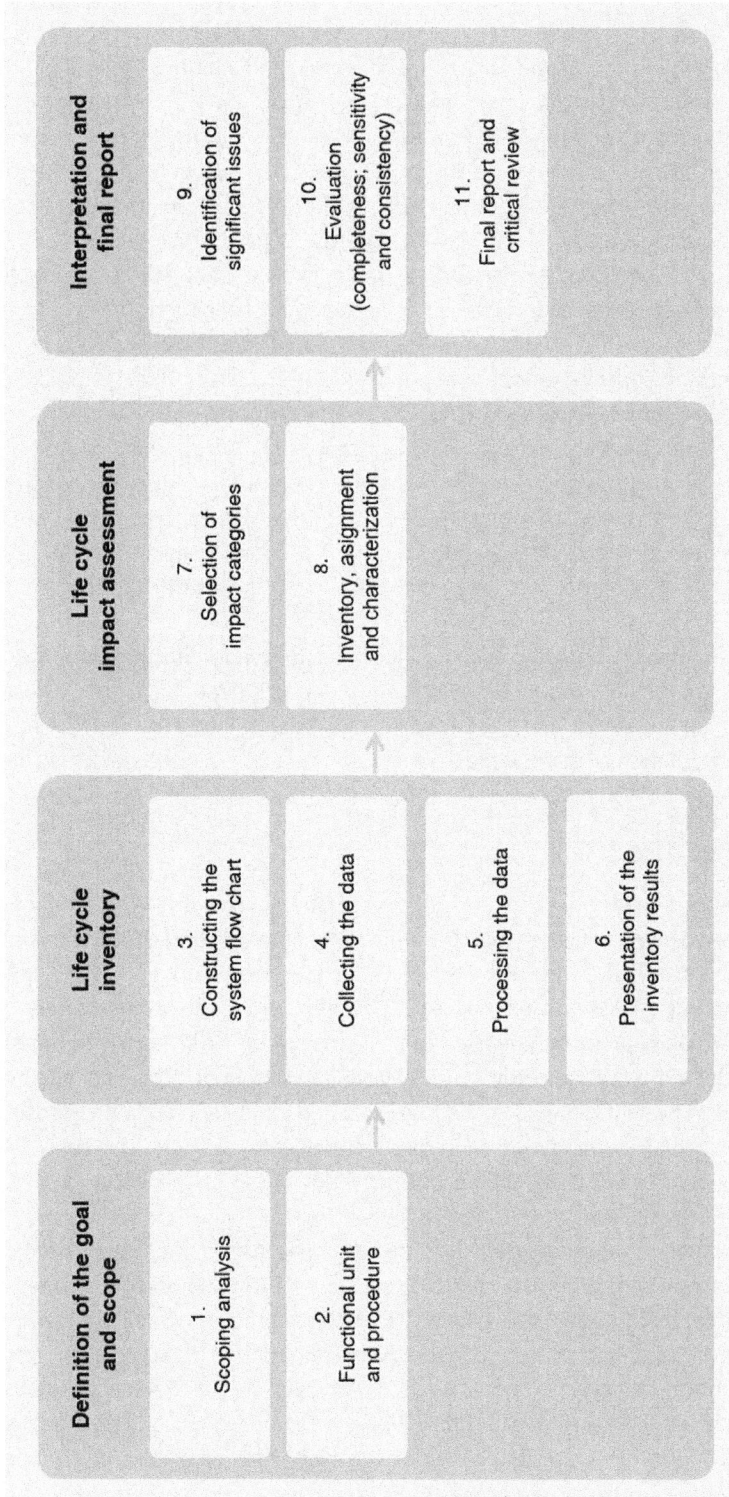

Figure 6.2 LCA basic structure.
Source: authors' elaboration.

Life cycle inventory

The stage of the inventory basically consists in collecting and quantifying inputs and outputs needed for the product or service under analysis throughout its entire life cycle. In the given cycle, each of the stages or the entire process will be considered sub-processes and evaluated as such. For each sub-process, we will collect input and output data and we will assign such data to the selected "functional unit" delimited in the previous phase. The analysis finally will create an inventory of flows (water, energy, raw materials, releases to air, water and land, etc.) for the product and/or service under analysis that will be presented as the final inventory output. The results obtained need to be consistent with the proven scientific evidence of the conservation of mass, energy and its relations, as well as the principles of thermodynamics and stoichiometry.

In the establishment of inputs and outputs of the different subsystems, it is possible to end up with co-products or by-products, substances or objects that result from a particular part of a production process that was not designed to produce such items. In any case, the assessment needs to precisely define until which part of the process these by-products influence the principal unit identified for the analysis and in which way. Once the limit has been allocated, we also need to assign these impacts for the final valuation.

The final presentation of results must include the following information: (a) A detailed description of the examined product system; (b) An analysis of all the processes used in the product system; (c) A system flow chart; (d) Data allocation rules; (e) A model of the system; and (e) A final calculation of the inventory.

Life cycle impact assessment

The inventory analysis is followed by an impact assessment. This stage of the LCA evaluates the magnitude and significance of the potential environmental impacts of undesirable outputs listed in the previous phase of the LCA. To facilitate the methodological aspects, the structure of the impact assessment follows three mandatory requirements based on ISO 14040: The selection of impact categories; the assignment of the inventory to this selection; and the characterization of indicators as a result. At this stage it is convenient to consider the distinction we saw in the previous chapters between impacts, effects and consequences. This process is usually developed in four steps: classification of impacts; characterization of the inventory in its associated impacts; normalization based on the reference values of its relevance; and final valuation.

There is a long list of different classifications of impact categories (e.g., climate change, acidification, eutrophication, human toxicity, etc.); see Klöpffer and Grahl (2014) for a review. After selection, the results obtained in the previous phase are associated and their relevance assessed. The final valuation, usually conducted by weighting factors, has been always controversial (it is not considered in the ISO14040); however, it is still widely used in practice. Different impact categories are very difficult to use for comparisons, and on many occasions these data respond to other social, political and ethical considerations. In any case, when applicable, weighting factors are to be used cautiously.

Interpretation

This is the phase of the LCA where the information obtained in the inventory and the impact assessment are combined and the results are interpreted in accordance with the objectives and the scope of the study. Based on the results obtained in the previous phases, an identification of significant issues and an evaluation check is done before to produce a series of conclusions to explain the limitations and provide recommendations.

The identification of significant issues is intended to highlight those parameters for which the obtained results in the previous phases showed a significant quantitative difference with the involved uncertainties. Following that, the obtained results are scrutinized through three different completeness, sensitivity and consistency checks. The completeness check relates to all relevant information, especially to the provision of significant parameters. The sensitivity checks estimate the uncertainties implicit in the results due to data quality, criteria, allocation rules and impact categories. Finally, the consistency checks determine whether the assumptions, methods and data are consistent with the goal and scope of the assessment. Final conclusions and recommendations will establish a general assessment of the product/service to reduce its environmental impacts in the different parts of its life cycle.

Final report and critical review

Reporting and critical review have been introduced as formal parts of the entire assessment to be sure that LCAs are done precisely. The issue here is to ensure that results, conclusions and recommendations are reported in an adequate form to the intended audience, addressing the data, methods and assumptions applied in the study and the limitations thereof. At the same time, a critical internal and/or external review is thought to improve technical and scientific quality and increase reliability.

Ecological key performance indicators

LCA is a tool to assess the environmental aspects and potential impacts associated with a product and/or service. The use of this tool – through simplified and more focused methodologies – can also enable us to come up with key ecological performance indicators, the three most relevant being the carbon footprint, the water footprint and the ecological footprint. Performance indicators can be calculated in the same way LCA does, from cradle-to-grave or even from cradle-to-cradle. These key performance indicators are for the most part used individually today, but may also be used in conjunction, for example as parts of the so-called "footprint family" (Galli et al., 2012). Carbon, water and ecological footprint indicators can reveal information about humans' individual behavior and/or about companies' sustainability strategies, and they can be used together to explore the pressures we put on the different earth systems (biosphere, atmosphere and hydrosphere) complementarily. The use of these key performance indicators is becoming more and more common in product declarations, as we will see in the next section.

The calculation of the key performance indicators follows a life cycle approach, particularly the group of resource flows and environmental loads associated with a product or organization from an extended supply-chain perspective. As was seen in the previous section, in its best scoping consideration, it includes all stages from raw material acquisition to end-of-life processes, and all relevant related environmental impacts, health effects, and resource-related threats and burdens to society.

Carbon footprint (CF)

A carbon footprint (CF) measures the total amount of greenhouse gas emissions that are directly caused by an activity or are accumulated over the life stages of a product or service (Galli et al., 2012). The CF can be used for multiple purposes and applied to products, services, companies, industry sectors, individuals and countries. When used for corporate applications, it measures the total amount of the six greenhouse gases or family of gases identified by the Kyoto Protocol (carbon dioxide (CO_2), methane (CH_4) and nitrous oxide (N_2O), together with families of gases including hydrofluorocarbons (HFCs) and perfluorocarbons (PFCs) – CO_2, CH_4, N_2O, HFC, PFC, and sulfur hexafluoride (SF6) expressed in kg CO_2-equivalent. A CF is calculated by multiplying the actual mass of each gas emitted with the global warming potential factor for this particular gas, and then adding these six gases together. In this way, the different GHGs emitted by a product system or by a company are comparable and additive, producing a unique number as a key performance indicator.

When related to products and services, there are different guides that can help in its calculation. In 2008 the British Standard Institute developed its PAS2050 guide for calculating CF; more recently, in 2013, the International Standard Organization came up with the ISO/TS 14067, which specifies principles, requirements and guidelines for the quantification and communication of the CF of a product (CFP). In both cases, they are based in the international LCA standards seen previously, and are only applicable to the impact category of climate change. Another way to calculate CF is the so-called Greenhouse Gas Protocol (GHGP) developed by the World Resource Institute and the World Business Council for Sustainable Development, which enables companies to measure, manage and report greenhouse gas emissions from their operations and value chains, and which is widely used for those companies reporting to the CDP in their reporting activities (see also Chapter 8 of this book).

In calculating the CF of a company or of a product/service, since the very beginning there has also been a need to establish the scope and boundaries of the analysis, as was seen when starting a global LCA. Box 6.1. presents a general way to make these delimitations following the GHGP.

Box 6.1   Operational boundaries in the case of GHG emissions and CF calculations

In order to develop a framework for CF calculation, it is necessary to better understand the question of operational boundaries. Companies have two types of GHG emissions:

- **Direct,** generated from sources owned or controlled by the focal company, such as manufacturing plants, vehicles, or maintenance activities. This is the source, for example, of emissions from the production of electricity, heat and steam; emissions resulting from chemical processing; or emissions from company-owned vehicles for the transportation of materials, products, waste, and employees (trucks, planes, ships and cars), and

- **Indirect,** generated from sources that are not owned or controlled by the company. For example, emissions from the production of electricity purchased by the company, emission from the use of products and from the product's end-of-life, emissions related to the manufacturing of components and other materials imported and used by the company, emissions from employee business travel, etc.

According to the international standards used to support companies in assessing their direct and indirect carbon emissions and in delineating their climate plan and goals, three types of "scopes" have been defined (WRI and WBCSD, 2004): Scopes 1, 2 and 3.

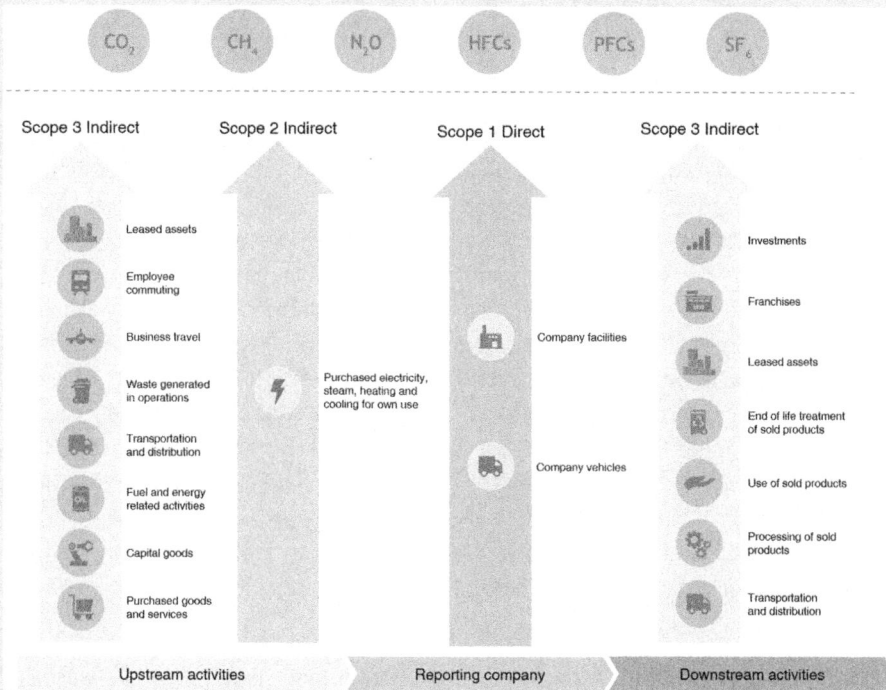

| CO_2 | CH_4 | N_2O | HFCs | PFCs | SF_6 |

Scope 3 Indirect

Scope 2 Indirect

Scope 1 Direct

Scope 3 Indirect

Leased assets

Employee commuting

Business travel

Waste generated in operations

Transportation and distribution

Fuel and energy related activities

Capital goods

Purchased goods and services

Purchased electricity, steam, heating and cooling for own use

Company facilities

Company vehicles

Investments

Franchises

Leased assets

End of life treatment of sold products

Use of sold products

Processing of sold products

Transportation and distribution

Upstream activities — Reporting company — Downstream activities

Figure B6.1 Scope 1, 2 and 3 emissions.
Source: Adapted from Puma.

As illustrated in Figure B6.1, Scope 1 reflects only the direct GHG emissions coming from sources that are owned or controlled by the company. Scope 2 refers to GHG emissions from purchased electricity utilized by the company. Scope 3 encompasses the other indirect GHG emissions, both upstream and downstream. GHG emissions include the six gases that were included in 1997 Kyoto Protocol (CO_2, CH_4, N_2O, HFCs, PFCs, SF_6).

When make calculations, companies must decide the scope of their strategy and set their operational boundaries. Today there is a clear trend to consider companies accountable for Scope 1, 2 and 3 GHG emissions. In other words, stakeholders now expect that large corporations address sustainability through comprehensive climate plans that incorporate the impact of the extended value chain, from carbon emissions associated with acquired components and materials to those occurring from product consumption, utilization and end-of-life treatment.

Source: authors' elaboration on multiple sources.

Water footprint (WF)

The water footprint (WF) measures the appropriation of natural capital in terms of the volume of freshwater required for human consumption. Departing from the concept of virtual water popularized by Allan (1998), Hoekstra and Hung (2002) and Hoekstra (2003) defined the concept of WF as being the "cumulative virtual water content of all goods and services consumed by one individual or by the individuals of one country." Virtual water is the water "embodied" in a product or used in a service, not in a real sense but in a virtual sense. It refers to the water needed for the production of the product or service.

Mirroring the previous CF, WF can also be used for multiple reasons, and for our purposes it can be calculated for an entire company or for a product and/or service. In particular, the water footprint of a product or service would be the amount of water that is consumed and polluted, analyzing all stages of its life cycle. This amount of water relates both to the direct use by a consumer or producer, but also to the associated indirect water usage. Similar to the CF, there are guides to develop WF assessments that are starting to be used (Hoekstra et al., 2011; see also, for example, the link in Annex 1 at the end of this chapter to the water footprint network).

The WF includes direct (controlled by the company) and indirect (uncontrolled by the company) usages of water. For each of these two aspects, three different components can be calculated (see also Figure 6.3).

- The **blue water footprint** refers to consumption of blue water resources (surface and groundwater) along the supply chain of a product. "Consumption" refers to the loss of water from the available ground-surface water body in a catchment area. Losses occur when water evaporates, returns to another catchment area or the sea, or is incorporated into a product.

• The **green water footprint** refers to consumption of rainwater resources that can be obtained directly or in the form of moisture (directly for the atmosphere) but that does not become run-off. In this case the WF excludes non-consumed water that is returned.

• The **grey water footprint** refers to pollution and is defined as the volume of freshwater that is required to assimilate the load of pollutants given natural background concentrations and existing ambient water quality standards.

A product's water footprint tells us how much pressure that product puts on freshwater resources. It can be measured in cubic meters of water per ton of production, liters per kilogram, or gallons per pound or per bottle (e.g., in the case of milk).

Ecological footprint (EF)

As we saw in Chapter 1, the ecological footprint (EF) (Wackernagel et al., 1999) is a resource and emission accounting tool designed to track demand on the biosphere's regenerative capacity of all human activities at different scales. It can be applied to firms, organizations, products, services and individuals.

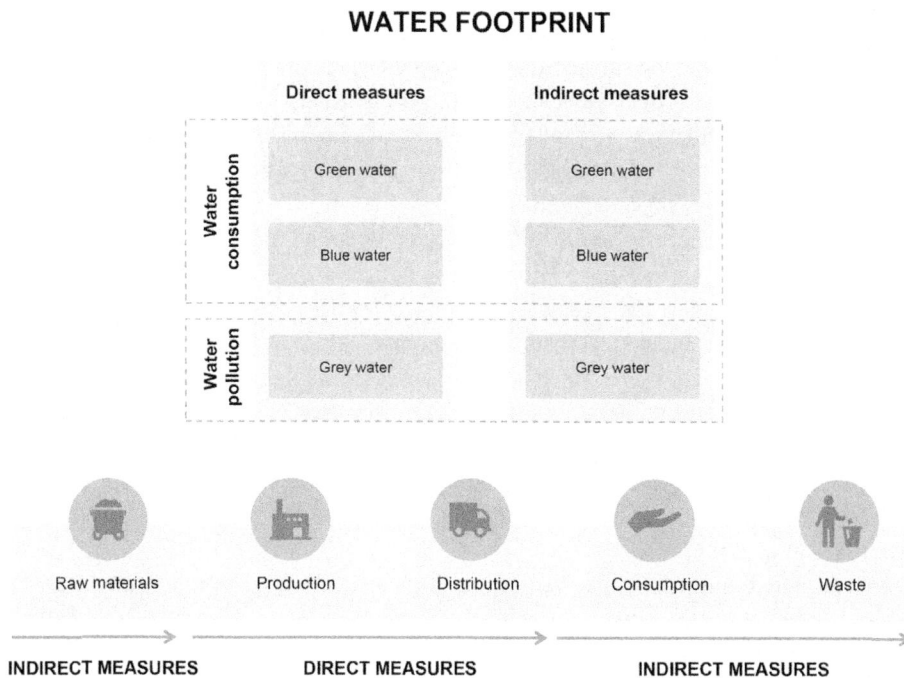

WATER FOOTPRINT

	Direct measures	Indirect measures
Water consumption	Green water	Green water
	Blue water	Blue water
Water pollution	Grey water	Grey water

Raw materials → Production → Distribution → Consumption → Waste

INDIRECT MEASURES | DIRECT MEASURES | INDIRECT MEASURES

Figure 6.3 Water footprint components and measures.
Source: authors' elaboration.

As was seen before for CF and WF, guidelines have been developed to establish common methodologies to calculate this footprint (e.g., the EU Product Environmental Footprint Guide, 2012; the Ecological Footprint Standards, 2009; and Wiedmann and Barrett, 2010). The computation of an EF for a product is recommended to be done within working areas of product categories. As was also indicated for CF and WF, both direct (process-based) and indirect (extended input-output) calculations are necessary in order to obtain an EF. Following the Global Footprint Network (GFH) (2009), the direct (process-based) method has the advantage of detail, as individual product types and even brands can be analyzed; however, its general disadvantage is its lack of complete upstream coverage of the production chain. The extended indirect (input-output) method has the advantage of full upstream coverage but entails the disadvantage of generality, as input-output tables typically do not disaggregate down to the level of individual product types.

As indicated in Chapter 1, the company and/or product/service EF is calculated as the area of ecologically productive land (and water) required to provide all the energy and resources it consumes and to absorb all the waste generated throughout its life cycle; and it is measured as a surface area (normally in hectares).

Environmental product declarations (EPD)

LCA and ecological performance indicators are important tools for internal decision making, and they can support and orient innovation processes with regard to the reduction of product and services environmental impacts. The main problem with these tools is that they are extremely complex and technical, and therefore they have limited power to orient purchasing towards sustainable products. Since the 1990s policy makers, business organizations and NGOs around the world have attempted to stimulate the demand for green consumption, but we know that consumers need accessible, clear, simple and credible environmental information to choose from when making purchases.

Environmental Product Declarations (EPDs) have emerged as a tool to convey sophisticated environmental performance information about the products and services to individual consumers through the business-to-business market. In other words, EPDs have become a helpful instrument to transfer and communicate LCA information through the value chain, although they are not designed to substitute for ecolabels in supporting green consumption patterns. Over the years, as in the case of the other environmental assessment tools previously illustrated, multiple national and sectorial schemes have been developed for EPDs, and after many years of attempts at harmonization ISO has provided a common ground for this instrument, defining a standardized methodology (EC, 2002; Ibáñez-Forés et al., 2016).

Today we define Environmental Product Declarations (EPDs) as third-party verified and registered documents (data sheets) based on the requirements of ISO14025 that communicate transparent and comparable information about the life cycle environmental impact of products. EPDs are also known as "Type III environmental declarations", to distinguish them from other types of communication tools such as ecolabels. In particular, having an EPD for a product does not imply that the declared product is environmentally superior to its alternatives. This differentiates this instrument form the so-called "Type I environmental label," which indicates the overall environmental preferability of a specific product within a particular product category based on its Life Cycle Assessment and independent verification (EC, 2002: 20).

The EPD® system has become the most important international program with regard to environmental product declarations. It operates in accordance with the ISO14025 and provides the necessary procedures for preparing declarations, developing consistent and comparable data sets, verifying declarations, registering and adding to the visibility of a product through the EPD® system website.

EPDs are useful tools to respond to different communication needs. They can support organizations and individuals in comparing the environmental performances of products and services, and they help in tracking the product's improvements over time. They favor green public procurement, providing valuable and accurate information to support the demand for low-impact environmental goods and services. In business-to-business transactions, the EPD is a valuable vehicle to collect and provide key environmental information along a value chain, thus supporting decision making. Finally, it can be used also in business-to-consumer communication to strengthen the credibility of a green value proposition, but with the warning that products that have an EPD should not automatically be considered superior to those products that do not.

The process for obtaining an EPD is articulated in a sequence of steps, as follows.

- The EDP starts with a LCA of the product, which must be carried out according to the ISO14040 and ISO14044 and the specific product category rules (EPD International, 2017).
- The LCA outcomes and the other mandatory information must be communicated compiling the EPD reporting format.
- An independent verification must be carried out according to the principles and procedures of the EPD® system.
- Finally, the EDP is registered and published on the www.envirodec.com website.

Several companies are using EPDs according to the international EPD® system and the ISO14025 standard, ranging from agro-food to detergents and washing products, from wood and paper to construction and electricity. One innovative example in the utilization of LCA and EPD both for strategic purposes and communication is provided by Barilla, the world's leading pasta company, which we already introduced in Chapter 5. For an explanation of how the Italian company applies these tools, see Box 6.2 below.

Box 6.2 Environmental product declarations at Barilla

Established in 1877 and with revenues exceeding €3,400 million (2017), Barilla is a family-owned firm and one of the largest Italian food producers with operations in more than 100 countries. Barilla's attention to sustainability is a core value, linked to other features of its identity such as a long-term approach to doing business and a strong sense of responsibility along the entire value chain. The company's statement of purpose articulates these features: "Only one way of doing business. Improving the wellbeing of people, the planet and the communities from field to fork" (Barilla, 2016). The strategy "Lighthouse for the future" also reflects the firm's commitment to environmental and social issues. It builds on the motto "Good for you. Good for the planet" and focuses on integrating nutrition and environmental protection, while also focusing on local communities (Pogutz and Winn, 2016).

Barilla's focus on environmental protection started two decades ago. The company was among the pioneers of applying the LCA methodology to measure environmental impacts along the entire pasta supply chain. At that time the LCA diffusion as a corporate tool was still very limited (Rex and Baumann, 2008). Over the years, the utilization of this methodology has become part of Barilla's strategy, and in 2009 the company started to use it to analyze and measure the environmental impacts associated with its many products (Ruini et al., 2013). Building on its LCA expertise, Barilla published its first EPD for semolina pasta in 2010. In line with the ISO14040:2006 standard, Barilla LCA covers all phases of the pasta life cycle – from durum wheat cultivation to the cooking of pasta in consumers' homes – and assesses the environmental impact based on three indicators (Barilla and EPD International, 2017):

- CF (greenhouse gases produced by the system under assessment),
- WF (water consumption of the system, measured as the volume of water evapotranspirated by plants, consumed or polluted), and
- Ecological Footprint (EF) (amount of biologically productive land and water surface required to produce all of the resources an activity consumes and to absorb the waste it generates).

The results of the application of the EPD is illustrated in Figure B6.2. In this case, the product included in the analysis is classic durum wheat semolina pasta pieces (such as spaghetti, penne or fusilli). The shape is the only differentiating feature, as the only ingredients used in all of these products are water and semolina. The data refers to 1 kg of product and the related packaging for each retail unit of 500 g. The analysis was based on the specific product rules published for the EPD System: "CPC code 2371 – Uncooked pasta, not stuffed or otherwise prepared."

With regard to the boundaries of the analysis, the methodology considers all the phases, from upstream (durum wheat cultivation, milling and packaging production), to core (pasta production), to downstream (distribution, cooking, packaging and end-of-life), in compliance with the requisites of the EPD system. Figure B6.2 illustrates environmental performances in the case of Italy. Barilla has carried out specific analyses in the most important countries where the company operates (i.e., the United States, Turkey and Greece), where different hypotheses need to

PRODUCT ENVIRONMENTAL PERFORMANCES

Dry Semolina pasta - Italy for local consumption

	Raw material production	Milling	Packaging production	Pasta production	Distribution	Primary packaging end of life	from field to distribution	Cooking phase
ECOLOGICAL FOOTPRINT	7.8	0.1	0.9	0.5	0.1	<0.1	9.4 global m²/kg	1.6 / 3.1
CARBON FOOTPRINT	564	48	21	198	65	17	913 gCO₂eq/kg	607 / 1 062
VIRTUAL WATER CONTENT	1 379	<1	100	2	<1	<1	1 481 liters/kg	11 / 16

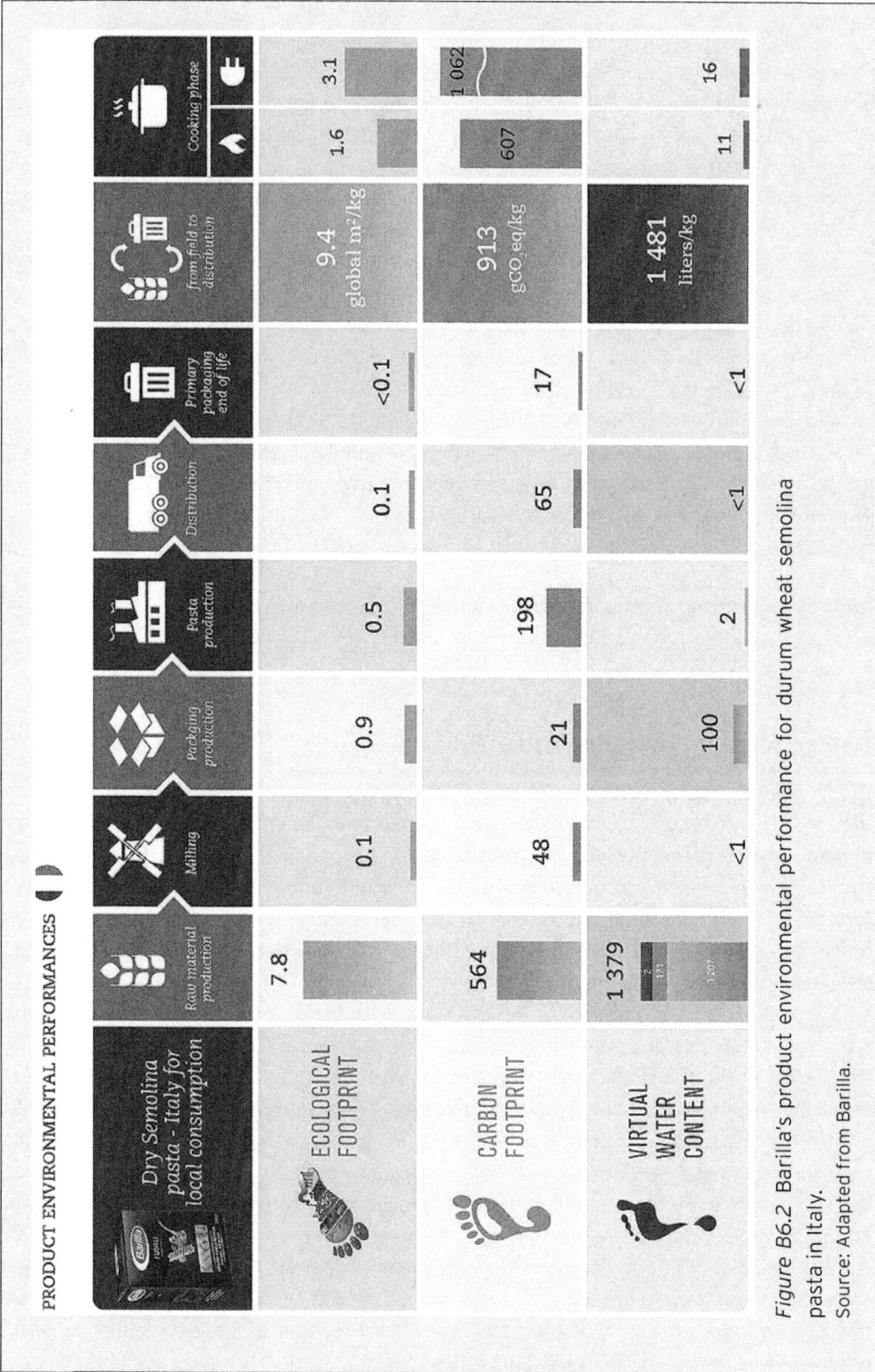

Figure B6.2 Barilla's product environmental performance for durum wheat semolina pasta in Italy.

Source: Adapted from Barilla.

be considered (e.g., variation in the yields for durum wheat cultivation, distribution, consumer behavior and cooking systems and waste treatment). The LCA/EPD clearly shows that the raw material, production, and cooking phases have the biggest impact on ecosystems (about 60% of the global warming potential of pasta is generated in the cultivation).

Based on the results of the first durum wheat semolina pasta EPD (2009), in 2010 Barilla developed a "sustainable agriculture" project designed to experiment with and implement several actions to reduce the impact of the pasta supply chain. The key pillars were: a handbook for farmers explaining sustainable cultivation techniques; a web decision support system (www.granoduro.net) designed by another company, Horta s.r.l., to support farmers in making cultivation decisions; and new types of contracts with farmers, with bonuses for durum wheat cultivation based on sustainable agricultural practices (Pogutz and Winn, 2016; Barilla and EPD International, 2017).

Over the years, Barilla has acquired knowledge and has developed competencies in the application of LCAs and EPDs. The company uses both tools for decision making, commercial purposes and communication with different stakeholders. In 2016, 61 of Barilla's products were EPD certified, equivalent to 67% of the overall production.

Source: authors' elaboration on multiple sources.

Design for the Environment (DfE)

Product innovation identifies a major area of intervention for companies that intend to reduce their load on the environment. The phase of consumption and utilization, in fact, in the case of many product categories (e.g., household appliances, personal care, cars, detergents, electronics) is responsible for about 80% of impact on ecosystems. For several companies, therefore, sustainability is about radically changing products and services, adopting new and more efficient materials, increasing end-of-life recycling, improving energy efficiency and reducing the packaging associated with products.

Design is a primary function for innovation in business, and companies know that more than 70% of the cost of a product is defined at early phases of the innovation process. Similarly, studies have shown that most of the decisions that influence the future environmental performance of a product are defined at this stage. The design decisions, in fact, condition a large array of material resource flows related to products. For example, the design specifications for a sport shoe in Europe can impact ecosystems both in South America (e.g., for rubber cultivation and leather) and South-East Asia, where the shoe is manufactured (e.g., air and water emissions, waste). The decision about the material, shape and size of packaging for a home care product influences the opportunity for recycling and the possibility to close the material cycle (closed loop systems and circularity). Similarly, choices about the energy and water efficiency of a washing machine directly affect the consumption of the natural capital in the product's use phase. It is therefore critical to weave sustainability and green issues within the process of product innovation, starting with the design phase to avoid future negative effects on the natural environment.

The engagement of ecological and sustainability discourse in design started in the late 1980s, with the first active interest of industry in environmental protection. In 1992, the US Environmental Protection Agency created the Design for the Environment (DfE) program. This was the first initiative driven by a governmental agency that focused on decreasing pollution and reducing human environmental risks (e.g., use of chemicals and toxic substances) associated with consumer and industrial products.

DfE (also ecodesign, green design, life cycle design, or design for sustainability) can be defined as "the systematic consideration of design performance with respect to environmental, health, and safety objectives over the full product and process life cycle" (Fiksel, 1996). DfE is therefore a strategic activity that supports firms in imagining and developing more sustainable products and services, lowering the environmental impact through re-designing or radically innovating the product concept. Ecodesign adopts a life cycle approach, considering all stages from natural resource extraction or cultivation to manufacturing, use and end of life. DfE is rooted in the Life Cycle Assessment (LCA) methodology. The LCA is useful for identifying environmental "hotspots": the life cycle stages with the greatest environmental load, which can be addressed using DfE.

Integrating DfE into the product innovation process

But how can a company incorporate DfE in its strategy? How can DfE be integrated in product innovation? Ecodesign combines three important aspects: business strategy, engineering design and focus on environmental aspects. For companies, it is therefore a crucial challenge to develop knowledge and capabilities that favor an effective integration of sustainability aspects into the innovation and design function. Effective DfE must combine lessening of environmental burden with maintenance or improvement of product quality, and with attention to consumers' needs and expectations. When designing a new product, traditionally, industrial designers have in mind several variables:

- branding and strategy,
- functionality and performance,
- reliability,
- aesthetics,
- manufacturability,
- logistics and reverse logistics,
- safety,
- cost structure,
- pricing decisions.

With DfE, analysis of the environmental impact is incorporated in the product development process, according to the life cycle thinking approach that extends across the entire supply chain. In other words, environmental issues are given the same importance as other industrial and marketing aspects that traditionally guided the innovation process (Ceschin and

Gaziulusoy, 2016). This approach, also known as integrative design, connects managers and experts from different areas with different competencies. Cross-functional teams can involve multiple organizational functions such as R&D, marketing, purchasing and supply chain, logistics, operations, technology and engineering, regulatory and sustainability/CSR. Moreover, they incorporate external stakeholders who provide important inputs for product innovation such as consumers, customers and retailers, experts from NGOs and the scientific community.

The DfE approach requires firms to adopt a new organizational logic for the product innovation process that involves the distribution of novel tasks across the company's many functions. To better investigate this new approach, we build on a typical model in product innovation, the "innovation funnel" (see Figure 6.3.). This framework provides a structured representation of the product innovation phases, based on the selection of product features through a series of stage-gates. As the design strategy narrows down, the product increases the specificities. The process starts with the first definition of the product concept and goes into several steps of progressive refinement and fine-tuning leading to product prototyping,

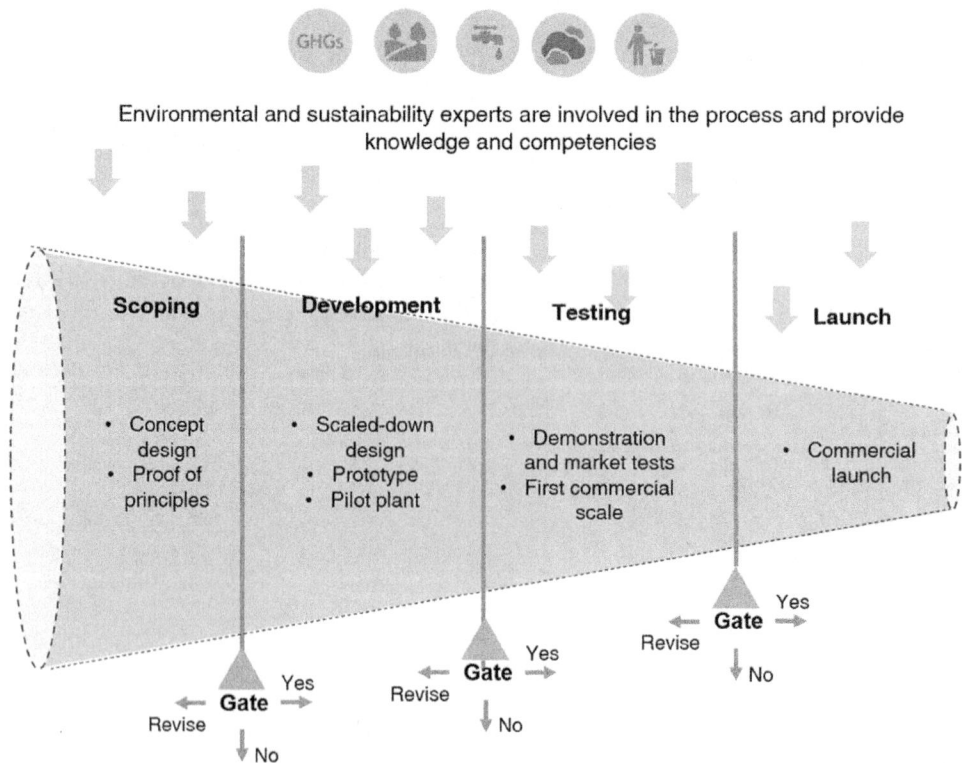

Figure 6.4 Sustainability and the innovation funnel.
Source: authors' elaboration.

demonstration, and first commercial scale; it culminates with the commercial launch of the new product. Innovation stages are separated by gates, where a decision is made to go to the following step, or stop the process. The basic idea of this model refers to the fact that stage-gates help in improving the efficiency and effectiveness of the innovation process, because the most critical aspects of the product are tested and if the tests fail, the process is interrupted, limiting the amount of money lost.

When using a DfE approach, a number of green and sustainability experts from inside and outside the company are involved in the process to provide the knowledge and competences required to draw attention to environmental issues from the early stages of product innovation. For example, the collaboration and alignment between procurement and supply chain managers, marketing managers and sustainability experts can favor a better understanding of future concerns related to the use of some environmentally sensitive raw materials. Anticipating these potential problems during the early stages of the product innovation process can favor the sourcing of alternative raw materials that fit with consumers' expectation with regard to product quality, aesthetics and functionality, and lessen the environmental impact. Likewise, collaboration among researchers and engineers can allow identifying a list of prohibited or restricted chemical substances that the organization should manage or eliminate to avoid future problems with product utilization and disposal.

Besides specific knowledge and competencies, incorporating sustainability in the innovation funnel also requires communication and coordination mechanisms among different teams of participants working on different typologies of products. Putting in place such organizational mechanisms helps firms bridge barriers to sustainable design, while favoring ongoing corporate learning processes among different teams and diffusing lessons learned among categories of products. For example, the results obtained from experimenting new and more sustainable materials in a specific product line can be extended to other product lines or other brands in the company portfolio, thus improving the overall environmental performance of the company.

Areas of intervention of DfE

We have defined DfE as a strategic process aiming at favoring a systematic reduction of product environmental impacts along the entire life cycle, starting from the early design phase of product innovation. The question we want to investigate now is what the main areas of intervention related to DfE are. It is evident that the use of the LCA approach extends from the analysis of environmental load generated from raw material to product recycling and disposal (cradle to cradle). Therefore, designers must consider environmental performance at the manufacturing and assembling phase, during product consumption and use, and at the end of its life. Another area that is becoming increasingly relevant for many product categories is raw material extraction and cultivation. DfE, therefore, extends the focus of product innovation to environmental performance of early phases of the product life cycle, seeking for more environmentally friendly materials, and favoring closed loop systems where virgin inputs are substituted with recycled ones. In the following box (Box 6.3.), we illustrate the example of Nike's sustainable innovation.

Box 6.3 Nike's sustainable innovation

Making sneakers has high environmental impacts because of the utilization of primary materials such as leather, synthetic leather and rubber. Leather is mainly cowhide and responsible for GHG emission and deforestation, while leather tanning is environmentally costly due to use of water and chemicals, besides waste generation. Synthetic leather is made from plastic and is non-biodegradable. Rubber is also made from petroleum and treated with chemicals (natural rubber, which could be a better solution for the environment, is much less used for sport shoes). Nike, the Oregon-based sport apparel company, has come a long way in its sustainability journey and is now a leading player in sustainable innovation investing in new low-impact materials, reducing emissions and waste generation through technological changes and product design. In particular, from concept to product creation, the company involves different teams and a set of skills following an integrated approach: "Working together to find new and better product solutions helps reduce the environmental and social impacts of our products significantly, while also adding value to our business. It also offers a more integrated way of thinking, moving teams beyond their own subject matter expertise to achieve greater efficiency and greater sustainability" (Nike, 2016: 24).

Among the initiatives introduced to increase product sustainability, attention to waste generation and recycling became a key strategy for Nike. The company started working on a new manufacturing technology that allowed creating footwear uppers directly from yarn rather than cutting them from fabric. This solution, called Flyknit, was first introduced in 2012. By fall 2015, 28 products in six categories were created using the technology, with consistent reduction in the generation of scrap waste. Moreover, Nike started using recycled polyester for all core yarn for Flyknit-based shoes, which meant diverting millions of plastic bottles from landfills.

Another important program developed by Nike to address waste generation at the product level was Grind. This initiative was started in the 1990s with the aim of closing the product life cycle loop favoring recycling. Grind is a materials palette obtained from the recycling and regeneration process of other materials and products. In 2016, 71% of Nike's products – footwear and apparel – was using Grind's materials. Waste streams, including old rubber, foam, plastic bottles and textile scraps were transformed into new materials with high performance. Between 2012 and 2016, Nike recycled more than 3 million plastic bottles into footwear and apparel. Nike Grind materials are also used for the preparation of sports surfaces such as running tracks, soccer and American football fields or gym room floors.

Source: authors' elaboration of multiple sources.

There are multiple strategies of DfE that combine different approaches such as designs for disassembling, resource recovering, recycling, disposal, energy efficiency, regulatory compliance, reducing human health and safety impact, or hazardous material minimization

(Fitzgerald et al., 2007). In order to systematize these multiple approaches, Simon et al., (2000) introduced the concept of Design for X, where X stands for any environmental criteria incorporated in the product design path, in accordance with the decision made by the company and the designers, and the phase of the product life cycle where it is applied. Table 6.2 shows some of the fundamental ideas behind DfE with some examples of initiatives realized by leading companies.

At the operational level, the implementation of the DfE framework can use a series of methodologies and tools (Fitzgerald et al., 2007; Fiksel, 2012; Lewis et al., 2017):

- guidelines and checklist documents that force managers to consider environmental issues from the product design phase,
- product design matrices that help to prioritize environmental impacts and link them with significant product features,
- environmental effect analysis (see also the previous chapters) to help in reviewing and understanding the effects of a product's life cycle on ecosystem services, and
- LCA, which can help in design and continuous improvement of products.

Table 6.2 Fundamental strategies behind the DfE framework.

Minimize the use of non-renewable or scarce resources and emphasize the use of renewables	This is probably the most notable example in the search for drastic product dematerialization, increase in energy efficiency pattern and the search for carbon neutrality. This strategy can be applied to products, production processes, raw material use and logistics. Google, the global technology leader, reached 100% renewable energy sources at the end of 2017 for its global operations (including both centers and field offices) by signing large-scale, long-term contracts to buy renewable energy directly. Google is a member of RE100 (see also Chapter 8), a collaborative global initiative uniting more than 120 influential businesses committed to 100 percent renewable electricity, working to massively increase demand for and delivery of renewable energy.
Increase product durability	Extend product life by better use of materials and improved design. Philips is the global leader in connected LED lighting systems and services. The Dutch company has made its products' durability and efficiency the selling point. The development of this technology has drastically improved energy efficiency, reducing environmental impact and increasing revenues.
Design products for re-use and material for recycling	Design for recycling favors material recovery, uses materials that can be recycled locally, and favors material identification for improving the condition of recycling and reducing associated costs. Sony, the global technological innovation and entertainment giant, subscribes to this principle through its producer responsibility philosophy over its products' entire life cycle, recycling-oriented product design, collecting and recycling used products, and building global recycling systems that suit the needs of individual countries and regions.
Design for disassembling	Design to favor reuse of components, quicker and cheaper product disassembly, and easy dismantling to recapture parts. This approach builds on a better product modularity that allows separation of the single parts. Acer, one of the world's leading IT companies, has established product design principles that emphasize modularization and components that are easy to dismantle and recycle, reducing product weight, providing better information and better labeling, reducing the use of plastic, and avoiding bonding, soldering and adhesive surface technologies.

(Continued)

Table 6.2 (Continued)

Minimize the environmental load of the production process	Design for manufacturability enables pollution prevention during manufacturing and at the same time uses fewer and safer materials. Hewlett Packard, the multinational IT company, has improved manufacturability by reducing the amount of material used in products and developing materials that: (1) are easier to manufacture, (2) lessen the environmental impact, and (3) have more end-of-use value.
Minimize the environmental impact of product use	Design to reduce the environmental burden of a product during the utilization phase of its life cycle. Electrolux, the Scandinavian global appliance company, is committed to improving its environmental performance in product use as well as in designing products for disposal. The company is aware that the maximum environmental impacts of products such as washing machines and refrigerators is in the consumption of energy and water and focuses on introducing up-to-date technologies to significantly reduce such impacts.
Use environmentally friendly packaging	Design for minimizing or changing packaging materials and rethinking the way the company sells its products. Seventh Generation, the US company specialized in selling cleaning, paper and personal care products has built a green reputation by the formulation of products made with plant-based ingredients. The company moreover carefully considers packaging materials (in terms of content, recyclability and performance) and strives for ease of use, so that its eco-conscious clients do not need to think about packaging at all.
Provide for environmentally friendly disposal of non-recyclable substances	At the end of the product life cycle, there should be no materials that cannot be reused or recycled. Every company should work hard to design new products that minimize non-recyclable substances. Residual and non-recyclable components must be treated through coherent waste management systems, minimizing the environmental impact.
Increase the logistics and reverse logistics efficiency	Design for efficient logistics across the whole product value chain. Cisco, the worldwide leader in IT networking, embraced a wide-ranging crusade to optimize reverse supply chain systems to enhance customer experience and increase firm profitability. While improving service to its customers, Cisco has favored relevant improvements in environmental efficiency, increasing product reuse opportunities and allowing recycling of much larger component parts.

Source: authors' elaboration on multiple sources.

The DfE is a useful framework guiding the process of product and service innovation. This approach needs to be integrated into sustainability strategy as well as support the relationship with the company's stakeholders. To sum up, the DfE can help managers in targeting sustainability at multiple levels:

(a) improving environmental stewardship and product performance through key performance indicators,
(b) ensuring compliance with formal and informal norms and regulations,
(c) responding to the awareness and concerns of clients and other stakeholders,
(d) mitigating possible risks that can translate into a reduction of finance liabilities, and
(e) improving product overall quality and profitability.

Implementing the DfE is key in helping companies build social-ecological system resilience through re-design and radical innovation at the product level. It is therefore part of pillars of the "Business In Nature" concept presented in this book. To conclude, in Box 6.4, we briefly illustrate another example of an innovative product based on the application of the DfE approach – the Setu™ multipurpose chair made by Herman Miller, one of the most important producers of home and office furniture in the world.

Box 6.4 Herman Miller's design for the environment

Based in Michigan, USA, Herman Miller is a global company that makes home and office furnishings and related technologies and services. With a revenue of about US$2.26 billion in 2016, and around 8,000 employees, it is acknowledged as one of the first companies to produce modern furniture. As indicated on its web page, Herman Miller aims to explore, research and develop insights that lead to innovative products and new ideas, solving problems for people, improving their health, well-being, productivity and enjoyment. One of its environmental principles for action is to work with the DfE framework: "As Herman Miller continues its 'Journey towards Sustainability,' designing our products with consideration for their environmental impact remains a central corporate strategy. Our long-term emphasis on product durability, innovation, and quality demonstrates that our company has effectively designed for the environment for decades" (www.hermanmiller.com).

The Setu™ multipurpose chair is a clear example of an environmentally friendly product that has been designed considering sustainability principles across the entire product life cycle (Herman Miller, 2017). The main results obtained are:

- environmentally friendly and non-toxic materials (41% aluminum, 41% polypropylene, 18% steel, by weight)
- use of recycled materials (44% by weight)
- less material content (20 lbs lighter than most task chairs)
- easy to disassemble (86% easily separable materials)
- recyclable (92% by weight)
- production line using 100% green power
- no air or water emissions released in production, and
- returnable and recyclable packaging.

Source: authors' elaboration based on multiple sources.

Another interesting case is provided by VF Corporation, the global leader in lifestyle apparel, footwear and accessories, with a portfolio of more than 30 brands (e.g. Timberland, The North Face, Vans, Wrangler). VF has started to integrate the principles of sustainability and circular economy in product design. In the following Box we briefly illustrate some of initiatives undertaken (Box 6.5.).

Box 6.5   Sustainable innovation and DfE at VF Corporation

VF Corporation is a global leader in apparel and footwear industry, controlling 30 brands such as Timberland, The North Face and Vans, some of which share a strong environmental and ethical orientation. VF has been a signatory of the Global Fashion Agenda's 2020 Circular Fashion System Commitment letter to accelerate the transition to sustainability and circular economy. In particular, the agreement requires that by 2020, all VF European designers have undergone training in circular economy in order to incorporate these principles in product innovation, designing accordingly. Educating designers is key to developing specific competencies to improve product environmental performance (e.g., reduction of waste, increased recyclability), studying and integrating sustainable solutions both in the manufacturing process and in the products. Another interesting initiative linked to the commitment letter is the collection of secondhand garments at Timberland and The North Face European owned and operated stores. Additionally, the company has entered into a partnership with the Dutch NGO Circle Economy (February 2018) with the goal to develop and adopt tools that can improve decision making for its brands with regard to end of life of products and circular opportunities.

Source: authors' elaboration based on multiple sources.

Ecolabelling

According to the Ecolabel Index global directory, a web-based organization that monitors and provides consulting on the world of eco-certification schemes, today (as of March 2018) there are 464 ecolabels in 199 countries, in 25 industry sectors. In 2009, the number of ecolabels tracked was 340, while in the 1990s there were only about a dozen (OECD, 1997; Delmas et al., 2013). What are ecolabels and why have they been spreading so much in the market?

Ecolabels are intended to informing consumers and individuals at large about the environmental impacts generated by a product or a service during the phases of its life cycle. On the one hand, they respond to the need to support responsible purchasing by providing the right environmental information to consumers, therefore facilitating the change towards more sustainable patterns of consumption. They are designed to simplify the process of identifying green products. Thanks to a simple and recognizable seal positioned on the packaging or on the product, ecolabels inform consumers that the labeled product has a better environmental performance than those without the label (Horne, 2009). They encourage firms to account for the impact of their products and services, and push companies to improve the performances with regard to environmental sustainability.

Ecolabels are also known as Type I labels. They are assigned as a result of a third-party certification process, and they are based on a standard under which the product is assessed against a set of criteria that relate to the (cradle-to-cradle) product life cycle. A product's criteria are a set of qualitative and quantitative technical requirements, usually identified by an independent body that an applicant (e.g., a firm) must meet to obtain the seal. Any applicant that applies to obtain an ecolabel for a product or a range of products undertakes a process of independent verification. After the review, in case of positive evaluation, the applicant is awarded the label: A license is granted to use it on the products in question for a specific period of time after paying a fee for the costs of the application.

According to the International Standardization Organization (ISO, 1999), there are other two kinds of labels: Type II and Type III. Type II labels are based on self-declarations of manufacturers, distributors, retailers, or importers; they thus have much less credibility than Type I because the standards and criteria might be less transparent and there is a conflict of interest in the evaluation process. Unlike Type III labels (like the EPDs previously illustrated), the assessment procedure for Type I labels is less extensive and does not consider all the impacts along the life cycle, usually focusing instead on some key parameters (e.g. toxic substances and chemicals, recyclable components and energy efficiency) that are considered the most important by the certifying organization. On the other hand, Type I labels are designated to apply to a small proportion of products and services (between 5% and 30%) in a specific product category and allow evaluation based on environmental performance, because they differentiate between those who have obtained the seal, those that have lessened adverse environmental impacts and others. Type III labels can be applied to any type of product, regardless of the ecological impact generated (OECD, 1997).

Ecolabels can be issued by different types of organizations: national and international governmental agencies, industry associations, NGOs and research institutions. Some of these bodies are multi-stakeholder platforms, which helps with increasing the credibility and legitimacy of the scheme guaranteeing that multiple experts' opinions are taken into consideration (Horne, 2009). Another distinction refers to the range of environmental issues covered along the product life cycle: Some ecolabels focus on single phases (e.g., the supply chain) or specific commodities, as in the case of the Marine Stewardship Council, the Forest Stewardship Council, or the Round Table for Sustainable Palm Oil (illustrated in Chapter 5), while others provide a broader view from cradle-to-cradle and include multiple categories of products and services, as in the case of the EU ecolabel scheme or the German Blue Angel. Table 6.3 offers some examples of national and international ecolabel schemes that have acquired widespread recognition, managed in accordance with ISO labeling standards.

Another important feature of ecolabels is that in many cases they are voluntary. Therefore, from a managerial standpoint, these types of labels can be used for strategic purposes, such as differentiating a product or a company, increasing the credibility of the green value proposition, and building trust and brand loyalty (Galarraga Gallastegui, 2002; Horne, 2009). As previously illustrated, one of the major barriers to sustainable consumption is consumer distrust and confusion over firms' environmental and social claims. We have illustrated in the first section of this chapter that green purchasing is a complex process and entails multiple

Table 6.3 Examples of ecolabels.

EU Ecolabel Award Scheme	The European ecolabel was established in March 1992 with the adoption of (EEC) Council Regulation n. 880/925. Over the years the European Commission has proposed certain revisions of the program in order to strengthen the efficiency and effectiveness of the scheme in promoting sustainable consumption patterns, and to facilitate its adoption by companies. More than 25 years after its introduction, over 54,000 products have received the EU seal (a flower) in many different areas of manufacturing and services, from cleaning products to textiles, from lubricants to holiday accommodation. Products are assessed in terms of their environmental impact during their entire life cycle.
Nordic Swan	In 1989, the Nordic Council of Ministers for Consumer Affairs introduced an ecolabel scheme for products in the Scandinavian region (Sweden, Norway, Finland and Iceland). It is the first multi-national ecolabeling program, and is intended to avoid competition between and confusion among national approaches. It covers a broad range of product typologies and the logo displays a swan.
Blue Angel	This label was introduced in 1977 in Germany as a voluntary third-party scheme. The process of certification is guaranteed by different bodies, including the Federal Environment Agency and the German Institute for Quality Assurance and Labeling. Blue Angel is known for having a high market penetration and high level of recognition among consumers. It covers 120 product groups, 12,000 products and services and 1,5000 firms. Products are assessed in terms of their environmental impact during their entire life cycle.
Eco Mark	The Japan Environment Association (JEA) administers the Eco Mark Program with the goal of promoting environmentally friendly lifestyles. A multi-stakeholder committee including academics, governments, consumer groups and experts from various industries defines the standards and awards the certification. Products are assessed in terms of their environmental impact during their entire life cycle.
Energy Star	Energy Star is a specific government-backed program created in 1992 by the US Department of Energy and the US Environmental Protection Agency focusing on energy efficiency and enabling consumers and business to identify energy-efficient products (e.g., household appliances, electronics, office equipment), homes (e.g., eating and cooling equipment), and commercial and industrial buildings. The efficiency criteria ensure that certified products prevent GHG emissions. The Energy Star scheme has now been extended to other countries such as Canada and New Zealand and international arrangements exist among countries that participate in the program. The EU agreed to take part in the Energy Star program in 2001.
EU Energy Label	The European Community Energy Label Council Directive 92/75/EEC (September 1992) introduced a non-voluntary program to label and display standard information regarding the consumption of energy and other resources on household appliances (e.g., refrigerators, washing machines, tires, electric motors and lighting). This information must be displayed on all new household products for sale, hire or hire purchase. Products are rated from A+++ for the most efficient to G. In July 2017 the Commission published a new Energy Labelling Regulation that will gradually replace the Directive with the goal of simplifying the labeling mechanism and allow consumers to more easily identify energy-efficient products.

EU organic label	Food and crops have generated high interest with regard to the introduction of programs for the organic farming certification in order to protect consumers from misleading business behavior, increase credibility, and favor the broad diffusion of products and cultivation techniques. The EU has introduced this regulatory framework to encourage improvement and progress in the sector. The logo (called the Euro leaf) is designed to make organic products easier for consumers to identify. It indicates that the products comply with the regulations for the organic farming sector in the EU. This means that for processed products, at least 95 percent of the agricultural ingredients are organic. Specific norms and rules govern the use of the logo for different types of products (e.g., pre-packaged and non-pre-packaged, different product categories, etc.).

Source: Authors' elaboration from multiple sources.

features that play an important role in guiding consumers behaviors such as price, product availability, environmental concern and knowledge, social values and norms, attitudes, brand reputation and credibility. Ecolabels work as a transparency and clarity mechanism, and offer a means of verifying accurate, non-misleading and scientifically sound information displaying a simple logo or a label. They thus protect consumers from uncertain or ambiguous environmental statements and greenwashing (Delmas et al., 2013).

Finally, the adoption of ecolabels can be used to gain access to markets, thereby allowing a company to increase its revenues and market share. For example, products certified with green labels can enter specialized retail channels (e.g., specialized retail stores that select products on the basis of their sustainability performance). Moreover, ecolabels can be used to determine a product's eligibility to enter in regional markets where consumers share high expectations with regard to sustainability and require credibility, or when regulations require firms to produce evidence of special product features. Another important area refers to the relation of ecolabels with green procurement policies. In many countries around the world, governments have associated public procurement with environmental and social protection, promoting the selection of more sustainable products and services through public purchasing policies. Public procurement counts for about 12% of GDP in industrialized countries and up to 30% in developing ones. Efforts to green this procurement can favor the development of novel generations of sustainable products and services, promoting innovation and technological changes. Ecolabels provide a credible, transparent and fair basis for setting the criteria for public procurement bids, therefore facilitating the selection of products and services. Companies that adopt these labels have the possibility to gain access to these markets and to be selected by public administrations.

To conclude, although we acknowledge that the efficiency of ecolabels in activating responsible consumption is complex to monitor and measure, they are important tools that favor the diffusion of green consumption patterns and support companies that innovate their products and services in order to lessen the associated environmental impacts. Choosing the right ecolabel can be helpful in supporting a company's journey towards sustainability, providing additional value to products and services, and giving legitimacy to the downstream innovations carried out to increase sustainability performance.

Summary

In order to implement the "Business In Nature" (BInN) view, a fundamental strategy is to focus on the design of new generations of products and services with reduced impact on the environment and that contribute to satisfying the needs of consumers. A growing number of companies are working on this trajectory, from textiles and apparel to food, from auto-motive and transport to retail, from home appliances to personal care. These companies are seeking new materials, new product concepts, new product architectures, increased resource efficiency and reduced carbon emissions. We have seen the examples of Nike Grind, a palette of premium recycled materials that the company uses in more than 70% of its footwear and apparel products, closing the loop and reducing the generation of waste. We have seen Barilla's product innovation approach, which builds on sustainable supply chain management in order to reduce the environmental, water and carbon footprints of its product portfolio. We have examined the case of Herman Miller, which has reinvented the idea of furniture by designing products based on recycled materials, low-resource utilization, ease of disassem-bly and using only renewable energy for the manufacturing phase.

Nonetheless, all of these innovative solutions will be ineffective if they remain on the shelves of supermarkets, or if they stay unsold and abandoned in companies' warehouses. That is why consumers play a paramount role in driving the change of our social-economic system towards greater sustainability. We started this chapter highlighting that millions of middle-class consumers will add to traditional consumption patterns in the next decades, putting additional stress on social-ecological systems. It is therefore critical to analyze what is unsustainable in our lifestyles and market logics in order to favor the emergence of sus-tainable consumption and green purchasing. We have underlined that sustainable consump-tion is not a new concept, and that for decades scholars and managers in marketing and management studies have investigated the business opportunities related to green consum-erism and the barriers to activating more conscious purchasing. We have concluded in this section that, building on the promises of the LOHAS segment and on the role of the new generations of consumers such as Millennials and Generation Z - who are much more con-cerned with sustainable and healthy lifestyles - there is grounds for hope for a transition to more sustainable consumption. We also introduced the life cycle thinking perspective, linking innovation to the whole life cycle of a product or service and extending the boundaries of design from cradle to cradle. Much of this chapter has been focused on explaining the LCA, a methodology that over the years has probably become one of the most important tools in building corporate sustainability and environmental management. The LCA is a techni-cal approach that allows us to identify and quantify the environmental and social impacts related to a product or service, from raw materials to end use, recovery and recycling. The EPD is another important tool that we have discussed: It helps to convey information about the environmental sustainability of a product or service to interested stakeholders. Both LCAs and EPDs have become more popular among business organizations, and can be used to understand where impacts are generated, to set priorities for designing new products, and to supporting decision making, as in the case of the Italian food company Barilla. We also pre-sented other tools that can guide companies in understanding their impact on ecosystems: Environmental, carbon and water footprints.

Design for the environment (DfE) has become an important framework to orient innovation towards sustainability, connecting different perspectives such as business strategy, marketing, engineering and environmental management. DfE highlights the importance of embracing a holistic and systemic view, and favors the adoption of environmental concerns at the initial stages of the product innovation process, which carries the twofold benefit of reducing costs for nature as well as economic and financial costs. Finally, we have analyzed what an ecolabel is, how they can encourage the spread of more sustainable purchasing habits (contravening some of the obstacles to the diffusion of green consumerism such as information transparency and credibility of green claims), and contributing to product differentiation.

Chapter 6 annexes

Annex 1

A.1.1: Questions for discussion

- What is sustainable consumption? Is green consumerism a new market trend?
- What are the main barriers to green purchasing? Why are Millennials and Generation Z considered interesting opportunities for developing conscious consumption patterns?
- Explain the concept of life cycle thinking. How might this framework impact the product and service innovation process?
- Illustrate the idea behind the LCA methodology. What is the life cycle inventory? What is the life cycle analysis?
- Using the LCA approach, where do you think the major impact of these groups of products is generated: a bottle of shampoo, a cotton T-shirt, a laptop, a 0.5 kilogram box of dry pasta, a washing machine, a refrigerator, a pair of sneakers, a bottle of mineral water?
- How would you proceed to assess and calculate the carbon footprint of your company? What are Scope 1, 2 and 3 emissions? What types of GHG emission are included in the Kyoto Protocol and need to be taken in to account when you assess your carbon footprint?
- What is an environmental product declaration (EPD)? Why and how do EPDs and the LCA methodology relate to each other?
- Illustrate the DfE framework. Exemplify some the most important options for product and service innovation suggested by the DfE methodology.
- What is an ecolabel? What types of ecolabels are you familiar with? Why and how can ecolabels contribute to increasing sustainable consumptions?
- According to the ISO, what are Type I, Type II and Type III labels? What is the difference among them?

A.1.2: Assignments

Assignment 1: Environmental campaigns and behavioral change

The role of NGOs in promoting changes in firms and personal behavior has increased over the last few decades. We addressed these aspects in Chapter 2, investigating why companies address sustainability. We now focus on specific environmental communi-cation campaigns that have been launched to raise awareness of global environmental problems such as deforestation or oil drilling in the Arctic, particularly among multina-tional companies and consumers.

In order to promote a class discussion on this topic, we encourage instructors to show, in class, the videos mentioned below.

You can start by showing a box of tea or any other product with the Rainforest Alliance Certified seal and ask students whether they have ever noticed this ecolabel and what it stands for. You can then show the following YouTube video: https://www.youtube.com/watch?v=3ilkOi3srLo

Greenpeace's campaigns against the destruction of the Southeast Asian rainforest as a consequence of palm oil cultivation had important positive effects on the activ-ities of many companies such as Unilever, Nestlé and Ferrero. Show the two videos below and start a discussion with your students about the ability of NGOs to exert pressure on business organizations:

https://www.youtube.com/watch?v=1BCA8dQfGiO
https://www.youtube.com/watch?v=odl7pQFyjso

Greenpeace's campaign against drilling for oil in the Arctic targeted the promotional partnership between the toy producer LEGO and Shell. Show the video and encourage discussion in the classroom:

https://www.youtube.com/watch?v=qhbliUqO_r4
The following questions can help start and fuel discussions:

- What strategies are NGOs using to engage businesses in tackling environmental problems?
- What are the benefits associated with these environmental campaigns? What are the risks involved?
- What other types of strategies can NGOs implement?
- Should NGOs play only the role of watchdogs, or should they compromise and partner with business to design new products and shape novel strategies?
- How can NGOs support the diffusion of sustainable consumption?

Assignment 2: Life cycle thinking and LCA

In order to better understand the LCA methodology and the life cycle thinking approach, consider the case of Oikos, an organic food company that produces grass-fed cheese and dairy products.

For its organic fresh yogurt line, Oikos utilizes plastic (high impact polystyrene – PS-HI) packaging with an aluminum foil laminate and a paper seal. For its drinkable organic yogurt line, Oikos uses polyethylene high-density (PE-HD) bottles sealed with either aluminum foil laminate or polyethylene low-density (PE-LD) caps. Oikos managers were convinced that this solution was the best to protect the product and guaranteed a good shelf life. Are these materials the best option to pack Oikos yogurts? Why not use glass or Tetra Pak containers? Are there any other solutions? What methodology would you use to analyze the impact of packaging on the environment? What types of variables should you consider? At what stage of the life cycle do you think that the main impact is generated? What types of relationships do you see with regard to the other functions that packaging addresses: Communication, product protection, product identification, image and brand representation, etc.?

Assignment 3: Carbon footprint and GHG emission assessment

This assignment focuses on the application of the carbon footprint methodology and on the calculation of GHG emissions associated with a retail company. Sustainability has become a relevant trend in retail, and retail store formats are adapting to today's consumer expectations. Features such as convenience, authenticity, and personalization are integrated with sustainability. Several retail companies, such as Walmart, Marks & Spencer, Tesco, Carrefour and IKEA have developed their own sustainability strategies and have adopted specific actions to reduce their carbon footprint.

In order to prepare a climate plan, you need to understand your carbon footprint and create a carbon inventory. Consider a typical retail company with hundreds of stores in different regions, warehouses and distribution centers, offices, transport and logistics, a broad range of product categories (food, textile, home appliances, electronics, etc.), store brands and support activities. Identify the main activities such a company would need to conduct, and try to define direct and indirect carbon emissions. Apply the GHG Protocol and identify Scope 1, Scope 2 and Scope 3 emissions (WRI & WBCSD, 2004). You can also look at the additional resources included in Chapter 8.

A.1.3: Further reading and additional resources

- The Life Cycle Initiative: www.lifecycleinitiative.org/
  In order to build a deeper understanding of life cycle thinking, use the web platform of the Life Cycle Initiative. It is a public-private, multi-stakeholder partnership enabling

the global use of credible life cycle knowledge by private and public decision-makers. Hosted by the UN's Environment division, the Life Cycle Initiative is an interface between users and experts on Life Cycle approaches. It provides a global forum to ensure a science-based, consensus-building process to support decisions and policies supporting the shared vision of sustainability as a public good. It delivers expert opinions on sound tools and approaches by engaging its multi-stakeholder partnership (including governments, businesses, and scientific and civil society organizations). Students can also take a course on Life Cycle Thinking using the e-Learning Module Kit. The course is designed to give participants an overview of LCA while developing an understanding of how to assess the impacts of any given sustainability issue that considers all of its life cycle stages. The module is also intended to serve as a guide to which kind of life cycle tools can be used, and for what purpose.

- If you are interested in EPDs, the International EPD® System (a global program for environmental declarations) maintains an online database of more than 800 EPDs, covering a wide range of product categories and companies in 39 countries (as of March 2018). The website contains many different resources and information that allow users to better understand the opportunities linked to the utilization of this tool. (www.envi rondec.com)

- With regard to environmental labels, the Ecolabel Index provides a repository of more than 460 ecolabels. It includes links to the webpages, a short description and additional resources. (www.ecolabelindex.com/)

- Other interesting webpages with regard to Type I labels are:
  http://ec.europa.eu/environment/ecolabel/
  www.epa.gov/greenerproducts/introduction-ecolabels-and-standards-greener-products

Annex 2

A.2.1: References

Allan, J. A. (1998). Virtual water: A strategic resource global solutions to regional deficits. *Groundwater*, 36(4), 545-546.

Anderson Jr, W. T. and Cunningham, W. H. (1972). The socially conscious consumer. *The Journal of Marketing*, 36(7), 23-31.

Armstrong, P. (2010, March 20). Greenpeace, Nestlé in battle over Kit Kat viral. Retrieved from http://edition.cnn.com/2010/WORLD/asiapcf/03/19/indonesia.rainforests.orangutan.nestle/index.html.

Barilla. (2016). Good for you. Good for the planet. Sustainability Report 2016. Retrieved from www.barillagroup.com/en/double-pyramid-our-daily-work

Barilla and EPD International (2017, 6 February). Durum wheat semolina pasta in paperboard box. Environmental Product Declaration. Retrieved from http://gryphon.environdec.com/data/files/6/7968/epd217%20Barilla%20Durum%20wheat%20semolina%20pasta%20in%20paperboard%20box.pdf.

Berkowitz, L. and Lutterman, K. G. (1968). The traditional socially responsible personality. *Public Opinion Quarterly*, 32(2), 169-185.

Business for Social Responsibility BSR and IDEO, (2008, 1 May). Aligned for Sustainable Design: An A-B-C-D Approach to Making Better Products. Retrieved from www.bsr.org/reports/BSR_Sustainable_Design_Report_0508.pdf.

Ceschin, F. and Gaziulusoy, I. (2016). Evolution of design for sustainability: From product design to design for system innovations and transitions. *Design Studies*, 47, 118-163.

Dangelico, R. M. and Vocalelli, D. (2017). "Green Marketing": An analysis of definitions, strategy steps, and tools through a systematic review of the literature. *Journal of Cleaner Production*, 165, 1263-1279.

de Leeuw, B. (2005). The world behind the product. *Journal of Industrial Ecology*, 9(1): 7-9.

Delmas, M. A., Nairn-Birch, N. and Balzarova, M. (2013). Choosing the right eco-label for your product. *MIT Sloan Management Review*, 54(4), 10.

Drew, D. and Yehounme, G. (2017, 5 July). The apparel industry's environmental impact in 6 graphics. Retrieved from www.wri.org/blog/2017/07/apparel-industrys-environmental-impact-6-graphics.

Elkington, J., Hailes, J. and Makower, J. (1988). The green consumer. New York: Penguin Books.

EPD International. (2017, 11 December). General programme instructions for the international EPD® system (Version 3.0). Retrieved from www.environdec.com/Documents/GPI/General%20Programme%20Instructions%20v3.0.pdf.

European Commission DG Environment. (2002, September). Evaluation of Environmental Product Declaration Schemes (Report No. B4-3040/2001/326493/MAR/A2). Retrieved from http://ec.europa.eu/environment/ipp/pdf/epdstudy.pdf.

Fiksel, J. (1996). *Design for environment: Creating eco-efficient products and processes*. New York: McGraw-Hill.

Fitzgerald, D. P., Hermann, J. W., Sandborn, P. A., Schmidt, L. C. and Gogol, T. H. (2007). Design for the Environment (DfE): Strategies, practices, guidelines, methods, and tools. In M. Kutz (Ed.) *Environmental conscious mechanical design*. London: Wiley Publications.

Gabler, C. B., Butler, T. D. and Adams, F. G. (2013). The environmental belief-behaviour gap: Exploring barriers to green consumerism. *Journal of Customer Behaviour*, 12(2-3), 159-176.

Galarraga Gallastegui, I. (2002). The use of eco-labels: a review of the literature. *European Environment*, 12(6), 316-331.

Galli, A., Wiedmann, T., Ercin, E., Knoblauch, D., Ewing, B. and Giljum, S. (2012). Integrating ecological, carbon and water footprint into a "Footprint Family" of indicators: Definition and role in tracking human pressure on the planet. *Ecological Indicators*, 16, 100-112.

Gilg, A., Barr, S. and Ford, N. (2005). Green consumption or sustainable lifestyles? Identifying the sustainable consumer. *Futures*, 37, 481-534.

Global Footprint Network (GNF). (2009). Ecological Footprint Standards 2009. Oakland: Global Footprint Network. Available at www.footprintstandards.org.

Gunther, M. (2015, 9 February). Under pressure: Campaigns that persuaded companies to change the world. *The Guardian*. Retrieved from www.theguardian.com/sustainable-business/2015/feb/09/corporate-ngo-campaign-environment-climate-change.

Herman Miller. (2017). Environmental Product Summary. Setu™ Chair. Earthright: Herman Miller's Sustainability Goals. Retrieved from www.hermanmiller.com/content/dam/hermanmiller/documents/environmental/eps/EPS_SET.pdf.

Hoekstra, A. Y. (Ed.) (2003). Virtual Water Trade. Proceedings of the International Expert meeting on Virtual Water Trade. Value of Water Research Report Series no. 12. UNESCO-IHE. Delft, The Netherlands. Retrieved from www.waterfootprint.otg/Reports/Report12.pdf.

Hoekstra, A. Y. (2009). Human appropriation of natural capital: A comparison of ecological footprint and water footprint analysis. *Ecological Economics*, 68: 1963-1974.

Hoekstra, A. Y. and Hung, P. Q. (2002). Virtual Water Trade: A quantification of virtual water flows between nations in relation to international crop trade. *Value of Water Research Report Series* no. 11 UNESCO-IHE. Delft, The Netherlands.

Hoekstra, A. Y., Chapagain, A. K., Aldaya, M. M. and Mekonnen, M. M. (2009). *The water footprint assessment manual: Setting the global standard*. Water Footpriny Network. Earthscan Publications. London, UK.

Horne, R. E. (2009). Limits to labels: The role of eco-labels in the assessment of product sustainability and routes to sustainable consumption. *International Journal of Consumer Studies*, 33(2), 175-182.

Ibáñez-Forés, V., Pacheco-Blanco, B., Capuz-Rizo, S. F. and Bovea, M. D. (2016). Environmental product declarations: Exploring their evolution and the factors affecting their demand in Europe. *Journal of Cleaner Production*, 116, 157-169.

Jackson, T. (2008). The challenge of sustainable lifestyles. (pp 45-60) In The Worldwatch Institute. *State of the world. Innovation for a sustainable economy*. New York – London: W.W Norton & Company. Retrieved from www.worldwatch.org/files/pdf/State%20of%20the%20World%202008.pdf.

Klöpffer, W. and Grahl, B. (2014). *Life Cycle Assessment (LCA): A guide to best practice*. Weinhein, Germany: Wiley-VCH Verlag GmbH & Co. KGaA.

Lewis, H., Gertsakis, J., Grant, T., Morelli, N. and Sweatman, A. (2017). *Design and environment: A global guide to design green goods*. New York: Routledge.

Makower, J., Elkington, J. and Hailes, J. (1991) *The green consumer supermarket shopping guide*. New York: Penguin Books.

McDonough, W. and Braungart, M. (2003). *Cradle to cradle: Remaking the way we make things*. London: North Print Press.

Nielsen Company. (2016). The Sustainability Imperative. Retrieved from www.nielsen.com/us/en/insights/webinars/2015/the-sustainability-imperative.html.

Nike. (2016). Sustainable innovation is a powerful engine for growth (report no. FY14/15). NIKE, Inc. Sustainable business report. Retrieved from http://nikeresponsibility@nike.com/

OECD. (1997). Eco-labelling: Actual effects of selected programmes. Retrieved from www.oecd.org/officialdocuments/publicdisplaydocumentpdf/?cote=OCDE/GD(97)105&docLanguage=En.

Ottman, J. A., Stafford, E. R. and Hartman, C. L. (2006). Avoiding green marketing myopia: Ways to improve consumer appeal for environmentally preferable products. *Environment: Science and Policy for Sustainable Development*, 48(5), 22-36.

PCF PilotProjekt Deutschland. (2008). Case study: Shampoo by Henkel AG & Co. KGAA.

Peattie, K. and Crane, A. (2005). Green marketing: Legend, myth, farce or prophesy? *Qualitative Market Research: An International Journal*, 8(4), 357-370.

Pogutz, S. and Winn, M. I. (2016). Cultivating ecological knowledge for corporate sustainability: Barilla's innovative approach to sustainable farming. *Business Strategy and the Environment*, 25(6), 435-448.

Remy, N., Speelman, E. and Swartz, S. (2016). Style that's sustainable: A new fast-fashion formula. Retrieved from www.mckinsey.com/business-functions/sustainability-and-resource-productivity/our-insights/style-thats-sustainabl-a-new-fast-fashion-formula.

Rex, E. and Baumann, H. (2008). Implications of an interpretive understanding of LCA practice. *Business Strategy and the Environment*, 17(7), 420-430.

Ruini, L., Ferrari, E., Meriggi, P., Marino, M. and Sessa, F. (2013). Increasing the sustainability of pasta production through a life cycle assessment approach. In V. Prabhu, M. Taisch and D. Kiritsis (Eds). *Advances in production management systems. Sustainable production and service supply chains* (pp. 383-392). Berlin, Germany: Springer.

Simon, M., Poole, S., Sweatman, A., Evans, S., Bhamra, T. and Mcaloone, T. (2000). Environmental priorities in strategic product development. *Business Strategy and the Environment*, 9(6), 367-377.

Srinivas H. (2011). Sustainability concepts: Sustainable consumption. Retrieved from www.gdrc.org/sustdev/concepts/22-s-consume.html.

Straughan, R. D. and Roberts, J. A. (1999). Environmental segmentation alternatives: a look at green consumer behavior in the new millennium. *Journal of Consumer Marketing*, 16(2): 558-575.

UNEP/SETAC Life Cycle Initiative at UNEP. (2015). Guidance on organizational life cycle assessment. UNEP Publications. Retrieved from www.lifecycleinitiative.org/wp-content/uploads/2015/04/o-lca_24.4.15-web.pdf.

UNEP/SETAC Life Cycle Initiative at UNEP. (2009). Guidelines for Social Life Cycle Assessment of Products. Retrieved from www.unep.fr/shared/publications/pdf/dtix1164xpa-guidelines_slca.pdf.

Vaughan, A. (2014, 9 October). LEGO ends Shell partnership following Greenpeace campaign. *The Guardian*. Retrieved from www.theguardian.com.

Wackernagel, M., Onisto, L., Bello, P., Linares, A. C., Falfán, I. S. L., Garcia, J. M., . . . and Guerrero, M. G. S. (1999). National natural capital accounting with the ecological footprint concept. *Ecological Economics*, 29(3), 375-390.

WBCSD. (2008). Sustainable consumption fact and trends from a business perspective. Retrieved from www.wbcsd.org/Programs/People/Sustainable-Lifestyles/Resources/Sustainable-consumption-facts-trends

Wiedmann, T. and Barrett, J. (2010). A review of the ecological footprint indicator – perceptions and methods. *Sustainability*, 2(6), 1645-1693.

WRI and WBCSD. (2004). The greenhouse protocol for project accounting. Retrieved from https://ghgprotocol.org/sites/default/files/standards/ghg_project_accounting.pdf.

WWF and SustainAbility. (2007). One planet business – creating value within planetary limits. Retrieved from http://assets.wwf.org.uk/downloads/one_planet_business_first_report.pdf

Annex 3

A.3: Case study resources

Feitag is a Swiss venture founded in 1993; it has pioneered the idea of product recycling by producing fashionable bags (e.g. bike messenger bags) utilizing truck tarpaulins. Over the years Freitag has developed a strong brand identity, extending its product portfolio to other accessories and new product categories. Sustainability for Freitag is a core value that influences branding and functional strategies (e.g. innovation, marketing and manufacturing). The case allows exploration of the interconnection between sustainability values and design-driven strategies. Moreover it allows investigation of the market segmentation and consumer purchasing pattern in the fashion industry.

- von Wittken, R. (2017). The venture Freitag: From recycled bags to sustainable fashion. Oikos Case Writing Competition 2017, Corporate Sustainability Track. Available at Oikos International Homepage website. Published by TUM School of Management (TU Munich), Germany. Prize winner.

In this chapter we introduced Herman Miller, the office furniture manufacturer who, in the late 1990s, started applying the cradle to cradle (C2C) system approach to product innovation, introducing the Design for Environment (DfE) methodology. The following case illustrates how a company can implement the C2C protocol. Moreover, it helps investigating the process of design of a product, the Mirra chair, as well as the impact of the new protocol on the firm's internal processes: product innovation, manufacturing and supply chain management.

- Lee, D and Bony, L. (2007), *Cradle-to-cradle design at Herman Miller: Moving toward environmental sustainability*. Reference n. 9-607-003. Available at The Case Center. Published by: Harvard Business Publishing, United States.

Another case that can be used to analyze the role of Life Cycle Assessment and Environmental Product Declaration in driving the development of a sustainability strategy and product innovation is Barilla:

- Pogutz, S. (2015). *When supply chain management drives environmental sustainability: The case of Barilla*. Published by: Università Bocconi, Bocconi Graduate School, Italy.

7 Innovating business models for sustainability

Learning objectives

- Illustrate the concept of business model and business model innovation.
- Link business model innovation to the sustainability challenges.
- Explain how business model innovation can contribute to reduce dramatically corporate environmental impacts and benefit society.
- Explore how to design and manage a sustainable business model.
- Examine the notion of circular economy.
- Provide a typology of circular business models.
- Illustrate other business models such as those of industrial symbiosis and industrial ecology.

Chapter in brief

This chapter investigates how sustainable business models allow companies to act as a game changer, modifying the competitive arena and finding innovative ways to meet demand while addressing the environmental and social challenges of our century. Business model transformation represents the most important response to the "Business In Nature" (BInN) view since it affects the business logic through organizational and technological innovations, stretching the boundaries of action to the entire value chain and engaging multiple stakeholders.

First of all, we focus on the concept of business model and provide a brief illustration from the view of management theory. We then analyze the ontology of the business model and the main components and their relationships. Second, the chapter investigates in detail the notion of business models for sustainability (or sustainability-oriented business models). Adopting the perspective of innovation, we explore how breakthrough business models can create positive socio-economic benefits while dramatically reducing the environmental impact and adding value to society. Further, we introduce and discuss the concept of Circular Economy (CE), its origins, characteristics and criticism. Finally, we provide a typology of circular business models illustrating how they contribute to slow down resource consumption and close the loop of material flow.

Introduction

In the past decade, the concept of business model for sustainability (also sustainable business model or sustainability-oriented business model) has gained growing attention in the field of management studies and, more specifically, in the subfield of corporate sustainability (Stubbs and Cocklin, 2008; Boons and Lüdeke-Freund, 2013; Schaltegger et al., 2016b). Several reasons explain this surge of interest both by scholars and practitioners. First of all, there is empirical evidence that a number of companies are breaking the rules of the game in several sectors, proposing innovative ways to approaching consumers while trying to minimize their environmental and social impacts. Tesla is probably the paramount case. Founded in 2003 by a group of engineers in the Silicon Valley, the company is a pioneer in electric mobility and is trying to transform the car industry with the explicit aim to make it more sustainable working at the same time as a tech-company (see Box 7.1). The electricity sector is also facing a revolution: the rapid growth of renewables, digital technologies and decentralized production are paving the way for innovative companies that balance the request for a low carbon economy with new consumer needs. Enel, for example, a leading player with over €70.5 billion revenues (2016) and about 65 million customers around the world, is breaking with the tradition of large fossil-fuel plants and centralized production. The Italian company has approved an investment plan (2016-2019) of 21 billion largely focused on distributed electricity, smart grids and renewables (Enel, 2015). Other examples are Philips Lighting, which transformed the business from luminaires and bulbs manufacturer to energy saving solutions (pay for light not for lamps); or Fairphone, a Dutch-based social enterprise that has successfully developed the market for a collaborative, conflict-mineral free and designed to last longer, repair and dismantle smartphone (Akemu et al., 2016). Even in the financial industry, sustainability is leading to profound changes: microfinance (Yunus et al., 2010) and the rise of impact investing (GIIN, 2017) are challenging the traditional market approaches, balancing good performances and excellent capacity to attract investors. These novel business models are pioneered both by newcomers and incumbents, and they are beginning to threaten established market positions in industry, promising a real revolution just like information technologies and digitalization changed the media, music, entertainment and retail industries.

The second reason for looking into business models is that the sustainability challenge calls for transformative changes of our dominant production and consumptions patterns in order to preserve social-ecological resilience. Cleaner technologies (Chapter 4), sustainable supply chains (Chapter 5), and eco-friendly products and services (Chapter 6) contribute to reduce environmental impacts and increase social benefits, but these solutions have also proven to be not enough to change the course. In many cases, these novel processes and products have fallen short on their promises by remaining anchored to niche consumer segments (e.g., premium price products) and not scaling up the mass markets (Kiron et al., 2013), producing only limited results when it comes to the magnitude of the ecological crisis.

The concept of the business model and, more specifically, that of business model innovation provides a particularly suitable perspective to analyze and discuss the need for breakthrough transformations. A business model innovation can be defined as a "framework or recipe for creating and capturing value by doing things differently" (Afuah, 2014: 4); it embodies a profound change in the rules of the game. The power of business model innovation lies in the fact that it catalyzes multiple changes at organizational, technological, operational and financial levels and involves a system perspective, including collaborations with several stakeholders.

The third reason that has attracted scholars and practitioners to further investigate business model and sustainability refers to another important anecdotal evidence that has emerged from research on corporate sustainability (Kiron et al., 2017). In a nutshell, success-ful sustainable strategies, and the existence of a clear business case, seem to be strongly connected to business model innovation. In a landmark study by MIT Sloan Management and The Boston Consulting Group investigating thousands of managers and interviewing more than 150 executives and leaders, it was found that the companies that added more profit from sustainability-related activities were those that combined innovation in the value chain with changes in the value proposition and target segments (Kiron et al., 2017: 11). Relevant modifications in existing business models or entirely new ones are key conditions to fully capitalize from the opportunities offered by the Anthropocene challenges.

To conclude, the business model perspective provides a powerful tool to frame and ana-lyze corporate responses to sustainability. It links the firm-level to the system-level and helps to overcome the limits of approaches that focus only on green technologies, without explor-ing the organizational and market dimensions of sustainability. Moreover, it can stimulate visionary leaders and entrepreneurs in developing new approaches that can combine value creation processes while contributing to social-ecological resilience.

In the next paragraphs, we are going to investigate in depth the concept of business model for sustainability: What is a business model for sustainability or a sustainable business model? How does business model innovation link with sustainability? Why can sustainability-oriented opportunities be better captured through business model innovation? Before doing so, we illustrate what is a business model and what are its constituting elements.

Box 7.1   Tesla innovative business model: the challenge of electric mobility

2 August 2017. Tesla, the popular electric carmaker, has just released its financial results and shareholders letter for the Q2. The company has delivered about $2.8 billion in revenues with a loss of US$2.04 per share. Tesla has produced 25,708 vehi-cles in Q2, 40% more cars than in the same period in 2016. In the first six months of 2017, Tesla sold about 47,000 of its Model S and Model X vehicles worldwide (Lambert, 2017). Just to provide a comparison, in 2016 Toyota announced global sales of 10.1 mil-lion cars, Volkswagen 10.3 million, General Motors (GM) 9.9 million and Ford over 6.6 million. Despite the limited sales and the negative operating income, Tesla's market value in April 2017 has surpassed both GM and Ford, becoming, for the first time since its foundation in July 2003, America's most valuable automaker (La Monica, 2017).

How come? What drives investors toward Tesla stocks despite unsatisfying financial results? Tesla's usurping of GM and Ford has spurred a lively debate over the "real" value of Elon Musk's company in the financial community. According to several business analysts, while GM and Ford have solid financials and a vast portfolio of vehicles, Tesla has produced nothing but promises. If we take a cash flow perspective to assess the corporate value, Tesla has never been able to generate profit. On the other hand, investors look like they have bought Tesla's vision that electric vehicles will rule the world. Tesla's mission, in fact, states: "Tesla's goal is to accelerate the world's transition to electric mobility with a full range of increasingly affordable electric cars. We're catalyzing change in the industry." The capacity to attract shareholders and consumers is distinctive, and the company is the only novel carmaker that has acquired legitimacy in the industry since many years.

In order to try to provide an answer to our previous questions, we need to investigate Tesla's unique and innovative business model, very different from that of other automotive manufacturers. First, with regard to the company's value proposition, Tesla has approached the electric car market targeting high-end products that combine appealing design, amazing acceleration and fun to drive. This value proposition is captured by the words of Tesla CEO, Elon Musk: "If we could have [mass marketed] our first product, we would have, but that was simply impossible to achieve for a startup company that had never built a car and that had one technology iteration and no economies of scale. Our first product was going to be expensive no matter what it looked like, so we decided to build a sports car, as that seemed like it had the best chance of being competitive with its gasoline alternatives" (Zucchi, 2018). Starting with the Roadster model (launched in 2008), Tesla has promoted its aspirational brand selling luxury sport cars like Model S and the Model X to the early adopters. Tesla has then started to enter the mass-market with the introduction Model 3 and the pre-booking strategy based on the US$1,000 down payment that the company has asked from its clients. Another distinctive feature of Tesla's value proposition relates to the software and hardware applications embedded in the vehicle (Wharton, 2016). The car is fully connected in the IT ecosystem with regard to charging stations, service providers and maintenance. Tesla's self-driving technology (autopilot, sensors, computing power, vision processing tools, etc.) has further contributed to strengthen consumer perception about the uniqueness of this brand.

Second, looking at the customer interface and distribution channels, unlike the other carmakers that sell thorough dealer franchises, the business model of Tesla is based on direct sales. The company has created a network of owned showrooms and service centers placed in prestigious city locations that allow providing a unique experience to its customers. This approach has been compared to the Apple store strategy.

Third, when it comes to the infrastructures and activities managed by the company, Tesla has directly invested in the grid in order to develop its supercharger system around the world. Superchargers offer reliable and fast charging services to satisfy the customers' needs. They are located both in popular highly travelled routes and in urban areas. As of October 2017, a milestone of 1,000 stations worldwide has been reached, with over 6,930 supercharger stalls. The grid represents a proprietary investment by Tesla that has developed its own standard for the charging technology. The expansion of the supercharging stations is a fundamental piece of Tesla development strategy to speed up the adoption rate of electric cars. The financial model of supercharging stations has also been innovative. For Tesla clients who have ordered Model S and X before 15 January 2017, supercharging was free. After this date, new clients have a free-of-charge amount of power (400 kWh per year, equivalent to about 1,600 km), while they have to pay a fee for additional electricity consumption, but at a lower price than other fuels. Finally, the company has started to build efficient solar-based charging systems fully off-grid and has announced an ambitious plan to disconnect several supercharger stations. According to Elon Musk, all stations will be equipped with solar panels and Tesla batteries and will be connected to the commercial electricity grid only to back up the shortage of power in regions where sun availability cannot guarantee full coverage. The last component of Tesla vision of the future is the battery.

When the Powerwall new adventure was announced, a gigantic rechargeable lithium-ion battery that can be installed in any home or business (e.g., warehouses or even factories), many questioned if this was going to transform the Tesla business model. The Powerwall functions as a backup for energy storage from solar panels or from the grid (e.g., when prices are low) and represents another key component in the strategy for a low-carbon future (Hsu, 2015). Like in the case of the Model S and X, this technology is addressing a luxury market segment, but the price is likely to come down while scaling up the market, as had happened with the introduction of Model 3. Tesla has partnered with Panasonic, the Japanese conglomerate leader in electronics, to promote technological innovation in batteries and to match the growing demand. Panasonic was already the exclusive supplier for Tesla cars, and they consolidated the relationship with the construction of the so-called Gigafactory, a massive battery factory built in Nevada for making a new generation of efficient and reliable batteries. The Gigafactory has started production in January 2017. The aim is to make enough batteries for 500,000 cars a year, a huge target that responds to Tesla's goal to transform the automotive and electricity industries.

To conclude, Tesla is challenging the automotive industry; its innovative business model is triggering a competitive response from the incumbent players. Elon Musk has been able to demonstrate that the electric car is feasible and reliable, and Tesla is contributing to building the electric mobility market. Other drivers such as digitalization and connectivity, sharing mobility and changes in consumer preferences are contributing to this transformation, but Tesla is now a major force in changing the landscape of the sector, starting the phase-out of the old and inefficient internal combustion engine. Sustainability and social-ecological resilience require fewer cars on the street, but the cars we need must be fueled with electricity from renewables or other clean fuels, not with fossil fuels that generate carbon emissions and pollutants (Cohen, 2017). The transition toward electric mobility will take decades, but Tesla is a step in the right direction.

Source: authors' elaboration using multiple sources.

Business model

The term business model has become widely used by managers, consultants and scholars. Almost every business school offers courses focusing on business models, and the term is now common in strategy, innovation and organization studies. Of course, there are multiple types of business models. For example, inside the textile industry companies like H&M and Zara are well known to have totally different business models: the first does not directly own any factory and instead collaborates with hundreds of suppliers worldwide, while the second is a fully vertically integrated operator that controls every step of the supply chain. Recurring stories of novel business model in strategy and innovation textbooks are that of Amazon, e-Bay, Ryanair, Twitter, Instagram, Kickstarter, Netflix, Airbnb or Zipcar; all these newcomers are trying to transform their industries offering their customers brand new value propositions. To better understand the importance of this concept and the relation with innovation and sustainability, we first address the question: What is a business model?

We start with an analogy (see Casadesus-Masanell and Ricart, 2010: 197–198). There are many different cars on the streets that respond to very different logics: they are made of different components and parts (engine, wheels, seats, electronic systems), assembled in different ways, target different market segments. Traditional cars are made differently from hybrid ones and electric vehicles: they function with different fuels, have different transmission systems and generate different environmental impacts. A small city car responds to diverse needs when compared with a large and powerful SUV, or with a fast sport car. Different cars create different value for their clients. In our analogy, cars stand for business models. Just like in the streets where there are multiple car models, in the markets, there are multiple business models that respond to a set of specific requirements, generating value for consumers and shareholders in different ways. Like a car is made of parts, a business model consists of a system of elements and components that are linked together; like a car results from design processes made by engineers and marketing experts, business models are consequence of choices and decisions made by managers and entrepreneurs.

Business model ontology

There are many definitions of business models. Teece (2010: 173) emphasizes the value creation function: "A business model defines how the enterprise creates and delivers value to customers, and then converts payments received to profits." Casadessus-Masanell and Ricart (2010: 196) define this term as: "the logic of the firm, the way it operates and how it creates value for its stakeholders." Comparing to Teece (2010) previous definition, they extend the value creation process from customers to stakeholders in general. Zott and Amit (2010: 216) describe business models as "a system of interdependent activities that transcends the focal firm and spans its boundaries;" they emphasize the systemic notion and the interdependencies with other entities. Osterwalder et al. (2005: 10) provide a more comprehensive and detailed definition that includes all these elements and functions:

> A business model is a conceptual tool that contains a set of elements and their relationships and allows expressing the business logic of a specific firm. It is a description of the value a company offers to one or several segments of customers and of the architecture of the firm and its network of partners for creating, marketing, and delivering this value and relationship capital, to generate profitable and sustainable revenue streams.

This concept helps to understand how companies create and capture value through a series of activities, from raw materials to the final customers, and emphasizes the importance of specific and unique competencies to allow firms enjoying a competitive advantage (Chesbrough, 2007).

Building on the work of several scholars and authors (among others, Osterwalder et al. 2005: 10; Boons and Lüdeke-Freund, 2013: 10), although business models consist of a number of elements highly interconnected, we can identify the following major components that contribute to value generation.

- **Value proposition:** We refer to the bundle of products/services offered by the firm and to the value embedded for the customers.
- **Customer interface:** Here, the focus is on "how" the relations with the consumers are organized and managed (What are the market segments? What are the distribution channels that connect the company with the customers? What kind of relationship the company establishes with the customers?).
- **Infrastructure management:** In this case, we refer to the range of activities organized and managed by the company, including the supply chain and the set of partnerships, and to the capabilities necessary to perform successfully the business model.
- **Financial aspects:** The structure of cost and revenue flows associated with the business model.

In sum, a business model and the analysis of its parts offer a powerful tool to help managers and scholars with understanding how firms do business, providing a holistic representation of its logic. This perspective can be used for many different purposes: from analysis and comparison of different strategies, to management of firm's activities and performance assessment (Osterwalder and Pigneur, 2010).

Business model innovation

A specific approach to business models links with the field of innovation and looks at the processes of change (Chesbrough 2007, 2010; Amit and Zott, 2012). Business models are

not static and need revisions on a regular basis as a result of market forces and new customer needs, technological evolution and breakthrough, and societal transformations. Business model innovation is about "doing things differently;" it is about changing the rules of
the game (Afuah, 2014). In the words of Amit and Zott (2012), business model innovation is
changing how to do business, rather than what to do. Therefore, it goes beyond technology,
process or products innovation, combining changes in the major components of the business
models: value proposition, the market choice, the decision about the activities and how they
are organized, and how a firm captures value. In the Tesla case history (see Box 7.1), these
components are combined into an innovative business model that challenges the way in
which the dominant carmakers are organized and approach the market. This perspective
helps in understanding the attention that Tesla is generating among customers and investors
and can contribute to explaining the reasons for its current success.

Business model innovation has been given great attention by managers, entrepreneurs
and scholars for several reasons (Amit and Zott, 2012). First, because it can unravel new
opportunities for companies, unlocking the potential value of customers and technologies.
It is, therefore, an important and novel source of value. Second, it is more difficult to imitate.
Confronted with product or process innovation, it is much more complex to replicate a whole
new system of embedded activities than eroding the competitive advantage based on the
traditional forms of innovation. Finally, it can bring massive threats to market leaders from
outside the traditional industry boundaries, challenging the prevailing business model of the
incumbent companies. Once more, if we look at the case of mobility, automotive companies
so far have not been able to unlock the market potential of electric vehicles. Differently, Tesla
seems capable of threatening the big players and the oil companies pushing the electric
revolution. At the same time, the Tesla business model looks very complicated to imitate
because it combines multiple components: the electric car technologies, the supercharging
stations, the distribution network, the Powerwall battery, etc. We now turn the attention to
the relationship between business model and sustainability.

Business model for sustainability

Definitions, normative requirements and conditions

The business model perspective is particularly interesting when it comes to sustainability
because it offers the possibility to conceptualize in novel way business organizations, their
value creation logic, and their relation with the markets while considering the environmental
and social consequences of their activities. It is, therefore, not surprising that in the last
decade, several scholars and practitioners have analyzed corporate sustainability under the
lenses of business model innovation (Bocken et al., 2014; Boons and Lüdeke-Freund, 2013;
Schaltegger et al., 2012; Stubbs and Cocklin, 2008); but we can identify a specific focus on
business model transformation also in earlier work on corporate sustainability.

One of the most influential studies in the field was carried out by Paul Hawken, Amory
Lovins and Hunter Lovins who in 1999 introduced to the large business audience the concept
of Natural Capitalism, first in a book (Hawken et al., 1999) and later in a landmark article
published in Harvard Business Review (Lovins et al., 2007). This work already provided the

framework conditions to understand business models for sustainability and identified four practices for a shift toward what we have named in the previous chapters "true sustainability" (these practices are illustrated in Box 7.2). Natural capitalism acknowledges the fact that the global economy depends on natural resources and ecosystem services availability and offers a critique to the dominant "industrial paradigm," which neglects to recognize the value of natural and social capital (see Chapter 1). The authors' call for a profound transformation of the economy that they label "the next industrial revolution". This change is based on the principle that business and environmental interest overlap because long-term sustainability requires acknowledging the interdependence between production and consumption flows and natural capital flows.

Box 7.2 Natural capitalism business practices

According to Hawken, Lovins and Lovins (1999: 2-3), the journey toward natural capitalism calls for a major shift in four business practices, as follows.

- **Dramatically increase the productivity of natural resources.** The first step looks at the possibility of adopting product and process technologies that can radically increase the efficiency in the use of natural capital and ecosystem services.
- **Shift to biologically inspired production models.** This idea is inspired by the fact that nature produces no waste. Imitating natural cycles and shaping business models as closed-loop systems (see the section on the CE) where outputs become inputs for manufacturing other products will significantly reduce the harm to ecological systems.
- **Move to a solutions-based business model.** The authors suggest a transformation from traditional businesses based on selling goods to new models where value is generated through services. This approach can favor an alignment of the interest of providers and customers in a way that can foster the diffusion of innovative solution to increase efficiency and reduce waste.
- **Reinvest in natural capital.** The last practice suggested by Hawkens et al. advocates the idea of generating positive impacts on social-ecological systems investing in restoring, sustaining, or enhancing the natural capital and the ecosystem services.

Although the focus of this paper is not on the business model (the concept of business model was not yet legitimized in management literature), the authors are referring to the term in several cases, underlying that natural capitalism requires a substantial transformation in the way companies approach the market, organize the production and generate value. The paper illustrates several cases of companies that are transforming their practices in a way that we can nowadays define as the innovative business model for sustainability.

Source: authors' elaboration using multiple sources.

About ten years later, the concept of business model for sustainability (the authors pro-posed the label "sustainable business models") was conceptualized and discussed in another seminal paper by Stubbs and Cocklin (2008). These authors grounded their work on field-based case studies of two ideal-type innovative companies: Interface Inc., a world leading manufacturer in modular carpet that has developed a strategy based on the closed-loop system for recycling, zero waste and solution-based business model; Bendigo Bank, an Aus-tralian large bank that is implementing a business model focused on developing sustainable communities. The study allowed the authors to elaborate a set of normative principles that characterize the sustainable business model:

- a new definition of the business purpose,
- a stakeholder approach,
- a performance measurement system based on economic, social and environmental dimensions (Triple Bottom Line),
- a leadership style that acknowledges and drives structural and cultural changes in the organization to reach sustainability,
- an approach that promotes environmental stewardship and acknowledges the impor-tance of local communities.

Moreover, Stubbs and Cocklin (2008) introduce a system-based view to the analysis of sus-tainable business models. The authors illustrate this perspective as follows:

> Organizations can make significant progress towards achieving sustainability through their own internal capabilities, but ultimately organizations can only be sustainable when the whole system of which they are part is sustainable (. . .). An organization adopting an SBM [Sustainable Business Models] develops internal structural and cultural capabilities to achieve firm-level sustainability and collaborates with key stakeholders to achieve sustainability for the system that the organization is part of.
>
> (Stubbs and Cocklin, 2008: 122-123)

In other words, sustainable business models require the existence of system-level con-ditions such as sustainable infrastructures (e.g., waste treatment facilities, efficient and low-carbon transports and renewable energy) and a sustainable orientation in regulations (policies and taxes) and society (e.g., consumption models). Finally, planning a sustainable business model requires collaboration with multiple stakeholders including suppliers and customers, NGOs, competitors, communities, financial markets, governments and policy makers. The holistic view introduced by Stubbs and Cocklin is coherent with the former work by Hawking et al. (1999) and Lovins et al. (2007) and with other approaches such as "circular economy" and "industrial ecology" that have been developed under different scientific and theoretical perspectives (see the next sections of this chapter).

In more recent years, several contributions on business models and sustainability have been published in a variety of journals (for a recent literature, see the work by the Lüdeke-Freund et al. (2016)). Among these studies, we have selected three definitions (see Table 7.1).

Table 7.1 Some selected definitions of business model for sustainability.

Schaltegger et al. (2012: 112)	"[A]business model for sustainability can be defined as supporting voluntary, or mainly voluntary, activities which solve or moderate social and/or environmental problems. By doing so, it creates positive business effects, which can be measured or at least argued for. A business model for sustainability is actively managed in order to create customer and social value by integrating social, environmental, and business activities."
Schaltegger, et al. (2016b: 6)	"A business model for sustainability helps describing, analyzing, managing, and communicating (i) a company's sustainable value proposition to its customers, and all other stakeholders, (ii) how it creates and delivers this value, (iii) and how it captures economic value while maintaining or regenerating natural, social, and economic capital beyond its organizational boundaries."
Bocken et al. (2014: 44)	"Business model innovations for sustainability are defined as: Innovations that create significant positive and/or significantly reduced negative impacts for the environment and/or society, through changes in the way the organization and its value-network create, deliver value and capture value (i.e., create economic value) or change their value propositions."

Source: authors' elaboration using multiple sources.

Common features of business models for sustainability are:

- a process of value creation that goes beyond the economic dimension (e.g., shareholders and customers) and incorporates the environmental and social ones (all stakeholders),
- the idea, which we have broadly discussed in Chapters 2 and 3, that companies can create positive impacts on the environment and society.

Using the business model framework as baseline, we introduce a list of normative requirements and conditions (Boons and Lüdeke-Freund, 2013; Upward and Jones, 2016) that guide the design and implementation of a business model for sustainability.

- **Value proposition.** A sustainable value proposition must deliver both customer value and positive impact on the natural environment and society. For new companies, it is about designing and implementing a novel value proposition that balances the different stakeholder needs. For incumbents, a true sustainable value proposition requires a profound ex-post transformation of the status quo and of the strategy. Patagonia, the $800 million outdoor apparel company founded by Yvon Chouinard, one of the most popular leaders in sustainable business, has a clear mission focused on environmental protection and inspiring social change. The value proposition transcends the simple transaction and is based on the construction of a long-lasting relation with its customers and with the communities. At Patagonia sustainability is the core element of the value generated for its consumers, and positions the company as unique and differentiated from all other competitors on the basis. High quality, strong integrity, minimized impact on nature and the focus on innovation are all components of the Patagonia's brand value.
- **Customer interface.** Companies should extend their responsibilities downstream to consumers and other stakeholders (e.g., retail, recycling and waste management, etc.)

engaging them in responsible and sustainable consumption patterns. In other words, the companies should not pass the "burden" down to its clients. This practice overlaps with what we have illustrated in Chapter 6 about product sustainable innovation and design for the environment but must be integrated with other business model requirements.

- **Infrastructure system and management.** In this case, the company must acknowledge and integrate the principles and practices of sustainable supply chain and develop the resources and capabilities required to minimize the negative and increase the positive environmental and social impacts. This approach, extensively discussed in Chapter 5, incorporates the development of partnerships with other competitors and suppliers along the multiple phases of the supply chain, up to the phase of extraction and/or cultivation of raw materials.
- **Financial model.** The last component is the costs and revenues model and the ownership one. On the one hand, companies should try to develop a pricing model that allows them to address multiple market segments while responsibly redistributing value across the supply chain (e.g., in the case of agricultural commodities, a fair remuneration of farmers and cooperatives of small-holders). Moreover, they should find the right financial partners and the correct balance between equity and debt compatible with the structure of costs and benefits. For example, companies should orient themselves toward more "patient" investors, socially responsible investors, or impact investing funds that are more inclined to long-term benefits. Finally, the accounting system to control and manage the company should also include the measurement of the environmental and social performance.

Business models, to be successful and to generate value, must be consistent with the strategy elaborated by the firm (Figure 7.1). Therefore, when a company orients toward true sustainability (business sustainability 3.0 following Dyllick and Muff, 2015) and develops its sustainable strategy according to the BInN concept, its business model need to follow adapting and changing (Schaltegger et al., 2012).

Differently, in the case of start-up companies, new business models can be designed from scratch in order to incorporate the sustainability requirements illustrated earlier. Figure 7.1 helps in illustrating the relationship between the sustainability strategy and business model for sustainability.

Business model innovations: examples at industry level

Innovative business models involve transformations that go beyond the single firm boundaries, engaging multiple stakeholders in a broader value-network perspective. Like in the case of Amazon, Google, or Ryanair where the new business model has favored sudden and intense transformation in the industry, and also in the case of business models oriented to sustainability, the innovative combinations of resources and technologies can become a driving force for disrupting an industry structure, its dynamic and the competition. Some industries where novel business models are changing the landscape and opening new market opportunities are briefly illustrated and commented on below.

Sustainability strategy

Defines goals and drivers for
ecological, social and economic performance

GUIDES

Sustainable business models

Adapt, improve or redesign the value-creating activities and
components and the relations among them

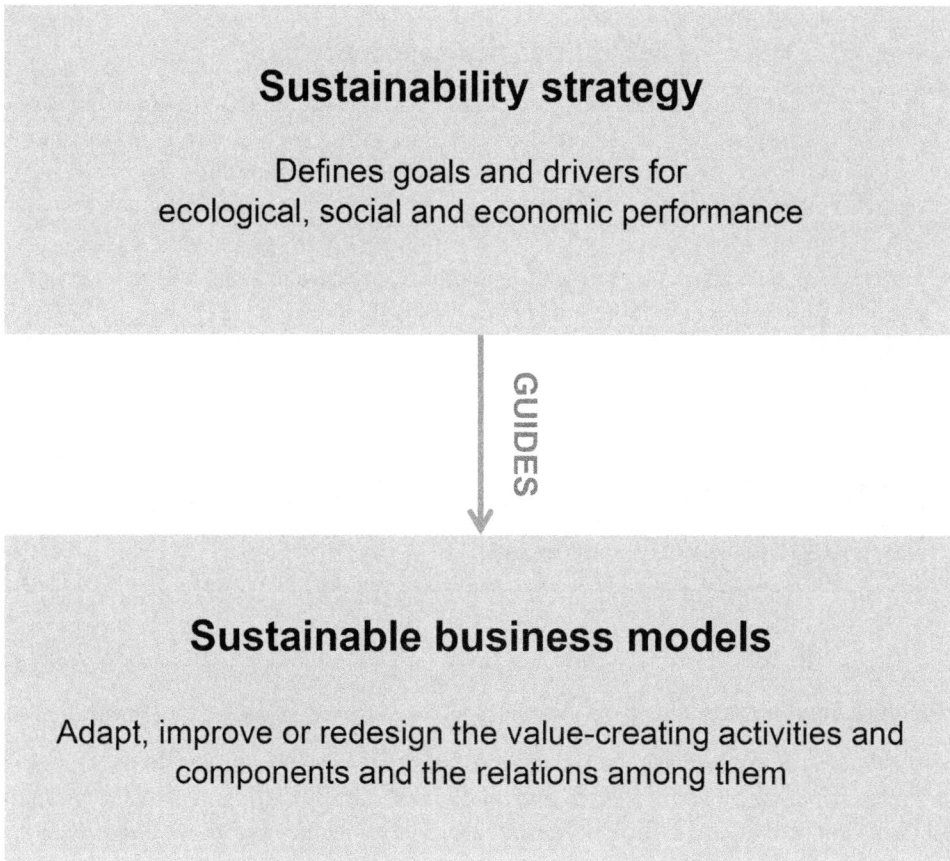

Figure 7.1 Sustainability strategy and business model for sustainability.
Source: authors' elaboration.

Energy and electricity

The process of digitalization (e.g., smart metering, remote control and smart sensors) combined with new technologies such as the diffusion of renewables (e.g., solar panels and electricity self-production), storage devices, microgrids and energy efficiency is changing the electricity production system from centralized to decentralized (World Economic Forum WEF with Bain & Company, 2017). These transformations are occurring at a pace unprecedented in the industry that for decades has been characterized by long-term investments and stable market positions. Utilities have faced severe losses in their market shares as a consequence of novel business models introduced by newcomers that have entered the industry such as private customers, independent project developers, investment funds and banks, new small and medium-sized companies (see Box 7.3). In this new competitive environment, utilities need to find new revenue sources, shifting their business model toward innovative services (e.g., services to balance demand and supply and to manage the load at wholesale level), providing new customer experiences, and integrating and exploiting distributed energy sources.

Box 7.3 The crisis of utilities business models

In the last five years, several incumbent utilities around Europe such as Centrica PLC's, British Gas, RWE, E.ON, EDF Energy and Scottish Power have suffered continuous market share losses as a consequence of the advent of renewables and new electricity distributors (Drozdiak, 2015; *The Economist*, 2013). These companies have failed in adapting to the market transformation spurred by low-carbon policies and incentives mechanisms, technological evolution and liberalization. In specific countries like Germany, where there are about 1.4 million PV users (*The Economist*, 2017), conventional power generators such as RWE, E.ON and Vattenfall have suffered several years of huge net losses (only in 2016 RWE reported a €5.7 billion loss and E.ON a €16 billion loss). Overall, as reported by numerous analysts, the value of the market capitalization of several electricity providers has been falling dramatically on a yearly basis since 2008.

Source: authors' elaboration using multiple sources.

Mobility and automotive

As previously discussed when illustrating the Tesla story, like in case of other sectors, the automotive industry might also face a revolutionary transformation as a result of digitalization and connectivity, electrification, low-carbon regulatory pressures and new forms of mobility (e.g., shared mobility). Focusing on sustainability, electrification is probably the major driver for a change since the decline in battery costs (according to Bloomberg data, from US$1,000 per kWh in 2010 to US$300 per kWh in 2016), the diffusion of charging stations, and an increased consumer acceptance of electric vehicles are creating new and strong momentum for penetration of novel technologies: hybrid, plug-in, battery electric and fuel cell (McKinsey & Company, 2016: 12). Forecasts for electric car adoptions are still extremely uncertain, but Bloomberg thinks they can count up to 25% of new vehicle world sales by 2030 and 35% by 2040 (World Economic Forum, 2017: 9). As a result, the industry might be affected by the uprising of novel business models, like in the case of Tesla. Mercedes, Audi and BMW are targeting Elon Musk's company with their own lines of electric vehicles. Jason Hoff, CEO of Mercedes-Benz USA, has recently declared US$1 billion investment in electric vehicle production since he agrees upon the prediction that soon 25% of new cars will be electric (Gurdus, 2017). At the same time, electric mobility is capturing the interest of utilities and electricity distributors that are looking at this technological convergence as a new market opportunity to deploy enabling infrastructure (e.g., charging stations, grid-edge technologies that allow interoperability and plug and play, and batteries).

Box 7.4   Examples of new business models in the agro-food industry

Cortilia

Cortilia is an Italian company launched in 2011 (www.cortilia.it). It was the first agri-cultural online market that linked consumers and farmers with the purpose of buying fresh and sustainable food from the countryside. The value proposition aims at responding to the request of consumers interested in tasty and quality food, authentic and traditional flavors, ethical purchases, and short and traceable supply chains. Cor-tilia offers a large variety of regional freshly picked and processed products combined with timely service (within a day after the order is placed via the web). Nurturing local farming and community relations is the other part of this distinctive business model that aims at supporting producers, processors, farmers, breeders, and their specific know-how and the preservation of cultural specificities.

Brooklyn Grange

One novel example of urban farming is Brooklyn Grange (www.brooklyngrangefarm. com), the leading rooftop farming and intensive green roofing business in the US. The company was established in 2010 with a mission "to create a fiscally sustainable model for urban agriculture and to produce healthy, delicious vegetables for our local com-munity while doing the ecosystem a few favors as well." Brooklyn Grange currently has more than 2 acres of rooftops under cultivation in New York and sells more than 50,000 lbs of food each year through farmers' markets, restaurants, subscriptions and wholesale accounts. The sustainability business model is that of zero mile and local food supply, but the company also captures and absorbs huge amounts of stormwater every year, which benefits the city water management system.

Source: authors' elaboration using multiple sources.

Agro-food industry

A traditional sector like agriculture is also undertaking unprecedented transformations driven by consumer pressures for sustainable and healthier products, the request for traceability and transparency, and new technologies that are challenging extensive mono-culture farming linked to global markets. Concepts such as organic farming, local farming (zero kilometers or zero miles), and slow farming are spreading around the world targeting new consumers' segments. The availability of open-source data and cutting-edge devices like sophisticated sensors that measure the soil characteristics (e.g., temperature, pressure, humidity, acidity), drones and integrated software solutions are helping the way farmers use chemicals and fertilizers and optimize resource efficiency, leading to the diffusion of what has been defined as "precision farming." Based on different technological solutions, another interesting pattern is urban farming, which is spreading in cities around the world. Ranging from vertical farming, where greens and herbs are cultivated in tanks or in roof gardens, to aquaponics operations that provide salad greens and fish through an integrated ecosystem with high resource efficiency, innovative start-ups are developing in this sector thanks to novel business models that are making agriculture an integral part of urban life. In Box 7.4 we illustrate some innovative cases.

Markets for ecosystem services

The conservation of ecosystem services, biodiversity and natural capital is of paramount importance for preserving our well-being. In the first chapter of this book, we illustrated the array of goods and services provided by ecosystems: nature-based goods like food, fibers, water and minerals along with other services that are often underappreciated and are of greater importance to our society like climate regulation, water purification, pollination, protection against natural and man-made hazards, and more. In order to preserve ecosystem services, natural capital market-based mechanisms have been introduced. Examples of these instruments are trading schemes in which the environmental damage is compensated by investing in nature in another place (e.g., the carbon markets), or the ecolabels, like in the case of certified and organic markets (e.g., timbers, fibers, or food) analyzed in the previous chapters, in which the value of the ecosystems is incorporated in the product price (MA, 2005; OECD, 2010). Another concept that has emerged in this area is that of payment for ecosystem services, also referred to as PES. PES describes those schemes in which the beneficiaries, or the users, of the ecosystem services pay to ensure its sustainability and timely provision. This includes paying for the management of the natural capital (e.g., ecosystem and land use management) in order to conserve or enhance the flow of services generated (e.g., the service of water purification and regulation provided by a forest and its conservation) above the amount that would have been provided in the absence of the payment (DEFRA, 2013; Sardá, 2015).

The last twenty years have witnessed a proliferation of market-based approaches for ecosystem services around the world, both regulated and voluntary, and at the international, national and local levels. We define "nature-based businesses" those entrepreneurial initiatives that have embraced novel business models aiming at taking advantage from the opportunities related to the development of these market-based mechanisms, marketing and promoting the protection, maintenance and the enhancement of ecosystem services, biodiversity and natural capital. Innovative companies have developed in this area targeting forest and carbon certifications, developing technologies and methodologies for accounting and monitoring natural capital, or specializing in providing the knowledge and the solutions as intermediaries between buyers (e.g., private organizations, individuals, NGOs, or governments) and sellers (e.g., farmers or landowners) of ecosystem services. Some examples are illustrated in Box 7.5.

Box 7.5 Examples of innovative business models in the market for ecosystem services

Carbon offset

To comply with climate regulations and cap-and-trade mechanisms implemented by several countries around the world (e.g., the European Union Emissions Trading Scheme or the Clean Development Mechanism under the Kyoto Protocol), markets for mitigation of business and individual carbon emissions through sequestration have developed. These markets include reforestation, afforestation and avoiding deforestation and maintaining ecosystem functionality. In other words, forests' carbon reserves can be monetized and sold as offsets to compensate for carbon emissions. My Climate, for example, is a Swiss climate protection foundation that emerged in Zurich in 2002. Today, the organization operates at global and local levels and connects the needs of buyers and sellers of carbon credits through projects that are providing quantifiable impacts over time. My Climate carbon offset initiatives (e.g., forest protection against logging and reforestation) must meet the highest international standards with regard to environmental and social protection such as CDM (Clean Development Mechanism) or the Gold Standard (www.golds tandard.org).

Habitat banking

Another innovative approach to ecosystem service, biodiversity and natural capital conservation is known as "habitat banking" (HB). In this case, a parcel of land (e.g., a wetland, a marine area or a habitat that supports the life of endangered species) is protected, restored and/or managed for its environmental and social values by a system of organizations and private companies. In exchange for permanently protecting the land, the bank owner is allowed to sell credits. These credits are then purchased by parties such as companies that impact the area through specific business activities as a compensation mechanism. For example, mining or oil companies that use natural resources for drilling and extracting might be required to mitigate their impact in order to maintain the "license to operate" among stakeholders. The money paid for the credits is then invested by the "bank" that ensures the respect of principles such as the maintenance – or the increase – in the ecosystem functionality. According to the type of HB, the compensation mechanism can be applied to the same location where the damage was generated or to other areas with the same type of environmental characteristics (e.g., species or hydrology). (Bishop et al., 2009; Bovarnick et al., 2010).

Source: authors' elaboration using multiple sources.

Energy and electricity, food, mobility, ecosystems' conservation provides examples of industries where innovative business models can reduce the environmental impact and contribute to address the planetary boundaries challenges. The diffusion of renewables, electric vehicles, or organic agriculture can generate positive benefits in terms of low-carbon emissions, decline of air pollutions and reduction of chemical compounds (e.g., fertilizers and pesticides), and therefore contributing to combating climate change, ocean acidification and health disease. Other innovative business models for sustainability have been developed with the purpose of responding to social challenges like reducing poverty, hunger, inequalities, or creating decent work. This is the case of social enterprises, companies that exist with the purpose to satisfy a social mission and where the core of the business model, its value proposition, is to generate positive externalities for society through a strong integration with local communities and stakeholders (Bocken et al., 2014). Micro-finance, impact investing, or crowd-sourcing are based on novel business models that are challenging traditional financial and investment approaches. Other new business models based on sharing schemes and collaborative consumption that generate value leveraging under-utilized assets are rapidly spreading.

In the next paragraphs, we will further investigate some of these solutions and the benefits they could bring, starting from the Circular Economy (CE), a pragmatic concept that has gained momentum in recent times and that can help with fixing some of the environmental woes of our production and consumption approach.

Circular economy

In each mobile phone we use, there are gold and precious metals that are probably lost once we decide to buy a new and fancier one. More precisely, from one ton of old mobile phones, we can obtain about 180-300g of gold, 8,000g of copper and various precious metals, while in a traditional goldmine, from 1 ton of earth, which involves mining several hundred kilometers in the ground, we can eventually get 2-5g of gold. Extracting gold from cell phones and other minerals from e-waste (electronic waste) should represent a source of sizable profit for everybody, with evident benefits in terms of consumption of environmental resources and social risks. Unfortunately, it is not a problem related only to mobile phones. In our production and consumption system, inefficiencies are extremely common and extend to the large majority of goods, as follows.

- According to some recent publications (Ellen MacArthur Foundation, 2016; WEF, 2015), in the food supply chain, around 31% of food is lost or wasted. In addition, 95% of the fertilizer used to grow crops does not provide any nutrients to the human body and is lost because it is not absorbed by the crops or because it disappears with inedible or wasted crops.
- When we look at mobility, a car is typically parked 92% of the time. An average car has five seats but carries an average of 1.5 people per trip. The "tank-to-wheel" efficiency rate is extremely low, and more than 86% of the fuel is not used to transport people but gets lost in engine and transmission losses and in rolling resistance (Ellen MacArthur Foundation and McKinsey Center for Business and Environment, 2015).

- In 2014, about 311 million tons of plastics were produced and about 26% of it was packaging. According to the data available, 14% of plastic packaging is recycled, and when it comes to Polyethylene terephthalate – PET used for plastic bottles, only 50% is collected. Moreover, a large percentage of plastic packaging leaks out and is lost, is mismanaged, or is illegally dumped. The consequences are dramatic: today the weight ratio of plastic to fish in the water is 1:5. If we do not introduce radical transformations in our production and consumption models, we expect this ratio could grow to 1:1 in 2050 (Ellen MacArthur Foundation, 2016).

Since the Industrial Revolution, our economy, despite evolving and diversifying in multiple ways, has been dominated by a linear model of production and consumption, where resources follow a "take-make-dispose" pattern: goods are manufactured from raw materials extraction and cultivation, are then consumed and used, and are finally disposed as waste. We clearly acknowledged that this model has favored exceptional improvements in our society, dramatically increasing our prosperity and our well-being. At the same time, this model has proven to be extremely inefficient when it comes to the utilization of natural resources and ecosystem services (see Chapters 1 and 2) especially when we consider the design and implementation of systems to capture, reuse and regenerate material leakage and disposal (Ellen MacArthur Foundation, 2013a: 6). In the last decade, the concept of CE has received increasing attention worldwide as a possible response to the planetary boundaries (Rockström et al., 2009), increased scarcity of natural resources, volatility of commodity prices and risks of sudden supply chain disruptions. Governments, the business communities and NGOs have promoted a large debate on CE in order to imagine a new economic model in which goods are designed as tomorrow's resources, activating virtuous cycles that can create more value by recovering and recycling when the goods have ended their service-life. As a result, the CE idea has popularized and gained momentum among policy makers (Rizos et al., 2017), academics (Ghisellini et al., 2016; Lieder and Rashid, 2016) and practitioners, making its way through journals, magazines and media, and becoming a key topic at international events like the World Economic Forum in Davos (WEF et al., 2014). See Figure 7.2. for a visual representation of the linear vs circular model.

Origin of the circular economy (CE)

The roots of CE can be traced back to environmental economics and to the work of Pearce and Turner (1989), who, in turn, were inspired by the previous studies of Boulding (1966) and Georgescu-Roegen (1971) on system theory and thermodynamics. Boulding and Georgescu-Roegen (1971) developed the idea of an economy as a closed-system, instead of an open one like in the neo-classical paradigm where natural resources are implicitly considered abundant and substitutable. Kenneth Boulding and Nicholas Georgescu-Roegen explore cross-disciplinary fertilization and apply the principles of thermodynamics to explain the physical and ecological constraints to the dominant economics of growth. Boulding offers a particular interesting metaphor to illustrate the concept of closed-system in his article "The economics of the coming spaceship earth" (1966). In this seminal paper, he wrote:

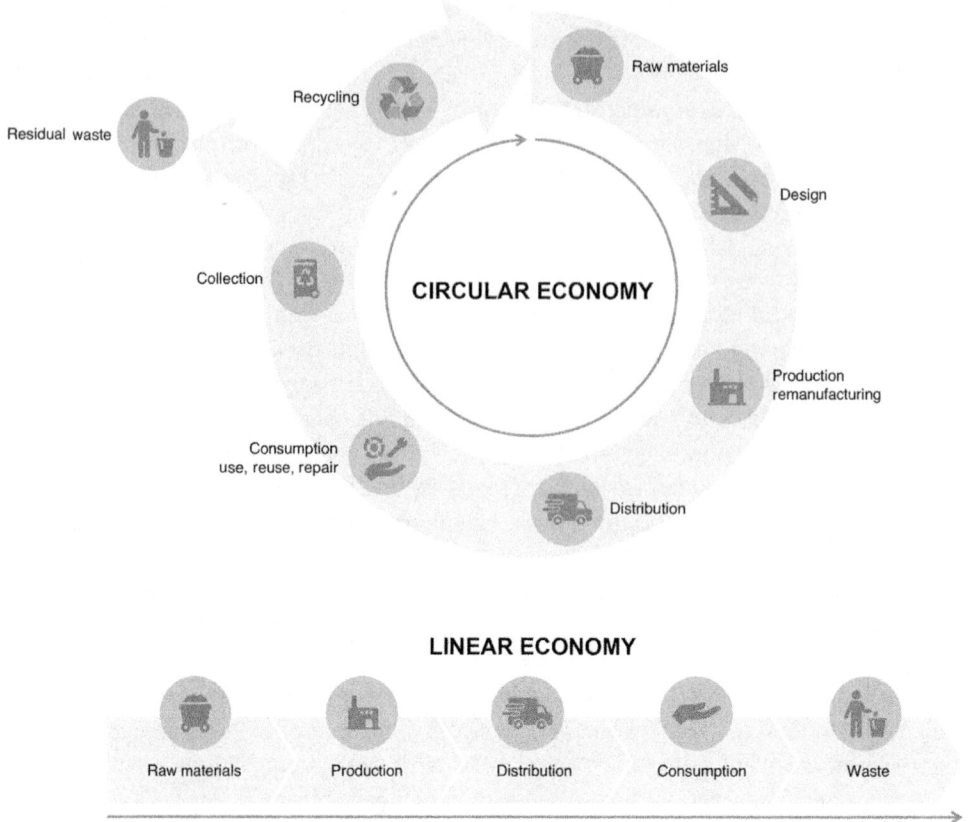

Figure 7.2 Linear vs circular economy.
Source: authors' elaboration.

The closed earth of the future requires economic principles which are somewhat dif-ferent from those of the open earth of the past. For the sake of picturesqueness, I am tempted to call the open economy the "cowboy economy," the cowboy being symbolic of the illimitable plains and also associated with reckless, exploitative, romantic and violent behavior, which is characteristic of open societies. The closed economy of the future might similarly be called the "spaceman" economy, in which the earth has become a single spaceship, without unlimited reservoirs of anything, either for extraction or for pollution, and in which, therefore, man must find his place in a cyclical ecological system which is capable of continuous reproduction of material form even though it cannot escape having inputs of energy.

(Boulding, 1966: 7–8)

System theory and the law of the thermodynamic explain the theories behind the degra-dation of "matter and energy" and calls for a re-conceptualization of our production and consumption patterns. This new circular economic model illustrated by Pearce and Turner (1989) acknowledges the interdependencies between the economy and the natural environment and assumes that the environment is both source of resources and receptor of waste. They also assume that waste is generated at all levels of economic activity – resource production, manufacturing and consumption – and that in order to maintain the limited stock of resources, waste should be recovered and recycled.

Another important literature stream that has contributed to conceptualize the CE is industrial ecology. First introduced by Frosh and Gallopulos in a paper published in 1989 by Scientific American, industrial ecology has emerged as a specific field of research, which adopts a system perspective and focuses on the analysis of material, energy and information flows that emerges from human activities and industrial societies (Boons and Howard-Grenville, 2009). A core element of industrial ecology is the idea that natural ecosystems and industrial systems operate according to similar rules. Industrial ecology attempts to understand how industrial systems work and how the exchange of flows with the biosphere is organized. Moreover, this approach adopts a normative standpoint: sustainability requires to close the material cycles related to man-made economic activities, therefore, minimizing the ecological impact generated by these activities on the natural environment. The implications of industrial ecology are broad and include introducing structural technological, cultural and economic changes. In this area, Frosch and Gallopoulos (1989) first highlighted the necessity of improving our production processes, minimizing waste (in particular, unrecyclable waste) as well as the consumption of scarce material and energy resources. This goal requires disruptive innovations in the manufacturing and design of processes and products and the development of waste management systems that must be integrated into the industrial production networks (Ghisellini et al., 2016).

Industrial ecology incorporates some of the basic elements of the circular ecology when it foresees the development of synergetic exchanges of materials (by-products and waste that become resources) and energy among firms. In both cases, the concept of value of resources is re-designed to account for the opportunities related to material savings, waste reuse and recycle, and reduction of environmental externalities. In this way, the value of resources is maximized throughout the extended supply chain, activating virtuous cycles of materials and energy.

Other theories and approaches that have been associated with CE are life cycle thinking, cradle to cradle and design for the environment (see Chapter 6). In our view, these approaches provide complementary and integrative elements to the concept of CE and are fundamental for its further development. At the same time, these approaches lack a strong holistic perspective, and the concept of innovation is still mainly focused on technological solutions based on the pivotal role of a specific "focal" company. Differently, the CE perspective responds to a strong systemic and collaborative view, which spans the macro (e.g., countries and regions), meso (e.g., industrial or organizational networks) and micro (e.g., firm to firm) levels, and where multiple types of innovations develop across industries, infrastructures, cultural and social environments. CE provides the new framework to favor the development of sustainable business models and the design and formulation of sustainable supply chains and products.

Definition, main features and limits

Once we have illustrated the limits of our "linear" economic system and the roots of CE, we can explore some of the most popular definitions and analyze the main components of a circular system. We start with what has probably become the most important conceptualization of this term, which has been provided by the Ellen MacArthur Foundation, a global organization established in 2010 with the aim of accelerating the transition to the CE. Ellen MacArthur Foundation describes this concept as:

> an industrial system that is restorative or regenerative by intention and design. It replaces the 'end-of-life' concept with restoration, shifts towards the use of renewable energy, eliminates the use of toxic chemicals, which impair reuse, and aims for the elimination of waste through the superior design of materials, products, systems, and, within this, business models.
>
> (Ellen MacArthur Foundation, 2013a: 7).

This definition is wide-ranging and goes beyond the requirements for designing reusable and recyclable goods and sustainable services; it is about designing a sustainable socioeconomical system. It is based on system-thinking and on the idea of non-linear and feedback-rich flows of energy and material, where non-living and living components interdepend and exchange. The perspective provided by the Ellen MacArthur Foundation builds on the concept of cradle to cradle (see Chapter 6) and on products that must be designed to be re-introduced in the biosphere without damaging the natural capital, or that must be circulated in the technosphere several times before losing their value. The ultimate goal of CE is to "enable effective flows of materials, energy, labor and information so that natural and social capital can be rebuilt" (Ellen MacArthur Foundation, 2013b: 26).

Among policy makers, the European Commission is probably the leading actor in this area. On 2 December 2015, a package that included several revised legislative proposals on waste management was adopted. In order to push forward the CE approach, this package also included an Action Plan aimed at supporting closed-loop product life cycles at every stage of the value chain: from production to consumption, repair and manufacture, waste management and secondary raw materials that become new resources for the economy. The EU Action Plan describes the CE as an economy: "where the value of products, materials and resources is maintained in the economy for as long as possible, and the generation of waste minimized" (European Commission, 2015: 2). CE is considered as "an essential contribution to the EU's efforts to develop a sustainable, low carbon, resource efficient and competitive economy" (European Commission, 2015: 2). The Action Plan is considered "to be instrumental in reaching the Sustainable Development Goals (SDGs) by 2030, in particular Goal 12 of ensuring sustainable consumption and production patterns" (European Commission, 2015: 3).

Another important definition has been provided by Peter Lacy and Jakob Rutqvis, two scholars and consultants from Accenture, a leading global professional services company, who have contributed to link the concept of CE to the business and financial community and to competitive advantage. In their bestselling book titled *Waste to wealth: The circular economy advantage* (Lacy and Rutqvist, 2015), they define the CE as an economy where "growth is decoupled from the use of scarce resources through disruptive technology and business models based on longevity, renewability, reuse, repair, upgrade, refurbishment, capacity sharing and dematerialization."

Building on these definitions, we can identify some key elements that can be considered founding components of the CE approach:

- a system perspective, based on material, energy and information flows as a central part of any economy, associated with the idea of system effectiveness,
- a normative approach that focus on a set of required actions to close the loops and enhance the natural capital. For example, favoring the reuse, repair, recycle of waste and increasing the resource efficiency (including the substitution of scarce and toxic resources with renewable ones),
- the link with the notions of competitive advantage and market opportunities, and the idea that a circular approach is beneficial to the environment, society and the economy (see Box 7.6),
- the link with design, innovation and new disruptive technologies, which can facilitate the transition to this system-based approach.

Box 7.6   The value of the CE

Several reports and publications have attempted to measure the potential impact in terms of environmental benefits, new job opportunities and contribution to GDP associated with the transition to the CE. The Ellen MacArthur Foundation and the McKinsey Center for Business and Environment think that in Europe, CE can favor an increase in resource productivity of 3% annually, translating to a total annual benefit of around €1.8 trillion per year, leading to a GDP increase of 7% (2015: 12). The adoption of this approach can also decrease GHG emissions up to 48% by 2030 and up to 83% by 2050. According to another study conducted by Cambridge Econometrics & BIO Intelligence Service (2014) focused on resource productivity and CE, an improvement of the EU's resource productivity of around 2% to 2.5% can generate a net impact on EU's GDP and could help create two million additional jobs in 2030. The same study has estimated that an increase in resource productivity by 3% can lead to a reduction in GHG emission by around 25%. CE can therefore become complementary to other policies in order to reach GHG reduction targets. Finally, the study by Accenture (Lacy and Rutqvis, 2015) has identified a US$4.5 trillion opportunity in introducing circular business models that transform waste from a problem to a resource.

Source: authors' elaboration using multiple sources.

The CE has been introduced almost as a magical solution to our environmental and sustainability problems and as an incredible business opportunity. But what, if any, are the limits and criticisms of this vision? Scholars have pointed at some possible problems that may require further research and that need to be considered when we design a future based on closed-loop production and consumption systems. First, the CE must consider the question of a limited planet and thresholds, including planetary boundaries as well as the finiteness of resources. Resource saving and resource productivity are extremely important for our goal of preserving social-ecological resilience, but our saving must be measured in absolute terms, not in relative ones. The risk of high increase of relative efficiency is well known (Jevon's paradox; see Chapter 4 of this book) and can lead to a backlash effect with an overall increase in consumption that partially or fully offsets the benefits of circularity. For example, more efficient use of resources can translate to cheaper products, increasing the overall demand. In other words, closing material and product loops do not necessary prevent the increase in primary production. Second, Europe and North America, despite being intrigued by the perspective of the CE, are very high in terms of per-capita consumption and waste production. Circularity, therefore, must be coupled by policies that aim at preventing waste generation targeting a reduction in our material consumption pattern and a decoupling of economic growth from the environmental impact (Tencati et al., 2016). Third, the possibility to reuse and recycle materials has technical limits and cannot be overpassed. As stated by the second law of thermodynamics, when the energy is transferred or transformed, more of it is wasted (and as a consequence, also matter is wasted). This makes a complete closure of loops hardly achievable because of a progressive loss of quality and quantity. For example, cellulose fibers brake and decrease in quality each time they get recycled, and this leads to a "down cycling" process that is complex to contrast and invert. Similarly, remelted glass reduces its workability after successive reprocessing. Even plastics lose flexibility and require virgin resins to be remixed in order to again become useful substances. Moreover, all these activities need energy to be performed (e.g., for melting, cleaning and sterilizing), with implication for GHG emissions and climate change. Other challenges relate to the fact that materials need to be collected and separated in order to be recovered and recycled, and this requires efficient take-back systems and the engagement of consumers, companies, NGOs and local authorities, among others. Furthermore, discarded materials are often mixed and contaminated with other materials or with hazardous substances thus limiting the possibility of activating efficient circular flows. Other challenging aspects related to the product design – e.g., cradle to cradle design, design for the environment, design for disassembling – have been discussed in Chapter 6. As a result, the possibility to recovery, reuse and recycle has to deal with physical/material limits and cannot be repeated indefinitely.

   In conclusion, CE is not the panacea to all environmental problems, and there are still several challenges to address and open criticisms. At the same time, it provides a novel view of production and consumption processes as inspired by natural cycles that open enormous potentialities to reach win-win solutions reducing environmental impact and promoting profitable business practices. In our view, circular principles, in order to deploy their full potential for a better world, need to be inscribed, embedded in a holistic framework that acknowledge the existence of the planetary boundaries and the idea that business operates in nature and under the laws driving earth-system processes. This includes the physical limitation to endless recycling and the necessity to reduce the use of ecosystem services and natural capital.

Circular business models

Two types of models

In the previous sections, we have illustrated and discussed sustainable business models and CE. In this final section, we bring together these two perspectives introducing the concept of circular business models. As previously discussed, the business model approach works at the level of an individual firm, although it involves a network of organizations along the value chain. Circular business models identify innovative strategies to transform the CE into business opportunities. Therefore, in this paragraph, we move from a macro perspective, where the utility and value of products, components, and materials and the environmental and social impacts associated with them are assessed at the system level, to the level of the individual firm where the notion of value and impacts are limited by the corporate boundaries. In particular, building on the works by Bocken *et al.* (2016) and Lacy and Rutqvis (2015), we identify two different types of sustainable business models that can contribute to the circularity challenge. On the one hand, we have business models that contribute to "slowing down" the resource consumption (Bocken et al., 2016: 312). These business models reduce the consumption of resources, improving the use of goods through different strategies. They offer to consumers an innovative value proposition, and they combine resources and capabilities along the supply chain in a different way when compared to traditional ones. The relation with the customers can also be profoundly different and even cost-revenues approach can be new. On the other hand, there are business models for closing the loop of materials. Here, the attention is mainly directed to innovative organizations that are extracting the value from what is considered waste in a "linear economy" model, focusing on closed material flows through recovery and recycling (Bocken et al., 2016).

Circular business models: examples at industry level

Some of these examples are presented below.

Product life extension

This business model allows companies to extend the life cycle of products and assets through actions such as repairing, upgrading, remanufacturing, and/or reselling. Thanks to this approach the company sells a different value proposition to its consumers (see Box 7.7.) and can exploit the residual value of products and components that would otherwise be lost through wasted materials. In terms of environmental benefits, this strategy reduces the amount of materials that need to be disposed and landfilled. Moreover, it provides a new potential stream of revenues for the company that can sell remanufactured components or upgraded products and the services connected.

Box 7.7   Fairphone case study

Fairphone was the name of a campaign against conflict minerals that started in the Netherlands in 2010 with the aim to raise awareness about the relations between smartphones and the conflict in the Democratic Republic of Congo. In 2013, the campaign converted into a social enterprise with a new objective "to produce a cool phone that puts human values first" (Akemu et al., 2016: 1-2), and in December 2013, the first generation of "fair phones" was delivered to the market. The company has now become inspirational for the entire market, raising attention and interest from customers, to designers, to large smartphone manufacturers and their stakeholders (Wernink and Strahl, 2015). Fairphone value proposition aims at making the supply chain transparent and "fair" with regard to human rights and sustainability issues. In the words of its founder and CEO Bas van Abel, the company ambitions are "to raise the bar in the electronics industry, we aim to increase our leverage with electronics suppliers to negotiate a healthier, more future-proof supply chain. This touches on a variety of issues, including the availability and life span of electronic components, the sourcing of Fairtrade gold and improving working conditions. By bringing these principles to the table, we can inspire an entire system change." Another core element of Fairphone strategy is a modular design that aims at extending the product life enhancing circularity of resources and materials. As a consequence, Fairphone has been designed in such a way that broken parts can be easily replaced or repaired for long lasting design. Moreover, modularity allows the phone to incorporate the pace of technological change. Finally, the company offers components that can be sold as spare parts and repair tutorials to help consumers in "do it yourself" repair and provides take-back programs to support recovery and recycling. Fairphone has developed a strategy for sustainability that embrace a product life extension business model and a sustainable supply chain.

Source: authors' elaboration using multiple sources.

Servicizing or product to service

This approach provides an alternative to the traditional model of "buy and own." Customers instead of purchasing a product can lease it or "pay per use" (see Box 7.8). The benefit can be particularly relevant for companies that are highly capital intensive, like business-to-business industrial equipment companies, and for firms that operate in business-to-consumer markets where the pace of innovation is limited and new products are not bringing high value for customers. In these cases, shifting from selling goods to servicizing can result in higher profits, including the possibility of extracting the residual value at the end of the life of the product. This approach radically transforms the firm's value proposition (instead of selling a product, the company sells a service), the financial model, and requires a different set of infrastructures and competencies (e.g., in order to offer the services, the company must manage the maintenance of the product). Examples include cars and clothes hiring, or tires (e.g., fleet solutions), phones or battery leasing. The benefits can be double: for manufacturers, it can increase the length and the depth of the relation with customers and it can expand the profitability when the product durability, efficiency and reusability are increased; for consumers, it can incentivize a more "responsible" use of the products (e.g., using the car only when necessary). Overall, this business model can reduce consumptions and contribute to slowing down the use of natural resources.

Box 7.8   Philips "Light as a service" case study

One of the most interesting case studies in the area of servicizing is Philips' strategy of offering its customers "Light as a service" or "Pay per Lux." Philips, a Dutch diversified company, focuses on improving people's lives through innovation and operates in two main business areas: health technologies and lighting. Philips is a frontrunner in sustainability with a long history that began in 1994 with its first sustainability program. In 2016, its CEO Frans van Houten launched a new five-year sustainability program "Healthy people, sustainable planet" that focused on sustainability through the supply chain, the operations and sustainable solutions (Philips, 2017). A core component of Philips' sustainability strategy is the attention focused on the CE and the transition to circular lighting as a new business model. In particular, the company, instead of selling bulbs that provide light, started to sell light itself. Customers do not need to invest in equipment, and the company takes care of management, maintenance and innovation. Moreover, the company has the incentive to maximize the level of energy efficiency and the recycling and re-use of components. In sum, this approach, based on a new value proposition and a revolutionary business model, increases the value for clients and ensures lower resource consumption and carbon emission. Philips is now extending this approach to the health-care business by refurbishing medical products.

Source: authors' elaboration using multiple sources.

Encourage sufficiency

This approach aims to unlock individuals from a logic of materialistic consumerism (Jackson, 2011; Bocken et al., 2016), introducing lifestyle changes and concepts such as down-shifting and frugality (Lorek and Fuchs, 2013). Business models that encourage sufficiency aim at favoring consumption behavior that searches for durability, quality and maintenance as core features of the product benefits. This is the case in some categories of goods, such as luxury or premium products (see the case of Patagonia in Box 7.9), where the value proposition focuses on long-lasting use experiences that preserve value over time, associated with long-term services and warranty. In the white goods industry, a well-known company Miele has built a brand value on high quality and durability and is slowing down consumption with a "classic" long life model. Its appliances have a reputation for lasting two times longer than their competitor models, they are highly efficient (about 80% of the energy consumption of a washing machine is during the utilization) and incorporate reusable materials. The benefits of these business models are two-fold: on the one hand, they guarantee high margins and favor customer loyalty while; on the other, they can contribute to reducing consumption with a positive impact on the environment.

Box 7.9   Patagonia "Don't buy this jacket" campaign

Around Thanksgiving 2011, Patagonia started the "Don't buy this jacket" campaign with a full-page advertisement in the *New York Times* (see Figure B7.9). This initiative, which asked consumers to rethink their lifestyle for a more sustainable one, can be considered one of the most interesting and effective examples in the area of green marketing. The company commented on its advertising saying: "We design and sell things made to last and be useful. But we ask our customers not to buy from us what you don't need or can't really use. Everything we make–everything anyone makes–costs the planet more than it gives back" (Rosenbloom, 2012).

   The campaign fully supported Patagonia's innovative business model that encourages sufficiency. Patagonia, as previously illustrated, is one of the leading and most admired brands in sustainability worldwide. Since its foundation in the 1970s, the company has been able to remain true to its eco-friendly ethos. Environmental protection is fully embedded in the business model, ranging from a supply chain that uses organic cotton and recycled polyester to a value proposition based on building a long-lasting relationship with its customers and communities. Patagonia's products are premium, and customers expect very high quality. The advertising was a part of the "Common Threads Initiative," a program based on four actions: reduce, inviting people to "buy only what they need;" repair, offering to repair what breaks; reuse, helping customers to find a home for gear no longer used; and recycle, taking back wasted products, keeping them out of landfills and incinerators. The final message anchored the initiative to Patagonia's core values and mission: "Reimagine. Together we imagine a world where we take only what nature can replace."

   The "Don't buy this jacket" campaign, despite being somewhat risky and, for many marketing experts, confusing, resulted in an increase in annual sales of almost 40% in the following years (Stock, 2013). It might seem paradoxical and controversial that an invitation to reduce consumption resulted in increased sales, but Patagonia's

Figure B7.9 "Don't buy this jacket" advertising campaign.
Source: Permission given by Patagonia®.

authenticity is at the core of its enthusiastic customers, and enthusiastic consumers become the best communicators and brand evangelists supporting sales growth and enlarging market segments (Rosenbloom, 2012).

Source: authors' elaboration using multiple sources.

Collaborative consumption and sharing platforms

These novel business models take advantage of the diffusion of digital technologies. They were proposed by Lacy and Rutqvis (2015) as a typology of circular solution because they favor the sharing of underutilized assets, thus increasing resource productivity. The idea is that platforms for collaboration among users, whether individuals or organizations, help increase the utilization rate of capital goods or durable products while reducing the necessity to buy, manufacture, or build new ones. It is probably the less "orthodox" of the circular models proposed, but it combines the possibility of saving materials and energy, already embedded in unutilized assets (in industrialized countries, around 80% of the objects we store at home are only used once a month), with the possibility of making money from it. BlaBlaCar, for example, operates as an aggregator and helps connect drivers who have empty seats in their cars and want to share fuel and travel costs with travelers who are interested in going to the same destination (or in the same direction). In this case, the environmental benefit is related to increasing the efficiency of the car by filling empty seats. Similarly, Airbnb is fast becoming the largest hotel company in the world in terms of available beds, but it has not constructed any new buildings or rooms. To conclude, sharing and collaborative business models are providing huge opportunities for new start-ups, and they can also contribute to a more sustainable social-ecological system by slowing down the production of new goods and increasing the utilization of existing ones. At the same time, this business model is probably more controversial since it can easily result in adding novel consumption patterns that, instead of optimizing material flows, induce the utilization of additional natural capital.

Resource recovery

With this typology, we refer to companies that transform and innovate their business model or position themselves at specific stages of the value chains in order to collect, utilize and exploit the materials contained in already disposed products. In other words, the value embedded in the products that reach the end of their life cycle is used to feed another economic activity and is turned into new forms of value thanks to technologies for recovery, recycling and up-cycling. This business model can be the business-as-usual approach in industries where materials at the end of their lives still have high economic value, like in the case of aluminum, cast-iron, steel, or paper. In these sectors, hundreds of firms have been profitably operating since decades managing, collecting and recycling waste and garbage from other companies, households, or individuals, closing material flows and reducing or eliminating leakages. In some cases, these companies are able to make profits both ways in its value chain, when they collect waste for others and when they sell recycled products made with this waste.

Markets for secondary raw materials can be created and shaped by the enforcement of environmental policies and regulations that introduce objectives and targets for waste collection, recovery and recycling. For example, in the European Union, the first Waste Framework Directive was introduced in 1975 to set recycling standards and to develop national waste prevention programs at the national level, while the first Directive on Packaging and Packaging Waste was introduced in 1994 (1994/62/EC). Today, another Waste Framework Directive (2008/98/EC) has been introduced to modernize the waste management principles and strengthen prevention, recovery and recycling of several material flows. Overall, only 36% of the waste produced in EU-28 is recycled (data as of 2014), but this percentage is increasing. When we specifically consider packaging waste, the EU-28 recycle rate in 2014 was above 60% of packaging materials, with countries like Germany (156.7 kg/inhabitant) and Italy (128.7 kg/inhabitant) reporting the highest amounts of packaging material recycled (EU statistics). When we consider specific material flows, the European packaging recycling rate for paper and board was above 80%; steel was 74%, glass 70% and aluminum 67% (data as of 2012, EU statistics).

The example of the EU shows that when regulations are enforced to close the loops, conditions for new markets and industries are established, and new companies can grow with the purpose of extracting value from wasted resources. Unfortunately, as illustrated in the introduction of this section on the CE, there are several products, materials and geographies (e.g., many developing countries) where policies and regulations are not enforced, or where markets for secondary raw materials have not developed due to different barriers: organizational complexity to activate waste collection and management systems, lack of technologies and know-how, economic reasons related to the market price for secondary raw materials. The case of food waste, textile waste, several types of electronic waste and hazardous waste, or the case of plastics are illustrative. In these situations, the introduction of circular and innovative business models based on novel competencies and resources, novel technological solutions, capable to target consumer segments interested in sustainability with new value propositions, can open interesting and profitable business opportunities. One well-known example of firms that have innovated their market approach to close the material loops is Interface (Lovins et al., 2007; Stubbs and Cocklin, 2008). This company, one of the world's leading manufacturers of modular carpet and a pioneer in sustainability, has developed a business model that offers to take its carpet back from its customers, recycle it and divert it from landfill in order to reduce the use of petroleum-based resources and extend the life span of materials. Two innovative companies in the textile sector that operate recovering and recycling wasted materials to produce new fabrics are illustrated in the Box 7.10.

Box 7.10   Aquafil and Orange Fibers

Recovering fishing nets to produce regenerated textile fibers–this is the business model developed by Aquafil, a medium-sized Italian company with about €480 million in revenues (in 2016). Aquafil specialized in nylon fibers for a varied range of industrial applications (e.g., carpet manufacturing for residential or automotive use, sportswear, clothing and fashion) and is now one of the leading players in the market. As a pioneer in circularity, the company started collecting used fishing nets around Europe and, thanks to innovative proprietary technological solutions, is able to regenerate these wasted materials into nylon and yarn (Econyl®). From 2010 to 2016, 20,000 tons of recovered fishing nets from aquaculture and the fishing industry have been transformed into Econyl® and re-commercialized at competitive costs and quality when compared to virgin oil-based materials. Among the company's clients are brands such as Adidas, Stella McCartney and Speedo.

Orange Fibers, founded in 2014, is a start-up that utilizes citrus juice by-products to manufacture sustainable fabrics. Every year, hundreds of thousands of tons of wastes are generated by this industry and they are usually thrown away. Orange Fibers has developed a novel technology that allows recycling of these materials, extracting the cellulose that is wasted from the industrial pressing and processing of the oranges, and transforming them into high-quality fabrics. Using nanotechnology technique, the fibers are also enriched with citrus fruit essential oil. The result is an innovative textile with specific features, and with the potential to contribute to the CE and to resource recovery. The company has started collaborating with fashion brands like Ferragamo and, with its innovation, contributes to introducing sustainability into the Italian-style tradition.

Source: authors' elaboration using multiple sources.

Industrial symbiosis

A specific typology of resource recovery is industrial symbiosis (Chertow, 2000), which scholars also consider a special form of industrial ecology; it builds on the notion of biological symbiotic relation in nature, a concept used to describe the exchange of materials, energy and information between two unrelated species in a mutually beneficial manner. The industrial symbiosis business model refers to the exchange of materials, by-products and waste among companies and other organizations so that the outputs of one process become the inputs of another process. Geographic proximity of beneficiaries, like in industrial parks, is another condition used to identify industrial symbiosis. The two most cited examples of this business model are Kalundborg symbiosis in Denmark where several public and private companies cooperate and exchange materials such as steam, ash, gas, heat, sludge and water (see www.symbiosis.dk/en) and British Sugar, a factory in Norfolk (UK) that, starting from processing sugar beet, obtains 12 different co-products from chemicals to food for animals and humans, which are sold and become a source of revenues for the company (Short et al., 2014).

To conclude, in this section we have illustrated several frameworks that can contribute to increase the circularity of our economy, slow down the production and consumption processes, improve resource productivity, and close the material and energy flows. The business models for sustainability require innovative strategic and organizational capabilities. The cases we have discussed, along with many others, demonstrate that win-win solutions are possible and feasible. The CE and circular business model are essential to a resilient social-ecological system that prospers within the limits of the planet.

Summary

Contributing to social-ecological system resilience instead of taking from society and nature is not an easy challenge for companies. In the Anthropocene epoch, firms are required to develop new logics to respond to the markets, radically changing the way in which they do business along the entire value chain. This is the only pathway to be sustainable and to implement the "Business In Nature" (BInN) concept.

Business model for sustainability (or sustainable business model) often represents the most radical change; this approach offers the possibility to re-organize companies from sourcing of raw materials to the end of life of the product and allows business to dramatically reduce the negative effects of their activities on ecosystems and societies. It is the most transformative among the form of innovation examined, and it changes the way organizations and their value network create, deliver and capture economic value, or change the value proposition for their clients and stakeholders (Bocken et al., 2014).

We started by illustrating the case of Tesla, one of the pioneers in electric mobility, to provide the first flavor of the ideas embedded in this chapter: How is Tesla competing? How is the Tesla business model organized? In order to pave the way for a "deep dive" into this topic, we first introduced the managerial and organizational literature on business models, illustrated the ontology and the core features of the concept, and then bridged this approach with the innovation perspective. In the following section, we have examined links with sustainability, starting with the seminal works of Lovins et al. (1999) on *Natural capitalism* and Stubbs and Cocklin (2008) framework based on a set of normative principles.

Using the business model framework as a baseline, we have introduced a list of requirements and conditions to design and implement a sustainable business model. Further, we have described some examples of industries where novel business models are changing the landscape opening new market opportunities: energy and electricity, mobility and automotive, the agro-food, markets for ecosystem service.

The second part of the chapter has examined the CE concept. This approach contributes to the BInN concept offering a system perspective consistent with the four key principles analyzed in Chapter 3. Building on some examples, we have exemplified the inefficiencies and the unsustainability of our linear economic model, and we have introduced circularity as a constituent feature of resilient social-ecological system. After positioning the CE in an historical and disciplinary perspective, we have investigated in detail its characteristics and limits. In this final section, drawing on the business model perspective with circularity, we have focused on the concept of circular business models. In particular, we have discussed two types of sustainable business models: those that contribute to "slowing down" the resource consumption and the ones that favor the closed loop of materials. Several short case studies of companies pioneering servicizing, product life extension, resource recovery, sufficiency and collaborative consumption have enriched the section.

Chapter 7 annexes

Annex 1

A.1.1: Questions for discussion

- What are the core features of a business model for sustainability (or sustainable business model)?
- What is natural capitalism? How does it relate to business models for sustainability?
- What are the key components of the Tesla business model? Can it be considered sustainable? Do you see any contradictions between Tesla strategy and sustainability?
- Discuss the concept of CE: origins, characteristics, and limits.
- Explain the relationship between CE, design for the environment, and cradle to cradle.
- Exemplify and discuss the main typologies of circular business models.
- Can we consider collaborative consumption and sharing platforms an example of the circular business model? Why or why not?
- Apply the sustainable business model framework to the Patagonia case. How are the "Don't buy this jacket" campaign and the "Common Threads Initiative" contributing to Patagonia's business model and brand positioning?
- Using the internet and other websites identify examples of companies with circular business models that can contribute to slow down resource consumption and increase closed loop of materials.

A.1.2: Assignments

Assignment 1: Business model for electric car

The links between electric vehicles and new energy systems offer huge opportunities for novel business models that can reduce air pollution and carbon emission and improve mobility efficiency. This assignment aims at investigating the new business models archetypes that are emerging in automotive-energy industries to favor this transition. Using the business model framework analyzed in this chapter, investigate the different options today available on the market and illustrate how they are and competing each other.

- What are the main actors involved: automakers, utilities, local governments, and other service providers?
- What are the types of electric vehicles: Hybrid Electric Vehicles, Plug-in Hybrid Electric Vehicles, Battery Electric Vehicles, etc.?
- What are the technologies required: charging stations, batteries, renewables, mobile apps, etc.?
- What is the value proposition of these new business models?
- How is the relation with clients going to change?
- What is the financial structure – costs and revenues – of these new business models?

In order to carry out your analysis, have a look at the following website and companies: Tesla; Zoe Renault, Wanxiang, BMW, Nissan leaf, Autolib'/Bolloré, Better Place.

The outcome of this assignment can be an individual paper of about 6,000 words.

Assignment 2: CE and up-cycle

This assignment aims at analyzing the opportunities related to waste management in some specific industries: for example, textile, food and drinks, white goods, electronics, tires, chemicals). Organize your class into teams of students (five or six students per team) and assign one industry and one value chain phase to each team.

Each group must identify some of the most innovative "up-cycling" initiatives and then quantify the economic opportunities and environmental and social benefits associated with these initiatives.

The expected outcome is a report of 3,500 words and a presentation in ppt (max 15 slides). Steps suggested for the analysis are:

- introduce the problem,
- examine the main waste flows/materials of the industry and try to identify the most interesting opportunities for "up-cycling." Also look outside the industry value chain for opportunities in other industries (the rubber in the snikers can become a material to create courts, tracks, fields; the fishnets can become a raw material for textile or carpets),
- asses the size of the market and the market prize for the waste with "up-cycling" potential,
- estimate the revenues you can obtain from selling the waste,
- describe the environmental and social benefits related to this business,
- illustrate three best practices that are developing this innovative approach, and
- provide conclusion and learning points.

A.1.3: Further readings and additional resources

- MIT Sloan Management Review, in partnership with Boston Consulting Group, has conducted several studies on sustainability strategies and innovation. In their 2017 report titled "Sustainability's next frontier. Walking the talk on the sustainability issues that matter most," they provide a valuable synthesis of eight years of research about HOW corporations address sustainability. This document is available at: http://sloanreview.mit.edu/projects/sustainabilitys-next-frontier/
- The Network for Business Sustainability has published an extremely useful report on sustainable business models titled "Business models for shared value" (2016) that offers an extensive and systematic overview of research and practice in this area. Please see: www.nbs.net
- The Ellen McArthur Foundation, one of the world's leading organizations on CE, offers a website rich of contents and tools, including definitions and approaches, case studies, educational programs, videos and publications. Please visit: www.ellenmacarthurfoundation.org
- If you are interested in the policy makers' views on CE, the EU website provides several documents: http://ec.europa.eu/environment/circular-economy/index_en.htm.
- Kalundborg provides one of the most important experiences on industrial symbiosis. This website - www.symbiosis.dk/en/ - (launched in 2015) explains how this industrial ecosystem functions and illustrates how companies can match output and input flows in economically efficient and sustainable ways.

Annex 2

A.2.1: References

Afuah A. (2014). *Business model innovation: Concepts, analysis, and cases.* New York: Routledge.

Akemu, O., Whiteman, G. and Kennedy, S. (2016). Social enterprise emergence from social movement activism: The Fairphone case. *Journal of Management Studies,* 53(5), 846–877.

Amit, R. and Zott, C. (2012). Creating value through business model innovation. *MIT Sloan Management Review,* 53(3), 41–49.

Ayres, R. U. (1989). Industrial metabolism: Theory and policy. In R. U. Ayres, U. E. Simonis (Eds.), *Industrial metabolism: Restructuring for sustainable development.* Tokyo, Japan: United Nations University Press.

Baden-Fuller, C. and Morgan, M. S. (2010). Business models as models. *Long Range Planning,* 43(2–3), 156–171.

Bishop, J., Kapila, S., Hicks, F., Mitchell, P. and Vorhies, F. (2009). New business models for biodiversity conservation. *Journal of Sustainable Forestry,* 28(3–5), 285–303.

Bocken, N. M. P., Short, S., Rana, P. and Evans, S. (2014). A literature and practice review to develop sustainable business model archetypes. *Journal of Cleaner Production,* 65, 42–56.

Bocken, N. M. P., Rana, P. and Short, S. (2015). Value mapping for sustainable business thinking. *Journal of Industrial and Production Engineering,* 32(1), 67–81.

Bocken, N. M. P., de Pauw, I., Bakkera, C. and Van Der Grinten, B. (2016). Product design and business model strategies for a circular economy. *Journal of Industrial and Production Engineering,* 33(5), 308–320.

Boons, F. and Howard-Grenville, J. (2009). *The social embeddedness of industrial ecology*. Cheltenham, United Kingdom, and Northampton, MA: Edward Elgar.

Boons, F. A. A. and Lüdeke-Freund, F. (2013). Business models for sustainable innovation: State-of-the-art and steps towards a research agenda. *Journal of Cleaner Production, 45*, 9-19.

Boulding, K. E. (1966). The economics of the coming spaceship earth. In H. Jarret (Ed.), *Environmental quality in a growing economy* (pp. 3-14). Baltimore, MD: Johns Hopkins University Press.

Bovarnick, A., Knight, C. and Stephenson, J. (2010). *Habitat banking in Latin America and Caribbean: A feasibility assessment*. United Nations Development Programme. Retrieved from www.cbd.int/financial/offsets/g-offsethabitatbanklac-undp.pdf.

Cambridge Econometrics and BIO Intelligence Service. (2014). Study on modelling of the economics and environmental impacts of raw material consumption. European Commission Retrieved from http://ec.europa.eu/environment/enveco/resource_efficiency/pdf/RMC.pdf.

Casadesus-Masanell, R. and Ricart, J. E. (2010). From strategy to business models and on to tactics. *Long Range Planning, 43*, 195-215.

Chertow, M. R. (2000). Industrial symbiosis: Literature and taxonomy. *Annual Review of Energy and the Environment, 25*(1), 313-337.

Chesbrough, H. W. and Rosenbloom, R. S. (2002). The role of the business model in capturing value from innovation: Evidence from Xerox Corporation's technology spin-off companies. *Industrial and Corporate Change, 11*, 529-555.

Chesbrough, H. W. (2007). Why companies should have open business models. *MIT Sloan Management Review, 48*(2), 22-28.

Chesbrough, H. (2010). Business model innovation: opportunities and barriers. *Long Range Planning 43*(2-3), 354-363.

Cohen, S. [Steve Cohen]. (2017, 31 July). Tesla's Model 3 and the Transition to Sustainability. Earth Institute, Columbia University, Sustainability (blog post). Retrieved from http://blogs.ei.columbia.edu/2017/07/31/teslas-model-3-and-the-transition-to-sustainability/.

de Jong, M. and van Dijk, M. (2015). Disrupting beliefs: A new approach to business-model innovation. *McKinsey Quarterly*. Retrieved from www.mckinsey.com/business-functions/strategy-and-corporate-finance/our-insights/disrupting-beliefs-a-new-approach-to-business-model-innovation.

Department for Environment, Food and Rural Affairs (DEFRA). (2013, May). Payments for ecosystem services: A best practice guide. Retrieved from www.cbd.int/financial/pes/unitedkingdom-bestpractice.pdf.

Drozdiak N. (2015, Mars 11). Germany's top power utilities face dimmer prospects. *The Wall Street Journal*. Retrieved from www.wsj.com.

Dyllick, T. and Muff, K. (2015). Clarifying the meaning of sustainable business: Introducing a typology from business-as-usual to true business sustainability. *Organization & Environment, 29*(2), 156-174.

Ellen MacArthur Foundation. (2013a). *Towards the circular economy Vol 1: Economic and business rationale for an accelerated transition*. Retrieved from www.ellenmacarthurfoundation.org/publications.

Ellen MacArthur Foundation. (2013b). *Towards the circular economy Vol 2: Opportunities for the consumer goods sector*. Retrieved from www.ellenmacarthurfoundation.org/publications.

Ellen MacArthur Foundation and McKinsey Center for Business and Environment. (2015). *Growth within: A circular economy vision for a competitive Europe*. Retrieved from www.ellenmacarthurfoundation.org/publications.

Ellen MacArthur Foundation. (2016). *The new plastics economy: Rethinking the future of plastics*. Retrieved from www.ellenmacarthurfoundation.org/publications.

Enel. (2015, 18 November). Capital Markets Day – Strategic Plan 2016-2019 [PowerPoint slides]. Retrieved from http://strategy2015.enel.com/files/9M/Enel_Capital-Markets-Day-18Nov15.pdf.

EU (2015, December 2). COM(2015) 614 final. Closing the loop – An EU action plan for the Circular Economy. Communication from the Commission to the European Parliament, the Council, the European Economic and Social Committee and the Committee of the Regions. Retrieved from http://eur-lex.europa.eu/legal-content/EN/TXT/?uri=CELEX:52015DC0614.

Frosch, R. A. and Gallopoulos, N. E. (1989). Strategies for manufacturing. *Scientific American, 261*(3), 144-152.

Geels, F. W., McMeekin, A., Mylan, J. and Southerton, D. (2015). A critical appraisal of Sustainable Consumption and Production research: The reformist, revolutionary and reconfiguration positions. *Global Environmental Change, 34*, 1-12.

Georgescu-Roegen, N. (1971). *The entropy law and the economic process.* Cambridge, MA: Harvard University Press.

Ghisellini, P., Cialani, C. and Ulgiati, S. (2016). A review on circular economy: The expected transition to a balanced interplay of environmental and economic systems. *Journal of Cleaner Production*, 114, 11–32.

Global Impact Investing Network, GIIN. (2017). *2017*, Annual Impact Investor Survey. Retrieved from https://thegiin.org/assets/GIIN_AnnualImpactInvestorSurvey_2017_Web_Final.pdf.

Gurdus, E. (2017, September 22). Up to 25 percent of cars may soon be fully electric, says Mercedes-Benz USA CEO after automaker invests $1 billion in electric vehicle production. [video file]. Retrieved from CNBC website www.cnbc.com/2017/09/22/mercedes-benz-usa-ceo-talks-1-billion-electric-car-investment.html.

Hall, S., Shepherd, S. and Wadud, Z. (2017). The innovation interface: Business model innovation for electric vehicle futures. University of Leeds. Retrieved from http://homepages.see.leeds.ac.uk/~earshal/Files/11167_SEE_electrical_vehicles_report_WEB.pdf.

Hawken, P., Lovins, A. and Lovins, H. (1999). *Natural capitalism: Creating the next industrial revolution.* Boston, MA: Little Brown & Company.

Hsu, C. (2015, June 10). Is Tesla Changing Its Business Model? *Investopedia*. Retrieved from www.investopedia.com/articles/investing/061015/tesla-changing-its-business-model.asp.

Jackson, T. (2011). Societal transformations for a sustainable economy. *Natural Resources Forum*, 35(3), 155–164.

Kiron, D., Kruschwitz, N., Rubel, H., Reeves, M. and Fuisz-Kehrbach, S. K. (2013, 16 December). *Sustainability's next frontier. Walking the talk on the sustainability issues that matter most.* MIT Sloan Management Review and The Boston Consulting Group. Retrieved from http://sloanreview.mit.edu/projects/sustainabilitys-next-frontier/.

Kiron, D., Unruh, G., Kruschwitz, N., Reeves, M., Rubel, H. and Meyer Zum Felde, A. (2017). Corporate sustainability at crossroads: Progress toward our common future in uncertain mimes. MIT Sloan Management Review Research report #58480 (in collaboration with the Boston Consulting Group).

Lacy, P. and Rutqvist, J. (2015). *Waste to wealth: The circular economy advantage.* Basingstoke, United Kingdom: Palgrave Macmillan. Quote taken from www.accenture.com/gb-en/insight-circular-advantage-innovative-business-models-value-growth.

Lambert, F. (2017, 2 August). *Tesla (TSLA) announces Q2 2017 earnings: revenue of $2.8 billion and loss of $2.04 per share.* Retrieved from https://electrek.co/2017/08/02/tesla-tsla-q2-2017-earnings/.

La Monica, P. R. (2017, 3 April). Tesla is worth more than Ford and GM is in sight [video file]. CNN. Retrieved from http://money.cnn.com/2017/04/03/investing/tesla-ford-market-value-gm/index.html.

Lewandowski, M. (2016). Designing the business models for circular economy: Towards the conceptual framework. *Sustainability*, 8(1), 43.

Lieder, M. and Rashid, A. (2016). Towards circular economy implementation: A comprehensive review in context of manufacturing industry. *Journal of Cleaner Production*, 115, 36–51.

Lorek, S. and Fuchs, D. (2013). Strong sustainable consumption governance–precondition for a degrowth path? *Journal of Cleaner Production*, 38, 36–43.

Lovins, A., Lovins, H. and Hawken, P. (2007, July–August). A road map for natural capitalism. *Harvard Business Review*. Retrieved from https://hbr.org/2007/07/a-road-map-for-natural-capitalism.

Lüdeke-Freund, F., Massa, L., Bocken, N., Brent, A. and Musango, J. (2016, September 8). Main report: Business models for shared value. Network for Business Sustainability South Africa. Retrieved from https://nbs.net/p/main-report-business-models-for-shared-value-4122f859-2499-4439-824e-7535631a14ed.

Millennium Ecosystem Assessment, MA. (2005a). *Ecosystems and human Well-being: Biodiversity synthesis.* Washington, DC: Island Press.

Magretta, J. (2002). Why business models matter. *Harvard Business Review*, 80(5), 86–92.

McKinsey & Company. (2016, January). Automotive revolution – perspective towards 2030. How the convergence of disruptive technology-driven trends could transform the auto industry. Retrieved from www.mckinsey.com/~/media/mckinsey/industries.

OECD. (2010, March 25). Paying for biodiversity: enhancing the cost-effectiveness of payments for ecosystem services. OECD. Retrieved from www.oecd.org/environment/resources/45137415.pdf.

Osterwalder, A. and Pigneur, Y. (2010). *Business model generation: A handbook for visionaries, game changers, and challengers.* Hoboken, NJ: John Wiley & Sons.

Osterwalder, A., Pigneur, Y. and Tucci, C. L. (2005, July). Clarifying business models: Origins, present, and future of the concept. *Communications of the Association for Information Systems*, 16(1).

Pearce, D. W. and Turner, R. K. (1989). *Economics of natural resources and the environment*. London: Prentice Hall.

Philips. (2017). Annual Report 2016. A focused leader in health technology. Retrieved from www.results. philips.com/publications/ar16#/

Rizos, V., Tuokko, K. and Behrens, A. (2017). The circular economy: A review of definitions, processes and impacts (CEPS Research Report No. 2017/8, April 2017). Retrieved from http://aei.pitt.edu/85892/.

Rockström, J., Steffen, W., Noone, K., Persson, A., Chapin, F. S., Lambin, E. F., Lenton, T. M., Scheffer, M., Folke, C., Schellnhuber, H. J., Nykvist, B., de Wit, C., Hughes, T., van der Leeuw, S., Rodhe, H., Sörlin, S., Snyder, P. K., Costanza, R., Svedin, U., Falkenmark, M., Karlberg, L., Corell, R. W., Fabry, V. J., Hansen, J., Walker, B., Liverman, D., Richardson, K., Crutzen, P. and Foley, J. A. (2009). A safe operating space for humanity. *Nature*, 461: 472–475.

Rosenbloom, J. (2012, 12 June). How Patagonia makes more money by trying to make less [video file]. Fast Company. Retrieved from www.fastcompany.com/1681023/how-patagonia-makes-more-money-by-trying-to-make-less.

Sardá, R. (2015). Payment for ecosystem services: concept and examples. In S. Nuss-Girona and M. Castañer (Eds.), *Ecosystem services: Concepts, methodologies and instruments for research and applied use* (pp. 137–153). Girona, Spain: Documenta Universitaria Publ.

Schaltegger, S., Lüdeke-Freund, F. and Hansen, E. G. (2012). Business cases for sustainability: The role of business model innovation for corporate sustainability. *International Journal of Innovation & Sustainable Development*, 6(2), 95–119.

Schaltegger, S., Lüdeke-Freund, F. and Hansen, E. (2016a). Business models for sustainability: A co-evolutionary analysis of sustainable entrepreneurship, innovation, and transformation. *Organization & Environment*, 29(3), 264–289.

Schaltegger, S., Hansen, E. and Lüdeke-Freund, F. (2016b). Business models for sustainability: origins, present research, and future avenues. *Organization & Environment*, 29(1), 3–10.

Short, S. W., Bocken, N. M. P., Barlow, C. Y. and Chertow, M. R. (2014). From refining sugar to growing tomatoes. Industrial ecology and business model evolution. *Journal of Industrial Ecology*, 18(5), 603–618.

Steffen, W., Richardson, K., Rockström, J., Cornell, S. E., Fetzer, I., Bennett, E. M., Biggs, R., Carpenter, S. R., de Vries, W., de Wit, C., Folke, C., Gerten, D., Heinke, J., Mace, G., Persson, L. M., Ramanathan, V., Reyers, B. and Sörlin, S. (2015). Planetary boundaries: Guiding human development on a changing planet. *Science*, 347.

Stock, K. (2013, August 28). Patagonia's "Buy less" plea spurs more buying. Bloomberg. Retrieved from www.bloomberg.com/news/articles/2013-08-28-patagonias-buy-less-plea-spurs-more-buying.

Stubbs, W. and Cocklin, C. (2008). Conceptualizing a "sustainability business model." *Organization & Environment*, 21, 103–127.

Teece, D. (2010). Business models, business strategy and innovation. *Long Range Planning*, 43, 172–194.

Tencati, A., Pogutz, S., Moda, B., Brambilla, M. and Cacia, C. (2016). Prevention policies addressing packaging and packaging waste: Some emerging trends. *Waste Management*, 56, 35–45.

The Economist. (2013). How to lose half a trillion euros. Europe's electricity providers face an existential threat. *The Economist, Print edition*, 409(8857), 27–30.

The Economist. (2017). A world turned upside down. *The Economist, Print edition*, 422(9029), 16–18.

Upward, A. and Jones, P. (2016). An ontology for strongly sustainable business models: Defining an enterprise framework compatible with natural and social science. *Organization & Environment*, 29(1), 97–123.

Wernink, T., Strahl, C. (2015). Fairphone: Sustainability from the inside-out and outside-in. In M. D'heur (Ed.), *Sustainable value chain management. CSR, sustainability, ethics & governance* (pp. 123–139). Cham, Switzerland: Springer.

Wharton. (2016, 11 April). Tesla speeds ahead in the electric vehicle market. *Knowledge & Wharton*. Retrieved from http://knowledge.wharton.upenn.edu.

World Economic Forum WEF, Ellen MacArthur Foundation & McKinsey & Company. (2014, January). Towards the circular economy: Accelerating the scale-up across global supply chains. World Economic Forum. Retrieved from www.weforum.org/reports/towards-circular-economy-accelerating-scale-across-global-supply-chains.

World Economic Forum, WEF with Bain & Company. (2017, Mars). The future of electricity new technologies transforming the grid edge. World Economic Forum. Retrieved from www3.weforum.org/docs/WEF_Future_of_Electricity_2017.pdf.

Yunus, M., Moingeon, B. and Lehmann-Ortega, L. (2010). Building social business models: Lessons from the Grameen experience. *Long Range Planning*, 43(2-3), 308-325.

Zott, C. and Amit, R. (2010). Business model design: An activity system perspective. *Long Range Planning*, 43(2-3), 216-226.

Zott, C., Amit, R. H. and Massa, L. (2011). The business model: Recent developments and future research. *Journal of Management*, 37(4), 1019-1042.

Zucchi, K. (2018, 25 January 25). What makes Tesla's business model different?, *Investopedia*. Retrieved from www.investopedia.com/articles/active-trading/072115/what-makes-teslas-business-model-different.asp#ixzz4v3Py3Sub.

Annex 3

A.3.1: Case study resources

In order to complement the learning experience, we suggest the following case studies that address the topics of business model transformation for sustainability and CE.

- Bohnsack, R. and van Heemstra, P. (2016). *The rise of a new industry: Business model innovation at the intersection of energy and mobility*. Published by Catolica Lisbon and Amsterdam University of Applier Science. Reference no. 316-0196-1. Available at The Case Center. Prize Winner.

  A couple of interesting cases on Tesla Inc.:

- Hettich, E. and Mueller-Stewens, G. (2014). *Tesla Motors: Business model configuration*. Reference no. 314-132-1. Published by University of St. Gallen. Available at The Case Center. Prize Winner.
- Heal, G. M. and Usher, B. (2017). *Architects of the future? Tesla, Inc, Energy, Transportation, and the Climate*. Published by Columbia CaseWorks, Columbia Business School. Reference no. CCW170306. Available at The Case Center.

On CE and business models:

- Rajakumari, D. J. and Saravanan, A. N. (2016). *Philips transition to circular economy: Can the innovation sustain?* Reference no. 316-0066-1. Published by Amity Research Center. Available at The Case Center.

A.3.2: Case study from this book

We suggest the case of Enel's transformation toward a sustainable business model. In the case study on page 301, we provide a short version and a list of questions for discussion. For the extended version and the teaching notes, contact the authors:

- Pogutz, S. and Vurro, C. (2016), *Sustainability and innovation at ENEL: Addressing the challenges of the energy industry.* Published by Università Bocconi, Bocconi Graduate School.

Case study: Enel's sustainable business model transformation

Following the nationalization of the electricity system, Enel, the Italian National Agency for Electric Energy, was founded in the early 1960s with the purpose of unifying the national system and boosting Italian post-war development. From the moment of its foundation, many things have changed for Enel, both internally and in the energy industry. Today, Enel is a global company with €70.5 billion revenues, over 61 million end users, 62,000 employees, and 46% of its production comes from zero-emission sources, a number that is meant to grow (Enel, 2017a). In addition, it is the only utility and the only Italian company included in the Fortune 50 Change the World ranking of 2017 (and 2015) and has been included in the Dow Jones Sustainability World Index for the 14th year in a row, proving its strong commitment to sustainability.

Living through change

How did Enel get to where it is today? As previously mentioned, it was through a combination of internal and external transformations. As a matter of fact, the electricity industry went through enormous changes in a relatively short period of time. What started off as a safe and reliable sector ruled by regional and national monopolies became extremely competitive thanks to the liberalization process and privatization of state-owned utilities that started in the 1990s. In addition to this, in the last decade, a number of changes in the institutional environment and external shocks affected the competitive landscape, calling for new strategies and innovative business models. Among these, the three greatest challenges are the development of renewables, the de-carbonization challenge, and the shift of energy demand from industrialized to emerging economies (CEC, 2008; IPCC, 2014). All of these may be seen both as challenges and as opportunities, but the question remains the same: how does Enel adapt its strategy to these transformations?

Enel's response to change

The launch of the subsidiary Enel Green Power in 2008 (EGP comprised all renewables: wind, geothermal, photovoltaic, and hydro) was a turning point in the direction of sustainability. In 2010, Enel pushed ahead EGP and renewables with an Initial Public Offering (IPO) (30% of the shares were sold), and thanks to a forward-looking strategy based on innovation and geographic diversification (e.g., in Latin America, North America, India and Africa) EGP grew steadily over the years. At the core of EGP business development was the firm belief that innovation and sustainability had to be pursued jointly and through openness to stakeholders and through active engagement of communities. This bottom-up approach, based on the adoption of the Creating Shared Value (CSV) framework by Porter and Kramer (2011), favored the integration of environmental, social, and governance indicators throughout the value chain and the inclusion of ex-ante sustainability analyses in determining whether or not to invest in a new project.

A deeper change was needed: EGP's approach had to be transferred to the larger Enel group in order to address the massive challenges that the industry was facing. The direction was clear: Enel had to find the right way to further push innovation on renewables and, at the same time, embed sustainability in every aspect of the organization. The new approach was fully captured in this statement by Enel's CEO and General Manager Francesco Starace: "A focus on the environment, social development and economic sustainability are three key factors for the growth of a global player in the energy sector. In Enel we believe in innovation to meet the need of our customers, offering new products and energy services to promote the social and environmental development of communities and generate shared long-term value" (Enel, 2015a). The reshaping of the organizational structure started with the decision to create a new "Innovation and Sustainability" function that, for the first time, reported directly to the CEO.

Enel's new strategic plan

Another milestone toward a transformative change was the strategic plan (2015-2019), released in March 2015. The focus was on new growth objectives and huge investments covering four key areas: renewables, the digitalization and modernization of grids, the improvement of operational efficiency and the rationalization of thermoelectric capacity. These goals had to be achieved through a flexible portfolio of small and medium-sized projects, which offered a quicker return on investment, and had to be developed with the full support of the communities that hosted them.

In November 2015, Enel presented the update of the strategic plan (2016-2019) to London's financial community. The company's new target was to invest in new capacity to allow the generation mix to reach over 50% from clean sources by 2019 (Enel, 2015b). Moreover, the plan proposed a simplification of the group structure, including the integration of EGP (in 2016, EGP was repurchased by Enel for €3.1 billion). The evidence of this new approach was the decision to dismiss 23 power generation plants in Italy (project Futur-e) and the termination of controversial projects such as HydroAysén plant in Chile, which represented a growth model no longer sustainable from an economic, social and environmental point of view.

Another major step taken by the company was the dissemination of the CSV approach into the entire group in order to ensure the integration of sustainability in decision-making across company functions and business units. In June 2015, the CSV

IN Program was launched engaging Sustainability Managers and Business Developers from the countries in which Enel operates to define customized tools and models, flexible enough to meet business peculiarities while fostering a proactive sustainable approach. By the end of the program, in September 2015, 14 sustainability tools fully integrated within the decision-making process were created and tested on real projects in pipeline.

Future outlook

In general, Enel took a proactive and dynamic approach to the industry challenges. As a result, the Italian utility shifted from centralized power generation to dispersed generation with investments in power plants all over the world. It transformed its business model to the very core in order to integrate sustainability in the strategy, culture, and operations of the company.

The situation in 2017 is positive, with large investments made in emerging countries and measurable increases in renewables. In addition, Enel has undertaken a new mission that goes hand in hand with sustainability and innovation: digitalization. In the now three-year strategic plan 2018–2020, presented in November 2017, digitalization is one of the fundamental keystones, involving more than a €5 billion investment (Enel, 2017b). Through the creation of the new brand Enel-X, which approaches energy as a service rather than a raw material, the company plans to respond to the new requirements coming from the shift from a centralized energy model to a distributed one. To conclude, Enel today is much more efficient and more sustainable than it has ever been, and, thanks to its dynamic business model, it is able to derive value and grasp the opportunities created from the radical changes that continue occurring in the energy sector.

Link to sustainable business models

This case illustrates the importance of dynamic business models when dealing with an industry characterized by a strong and frequent change. By going to the very core of the challenges and integrating sustainability in the company's strategy, Enel was able to innovate its business model in the right direction.

Source: authors' elaboration.

Questions for discussion

(1) Looking at the Enel's case, please analyze the relation between the strategy, business model transformation and sustainability.
(2) What is the role of EGP in the development of Enel strategy?
(3) How are innovation and sustainability interlinked? Is this always the case?

(4) Was creating the function "Innovation and sustainability" necessary to respond to industry challenges? What do you think was the rationale behind it?

(5) How can a company such as Enel keep up with the future challenge of the power and energy industry (e.g., renewable growth, electric mobility, service-oriented demand, etc.)? What are the necessary actions it has to take?

References

Commission of the European Communities. (2008). 20 20 by 2020 Europe's climate change opportunity, Brussels, 23.1.2008 COM(2008) 30 final.

Enel. (2015a). Sustainability report 2014. Retrieved from www.enel.com/content/dam/enel-com/gover nance_pdf/reports/annual-financial-report/2014/enel_sustainability_report_2014.pdf.

Enel. (2015b, 18 November). Strategic plan 2016–2019, Capital Market Day [PowerPoint slides]. Retrieved from http://strategy2015.enel.com/files/9M/Enel_Capital-Markets-Day-18Nov15.pdf.

Enel. (2017b). Strategic plan 2018–2020. Full speed ahead on digitalization and customers. Retrieved from http://strategy2017.enel.com/files/Strategic_Plan_2018_2020_ENG.pdf.

Enel. (2017a). Sustainability report 2016. Retrieved from http://sustainabilityreport2016.enel.com/en/ sustainability-report-2016/getting-know-enel/identity-card.

IPCC. (2014). *Climate change 2014: Mitigation of climate change. Contribution of Working Group III to the Fifth Assessment Report of the Intergovernmental Panel on Climate Change*. Cambridge, United Kingdom and New York: Cambridge University Press.

Porter, M. E. and Kramer, M. R. (2011). The big idea: Creating shared value. How to reinvent capitalism – and unleash a wave of innovation and growth. *Harvard Business Review*, 89, 1–2.

Theme III
Corporate sustainability implementation

8 Designing and implementing the sustainability plan

Learning objectives

- Provide a solid framework to design and prepare the sustainability plan (or plan for sustainability).
- Illustrate the relations between corporate sustainability and the sustainability plan.
- Explain how to identify the main sustainability challenges (understand the signals).
- Learn how to define the scope, the boundaries and the priorities of your sustainability plan.
- Understand the meaning of stakeholder engagement and materiality.
- Discuss how to identify sustainability goal(s) and targets.
- Discuss how to engage the organization and identify the resources.
- Link the sustainability plan with actions and initiatives.
- Provide a first base of knowledge on monitoring, measuring, and reporting sustainability.

Chapter in brief

This chapter responds to the need for a solid framework that supports managers in implementing corporate sustainability. We focus on business organizations, but the same approach can be applied to any organization. We call this framework a sustainability plan, a tool that guides and supports companies in addressing the sustainability challenges and executing the actions and initiatives. First, we define the concept of the sustainability plan, we discuss the relations with the firm's overall strategy and more specifically with the sustainability strategy (see Chapters 2 and 3). Then, we introduce the position of the Chief Sustainability Officer (CSO) and discuss her/his role in leading the process of embedding sustainability in the organization. Third, we introduce the structure of the sustainability plan following a typical scheme for strategic planning. The first phase starts with the examination of the external context and provides a structure for "understanding the signals" that we have explored in Chapter 2. Then we focus on defining the concepts of scope and boundaries, and the question of materiality in order to set priorities for the sustainability strategy.

We also introduce the concept of stakeholder engagement, which is a key step to prioritize strategic actions. Further, we examine how to engage the organization and we briefly discuss the organizational changes necessary to address the sustainability challenges and implement the actions. The following paragraphs introduce another key step in the process of framing the strategy: the definition of the goals and the specific targets with regard to sustainability. The chapter then elaborates on the articulation of the strategy into actions and tasks and on the execution. We conclude by introducing the final two steps in the process of planning – the monitoring and measurement of results – thanks to the introduction of key performance indicators, and the reporting and disclosure of the results through adequate communication tools and channels.

Introduction to the sustainable plan

In the past decades, several organizations (companies, foundations, NGOs, cities, governmental agencies) have started integrating sustainability into strategy and have developed different types of corporate sustainability strategies and plans (see Box 8.1 for the example of Unilever). Companies have been required to increase the level of formalization of their sustainability approaches as stakeholders have become more careful and sophisticated about corporate responses to environmental and societal challenges. Sustainability planning has, therefore, emerged as a new requirement for companies in order to:

- incorporate sustainability issues into strategy and organization,
- figure out what sustainability issues are important, where firms need to focus and when,
- orchestrate multiple actions to reach long-term value creation goals and contribute to social-ecological resilience, and
- allocate economic and human resources to these actions.

If not formulated and executed through a well-organized sustainability plan, environmental or social initiatives might be useless or even turn out to be perceived as forms of greenwashing. But what should be incorporated into the sustainability plan? Where should a company set the boundaries? For example, let's imagine a fashion retail company that has hundreds of stores around the world and is starting to address sustainability from zero: What dimensions of sustainability should the company focus on and prioritize? Should it be energy and carbon emissions in stores and offices? Waste management and waste recycling? Introducing shopping bags with 100% recycled materials? Bioplastics? Should the company implement ambitious initiatives for sustainable supply chains and decide to purchase only from certified suppliers? Or should the company involve consumers in take-back programs to increase end-of-life product recovery and recycling?

In broader terms, and more importantly, the company should understand how much sustainability is central to the firm's mission, its vision, and strategy. Is sustainability relevant for the brand image and for its consumers? What novel opportunities can emerge from embracing a sustainability purpose? Last but not least: Where should it start?

It's not easy and, unfortunately, there is no shared and accepted framework for realizing a sustainability plan. Unlike the case of measuring the environmental impact of a product, or sustainability reporting, where there are recognized standards and tools such as the Life Cycle Assessment (LCA) methodology (see Chapter 6), Environmental Management System (EMS) (see Chapter 4) or the Global Reporting Initiative (GRI; see later in this chapter), there is no common rule for sustainability planning. Therefore, the aim of this section is to provide a template that can help organizations with the process of introducing sustainability into their strategy and operations.

Box 8.1   Unilever Sustainability Plan

On 10 November 2010, Unilever first announced its Sustainable Living Plan (SLP) to the press. This was a novel and ambitious blueprint setting out the Anglo-Dutch company's new sustainability commitments and targets for the next decade. It was a major turning point for Unilever since the SLP was a bold attempt to integrate sustainability into the company strategy. In the words of CEO Paul Polman, it was a change to a new business model, including all brands and all divisions, from product development to the supply chains. The idea was framed in a courageous statement that over the years has inspired other companies to develop new ways of doing business: "our vision is to double the size of business, whilst reducing our environmental footprint and increasing our positive social impact" (Unilever, 2013). With operations in 190 countries, €51 billion turnover (in 2012), 173,000 employees, and 2 billion consumers worldwide who daily use Unilever products, this new approach represented a massive undertaking.

With the SLP, the company identified three main goals to be achieved by 2020:

- to improve the **health and well-being** of more than a billion people with a focus on hygiene, health and nutrition,
- to halve the **environmental footprint** across the entire value chain as the company grows its business, and
- to enhance the **livelihoods of hundreds of thousands of people** including smallholders and farmers.

Unilever was probably among the first multinationals that explicitly introduced the term "decoupling" in their plans and strategy, separating growth and shareholder value creation from the impact generated on ecosystems. At the same time, the Anglo-Dutch company shaped the purpose to make "sustainable living commonplace," taking on the responsibility to contribute positively to society. A third aspect that was clearly underlined in the SLP was the centrality of the customers and then employees, suppliers, and communities. Fulfilling the needs of these stakeholders was considered the key to benefit shareholders.

With the SLP for each one of the three main goals, Unilever set several progressive targets with a clear timeline in order to track improvements. For example, looking at the environmental dimension, the company adopted a life cycle approach focusing on the following areas: **greenhouse gases** (GHGs), **water**, **waste** and **packaging**. For each issue, the company formulated the target of halving the footprint by 2020: halve the GHGs impact of the company's products, halve the water associated with the consumer use of products and halve the waste generated with the disposal of products. This was a challenging task as, for example, 5% of the carbon footprint is from manufacturing and transport and 68% is related to consumer behavior (taking showers, washing, doing laundry, etc.), but the company obviously has only a limited possibility of directly managing consumer behavior. Similarly, the waste and water footprints are mostly associated to the product utilization and the agriculture phases. Another key issue addressed in the SLP was **sustainable sourcing**. Unilever targeted to purchase 100% of raw materials from agriculture sources with sustainable and traceable practices (tea, palm oil, fruit, vegetable, cocoa, sugar, etc.).

To achieve these targets and embed sustainability into the corporate strategy, Unilever SLP focused on the following: brands and innovation, collaboration with retailers and suppliers, respecting employees, partnering with cross-sector stakeholders' groups (e.g., in the case of the Roundtable on Sustainable Palm Oil), and governments (e.g., the UN Sustainable Development Goals [SDGs]).

In 2014, revision of the SLP was necessary, and Unilever decided to expand its commitment and targets to include paying more attention to human rights, women's empowerment, inclusive business, and improving livelihoods. Moreover, the company decided to enlarge its approach to generate a more holistic and system-wide transformational change (e.g., focusing on eliminating deforestation and supporting sustainable agriculture).

After several years of implementation, Unilever reported that the SLP generated business value for the company and its shareholders. Results on the three major goals and targets that span the environmental, social, and economic performance over the extended value chain are reported and communicated on a yearly basis. With regard to the business case, four dimensions for sustainability have been identified, as follows.

- **More growth and new market opportunities.** Unilever's "sustainable living" brands, these are the brands with sustainability integrated into their purpose and products, have been growing 50% faster than the rest of the company (Earley, 2016).
- **Cost reduction.** Waste reduction and increased efficiency have increased productivity. The company reports that since 2008, savings of over €700 million have been generated in its factories (www.unilever.com).

- **Risk reduction.** Sustainability has favored risk reduction (e.g., climate change or business interruption risks) along the entire supply chains. In 2017, more than 50% of agriculture commodities were sourced sustainably.
- **More trust and reputation.** Sustainability has been a key to maintaining solid and trustful relationships with customers and employees.

According to the company, new areas that go beyond the traditional boundaries of sustainability plans will be addressed in the years to come. The journey to corporate sustainability is long and complex, but the SLP launched and implemented by Unilever under the leadership of its CEO Paul Polman has fueled the idea that it is a firm's responsibility to generate a positive social-ecological impact thus contributing to major transformation in the business community and pushing many other companies in the same direction.

Source: authors' elaboration using multiple sources.

Strategy, corporate sustainability and sustainability planning

The process of addressing the sustainability challenges at the corporate level starts with the development of a sustainability strategy. As illustrated in Chapter 2, and more in depth in Chapter 3, this requires us to understand the extent to which sustainability goals and actions will be tied to the firm's strategy, its mission, vision and values and to the organization's plan of action to reach long- and short-term goals. We have highlighted that only few companies are "born sustainable," or with the explicit purpose to pursue a broad-based commitment to sustainability (Eccles et al., 2012). The case of Patagonia, a company that addresses consumerism as a part of the environmental and social problems and that targets the defense of ecosystems as its main purpose, is still very rare and unique. Instead, for the large majority of firms, sustainability involves a conscious choice about the level of integration with the corporate strategy and organizational culture. We had discussed that for these companies, which probably represent 99% of the market, sustainability calls for a change in their mind-set and in the way they provide products and services. In a nutshell, sustainability is a major decision that according to external (e.g., the pressures from social-ecological system drivers, the megatrends, the stakeholders) and internal factors (the role of the CEO and of the top management in acknowledging the importance of these issues for the company and its business) can assume different level of commitment and intensity. The result is that we have multiple levels of integration between strategy and sustainability, as outlined below.

- On one extreme, we have firms that have changed their mission and vision and have powerfully integrated sustainability into the company DNA through profound transformations to the point that they intend to contribute positively to society, maintaining social-ecological system resilience (the BInN view, see Chapter 3).

- On the other extreme, we have firms where the attention to environmental and social issues is still at the "compliance" level, where the business approach responds only to the mainstream market logics (e.g., short-termism and shareholder value maximization are the only base for decision-making choices).
- In the middle, we have a mosaic of companies that are trying to modify some parts of the strategic process, innovating certain products and services and reporting their results, but these companies are still anchored to the TBL approach and to the pursuit of the "business case" to justify their actions and to commit to a real change.

Our questions are: Do these companies have a sustainability plan? Do they have clear goals and targets to address the environmental and social challenges? Is the top management involved in this process? Or is sustainability emerging without any clear idea of the journey that the company must undertake? In our view, the implementation of a framework to guide corporate sustainability into business processes and actions is a fundamental condition for the successful integration of these dimensions into the organization dynamics. Each company that addresses these major challenges is required to implement a long-term plan, and this plan must start from the head of the company, engaging the top management and challenging the corporate and business strategy. It is, therefore, paramount to start this process with an analysis of the strategy of the company in order to understand if sustainability is going to require major changes in its mission, vision, values and decision making in the business model. The ultimate goal is to connect the two processes: on the one hand, the strategy formulation and, on the other hand, the strategy for sustainability.

To conclude, many sustainability champions, when they launched their sustainability plan, started with a reorientation of the mission and vision in order to build a new and clear inspirational purpose, linking sustainability to the strategic thinking. That was the case of Unilever, H&M, IKEA, and the other examples illustrated in this book. Therefore, the sustainability plan, in our view, should start with questioning the fundamental components of the strategy in order to inspire new directions:

- What services is our company providing to our society and to our planet?
- What solutions can we offer to address the social-ecological system challenges and contribute to their resilience?

Integrating sustainability into the mission and vision and building a purpose is the first and major step that calls for a top-down engagement and action. Besides the importance of a committed and inspired Chief Executive Officer (CEO) and a supportive board, as discussed in Chapter 2, one of the most important factors for the successful implementation of a sustainability plan is to identify who is leading the company through this complex journey. We will comment on this in the next section.

The process owner: illustrating the role of the Chief Sustainability Officer

Every company that undertakes a relevant change needs to address multiple obstacles and resistance from internal and external forces. It is well known from the literature that companies have many layers of managers and employees that are recalcitrant to sustainability, manifest skepticism and indifference to social and environmental issues and simply "don't want to do it." We have previously illustrated and discussed some of these barriers, like political, cultural, and psychological motives, that get in the way of the sustainability strategy and plan. Therefore, the first – and essential – step in building a sustainability strategy is to identify the manager who will take the responsibility to address these challenges: the process owner.

During the last decade, the figure of the CSO has emerged to play a new and important role within corporations of a certain size that are implementing sustainability. The CSO is a C-level executive who is directly involved in the development of the sustainability strategy and in the activity of internal and external communication. The role and span of responsibilities of this executive can vary significantly from firm to firm, but we can say that the CSO is the person who drives the formulation and execution of the sustainability strategy (Miller and Serafeim, 2014). We have illustrated the examples of Steve Howards, IKEA, and Hannah Jones, Nike, who have guided their companies through bold commitment and innovative solutions, and we have highlighted the importance of leadership as a distinctive trait of this role. The CSO, in fact, has to engage the whole firm and motivate managers and employees toward environmental and social goals. Moreover, the CSO needs to master a broad set of skills and competencies that range from the technical, with a profound control over the sustainability agenda, to the managerial and organizational, in order to build efficient relations inside the organization and innovative collaborations outside, with multiple stakeholders that must be involved in the journey.

It has been acknowledged in Chapter 2 that integrating sustainability calls for a very strong top-down approach. The cases of companies like Unilever, GE, Patagonia or Enel where iconoclastic CEOs have disrupted the status quo and have promoted a new business approach based on contributing positively to society are still infrequent. Anyway, in the majority of organizations it is the CSO who assumes critical importance in raising attention to the new macrotrends, engaging the head of the company, and obtaining the CEO's commitment and support. Otherwise the obstacles highlighted before can become powerful enough to derail the company from any virtuous journey. Networking with the centers of decision making (e.g., the Board of Directors), enlightening the CEO on the sustainability agenda, and nurturing these relations with the correct communication flows are, therefore, key actions to start the sustainability planning in any company.

Designing and preparing the sustainability plan

We now turn to the examination of the main phases of the sustainability plan. As illustrated in Figure 8.1., the first step in the process of planning refers to the analysis of external megatrends that can significantly influence the industry and the organization competitiveness in the medium and long term. We call this activity of scanning and sensing the external dynamics in the ecological, social, and technological spheres: understanding the signals.

The following step, scope and priorities, is about the identification of the boundaries of the sustainability plan and the priorities that the company must address with proper actions. Firms operate in multiple business arenas, with different levels of vertical integration and diversification, and compete in different geographical areas. How should these strategic and organizational elements be considered in the sustainability plan? Similarly, every company is exposed to multiple sustainability challenges connected to environmental and social impacts. Focusing only on the relations with ecosystems, we can mention climate change, biodiversity loss, waste and micro-plastics, water consumption, air pollution, and soil preservation. A key phase in the process of planning is, therefore, to clearly define and share the types of challenges that will be addressed first and what comes later. In other words, answering the question: Where should we focus?

Engaging the organization is another key phase of the process. The CSO needs to establish a multi-department steering committee of talented managers who collaborate on the formulation of the plan and support the implementation of the strategy. A growing number of organizations have already created a specific Sustainability Unit that responds to the CSO and works across functions and business departments (e.g., from operations and supply chain, to finance and investor relations) to coordinate and facilitate the execution of the sustainability-oriented initiatives. Engaging the organization is fundamental to both legitimate sustainability inside the company and to obtain access to the financial resources for implementing the programs.

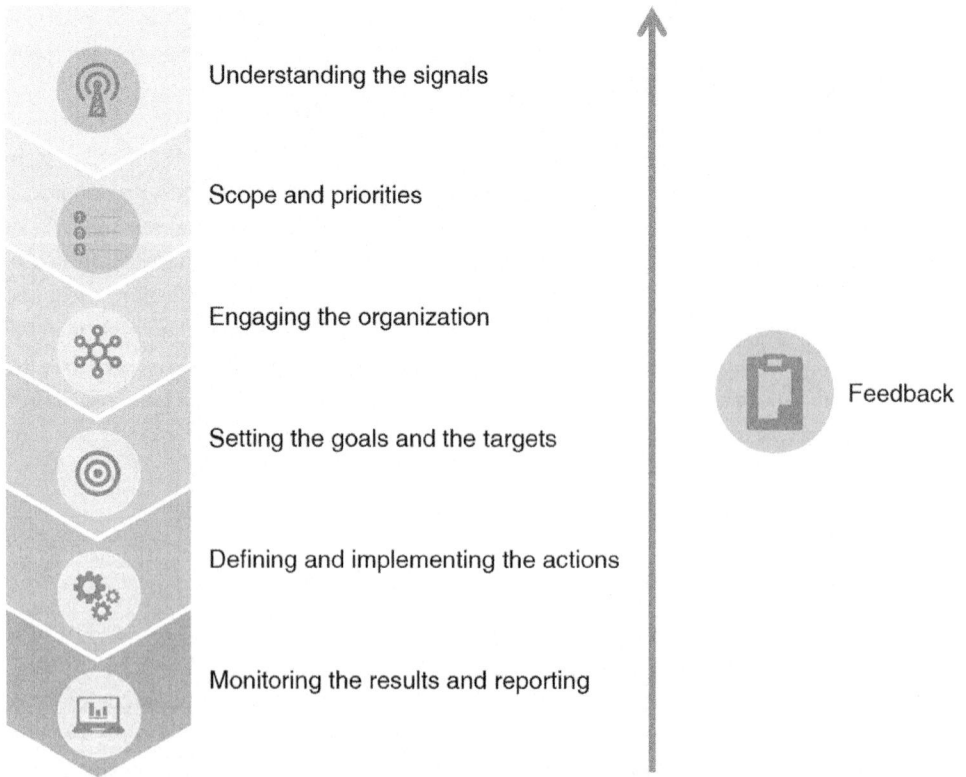

Figure 8.1 The sustainability plan.
Source: authors' elaboration.

Then next step is about setting the goals and the targets of the sustainability strategy. All companies that engage with sustainability planning must set ambitious goals that: (a) guide the company and stimulate innovations, (b) guide the engagement of managers and employees, and (c) orient the long-term performance. Nowadays, leading sustainability oriented companies are defining these goals on the basis of the Sustainability Development Goals of the United Nations, highlighting the contribution of the company to the global sustainability pattern. Goals should also be associated within more specific and short-term targets that can guide the day-to-day activities and can favor the monitoring, controlling and reporting of the results.

In the process of planning sustainability, defining and implementing the actions (e.g., initiatives, programs and projects) is, of course, a paramount step that requires a broad engagement of the organizational structure and external stakeholders. In this book, we have provided a framework to illustrate the multiple options that firms have made to walk the talk, putting sustainability into practice. These actions must be identified, selected, detailed, and combined with the strategic goals and targets previously illustrated.

The final step of every planning process refers to monitoring the results and reporting. This phase is important in order to check that the company is actually on track with regard to sustainability goals and targets. Functional to the monitoring process, and complementary to the sustainability planning activity, is the development of reporting systems. A sustainability reporting system allows the company to organize the data and the information related to its environmental and social actions and performances (Perrini and Tencati, 2006). Disclosure of information is becoming increasingly critical as several stakeholders are increasingly expecting transparency about how companies are addressing sustainability risks and opportunities (e.g., investors, consumers, policy makers, NGOs). Disclosure of information is becoming more and more critical. The "last mile" in the process is, therefore, about deciding to adopt a standard framework such as the GRI (see below in this chapter) and to communicate the sustainability strategy and performance to interested stakeholders.

The process of planning can be considered a cyclical one, like in the case of the management system analyzed in Chapter 4 (we refer to the Plan, Do, Check, Act cycle). The monitoring and reporting of the activities allows the company to analyze and understand the reasons for shortfalls in the accomplishment of sustainability goals and targets. Incorporating regular *feed-back* favors the revision of the sustainability strategy and the continuous alignment with the strategic and competitive goals.

Monitoring the external context: understanding the signals

As in the case of strategy, where the process of planning starts with the analysis of the so-called "external environment" to pinpoint the macro factors that can influence the organization, so also in the case of the sustainability plan we start with the analysis of global forces that can affect the company behavior. In Chapter 2, we identified a number of megatrends (e.g., increasing urbanization, aging of population, shifting of economic power) that are capable of shaping our future, impacting on the key variables of the IPAT equation. These megatrends are generating pressures on the resilience of social-ecological systems (the "I"), contributing to current globalization of the environmental crisis. At the same time, these megatrends are affecting many different industries and are generally not amenable to influence by a single company.

A key part of the process of planning is about monitoring, assessing, and forecasting the company's risks derived from these megatrends. In Box 8.2., we provide examples to illustrate this point. We focus on the case of palm oil, a raw material that can be found in about 50% of the products on supermarkets' shelves (e.g., food, cosmetics, detergents), with a growing demand in the years to come and several social and environmental challenges. How will these dynamics affect companies' sustainability strategies?

Box 8.2   Palm oil demand: scanning the risks

The global demand for palm oil has been sharply growing in the last decade (nearly doubling) to the point that today this raw material is one of the most widely traded vegetable oils globally. Governments of emerging countries in tropical regions have pushed palm oil cultivation as a tool to reduce poverty, reduce food hunger, and increase energy independency. The request for this raw material is projected to increase substantially in the future due to its versatility as an ingredient, the yields of the palm crop (almost four times that of other crops) and the low prices. On the negative side, the increase of the palm oil demand has led to a large expansion of the land used for its plantations, at the expense of tropical forests. The palm oil tree, in fact, grows only in humid tropical regions, where the rainforest is rich in native species and the soil is rich in carbon (the term used to indicate this type of soil is "peatland"). Nowadays, about 90% of palm oil plantations are in Indonesia and Malaysia. Therefore, the deforestation and the land conversion to this crop cultivation threaten the biodiversity and climate change. As the areas where palm oil is cultivated are extremely vulnerable, its production has been under global scrutiny since the 2000s when the UN and NGOs like Greenpeace and World Wide Fund for Nature (WWF) asked for adopting sustainable cultivation techniques and called for a moratorium on the use of land to avoid conversion of forests to land for cultivation.

The response to these requests was the creation of the Roundtable on Sustainable Palm Oil (RSPO). Established in 2004, this multi-stakeholder platform provides the standard and the certification system for the sustainable cultivation of this crop. The RSPO now covers about the 19% of the global demand for palm oil (data as of 2016) focusing on markets such as Europe and North America, where many large multinationals (Unilever, L'Oréal, Mars, Ferrero and Nestlé just to name a few) have committed to selling 100% RSPO palm oil and to have full supply chain traceability. These companies, through the Indonesian Palm Oil Pledge (IPOP), have also brought in some new perspectives to stop deforestation, attempting to engage governments on this target. However, countries such as China, India and the domestic markets (Indonesia, Vietnam, etc.) do not seem interested in certified palm oil.

Scanning and sensing future trends is paramount for companies that use this ingredient in their products. Land scarcity poses one major constraint to the production of sustainable palm oil in regions such as Indonesia and Malaysia that have already exceeded the sustainably planted areas (Pirker et al., 2016). Several future scenarios on palm oil are possible: ranging from a business-as-usual one, where expansion of production and deforestation continues at current rates, to a "zero deforestation" scenario, where no concession for additional palm oil plantation is allowed. Intensification of sustainable palm oil on "suitable land" is also an option, but, for many experts who are looking at its consequences on ecosystems services depletion, this is not the best solution (Pacheco, 2017). Other trends that companies need to consider when it comes to this vegetable oil refer to customer preferences. The growing demand toward sustainability-labeled products will also put pressure on palm oil producers to enlarge the size of sustainable cultivated land at the expenses of sensitive ecosystems in terms of biodiversity and peatland. Finally, another major controversy is about social issues, equity and human rights: several smallholders, due to low productivity, are being excluded from the market.

To conclude, these dynamics are going to affect future palm oil scarcity, quality, and prices. Therefore, the starting point for developing a specific sustainability strategy for this raw material is to understand these different trends and scenarios and to perform a solid assessment of the risks, analyzing the implications for the single company. Implications may range from finding alternative materials, when viable, to pledging a "zero-deforestation" supply chain and investing in actions to "walk the talk."

Source: authors' elaboration on multiple sources.

In order to understand these signals and how they are going to impact the firm, the CSO needs to perform an effective and efficient scanning, analyzing and selecting all information that can be relevant for the sustainability strategy. Several ecosystem goods and services are threatened by the pressures of demographic, economic, and technological drivers. Companies need to understand the relation between these drivers and the present and future availability of different raw materials (e.g., the quantity, the quality and the price). Similarly, it is extremely important to understand the implications of these trends at the level of governmental policies and regulations or in terms of future stakeholders' pressures.

If we look at the question of ocean protection, megatrends like the development of Asian economies, the increasing urbanization and the globalization of consumption patterns have determined the appearance of a relatively novel issue for companies: the challenge with plastics and micro-plastics. The urgent necessity to address this dramatic problem is pushing policy makers to develop specific regulations and laws on controlling or phasing-out the use of plastics for packaging or in product composition (e.g., microbeads from cosmetics). This is going to give rise to new challenges for many companies in sectors like beverages and cosmetics.

In order to understand these dynamics, the CSO has the responsibility of involving executives, managers and consultants. Gathering information from several data sources, both inside and outside the company, is essential to a successful planning activity: different departments and business units, NGOs, authorities, experts from the academia and research institutes, and industry associations. Table 8.1 provides some examples of data sources with regard to ecosystem trends.

The activity of monitoring and assessing the external context cannot be separated from the activity of scoping and prioritizing. With the sustainability plan, each company must set boundaries, focusing first on issues that require urgent responses. We are going to illustrate these phases in the next section.

Defining the scope and the priorities

Consider the case of global companies such as Coca-Cola, Nestlé, BMW, IKEA, Carrefour and ENEL. The question is: Where should they start with their sustainability plan? Some firms like Coca-Cola and Nestlé have multiple brands in their portfolios, and stakeholders might be concerned with supply chain transparency, product packaging and waste, or nutritional issues. Others such as IKEA or Carrefour have huge logistic systems and sell thousands of different products. They have large carbon footprints, extended supply chains, and impacts that their stores generate on traffic (customers and logistics) and waste. Differently, for a company such as ENEL, a lot of the sustainability attention should be about its contribution to climate change and its impact on local communities with the plants they install and operate. Finally, for a car maker like BMW, sustainability is about production efficiency, product innovation

Table 8.1 Example of sources of information on ecosystem drivers and trends.

Millennium Ecosystem Assessment	This Millennium Ecosystem Assessment by the UN provides information about the state and trends of several ecosystems and ecosystem services. It also helps with the examination of the drivers of social-ecological system changes. (www.millenniumassessment.org/en/index.html).
World Resource Institute (WRI)	The WRI is a global research organization that operates worldwide on sustainability and natural resource protection. It has identified six critical issues – climate change, energy, food, forests, water, and cities and transport – and offers charts, graphs, datasets and quality information on trends and forecasts.(www.wri.org)
Worldwatch Institute	The Worldwatch Institute, under the vital signs section, offers a complete collection of trends with concise analysis, detailed tables and clear graphs in five different categories: food and agriculture, energy and transportation, environment and climate, global economy and resources, and population and society. (www.worldwatch.org)
International Research Institute for Climate and Society	The International Research Institute for Climate and Society is a part of the Earth Institute, Columbia University, and focuses on issues to "understand, anticipate, and manage the impacts of climate" on society. It provides several resources such as forecast and prediction tools to support decision-making. (https://iri.columbia.edu/about-us/what-is-iri/)

Source: authors' elaboration using multiple sources.

(more efficient cars, recyclable parts), and more sustainable business models (electric mobility or new car sharing systems).

In order to help systematizing the matter, every organization needs to define the scope of its sustainability strategy, setting specific boundaries and priorities.

The scoping process

The ultimate aim is to decide "where" to start. In our approach, we identify three factors that can help with this analysis:

- the list of issues,
- the organizational and operational boundaries, and
- geographic specificities.

THE LIST OF ISSUES

The first step is the identification of the most important sustainability issues for the company. We have illustrated and discussed many of these issues in this book, and the number can be extremely broad, according to the level of detail and to the depth of the analysis. For example, if we consider the environmental dimension, we can focus on climate change, energy, resource scarcity, biodiversity protection, waste, air pollution, chemicals, soil contamination and water. But we can also incorporate animal welfare, ocean protection or address more specific challenges related to raw materials such as cotton, palm oil and coffee which can be critical for the company's supply chain and the conservation of the social-ecological systems where these commodities are grown. Similarly, for social issues at the workplace, the company can identify human rights, freedom of association, forced and child labor, gender and diversity. In order to address the global environmental, social and economic challenges, the UN SDGs identify a list of global priorities and aspirations for 2030. The 17 SDGs can support companies in aligning their sustainability strategies, helping with the identification of the key issues where they must focus and commit.

THE ORGANIZATIONAL AND OPERATIONAL BOUNDARIES

A second important factor is the definition of the organizational and operational boundaries of the sustainability plan. Companies often compete with portfolios of products, different business units, and dispersed value chains. Moreover, the organizational structure can encompass complex governance models with subsidiaries and multiple types of ownership and legal forms (e.g., joint ventures, alliance, franchising relations). What are the implications of these organizational solutions for sustainability?

First of all, in the case of large diversified firms, the choice should be made regarding what business units and brands the plan should focus on in the short term and in the medium to long term. Companies can decide to start with brands that are critical with regard to stakeholders' attention, or with business units where environmental and social impacts and risks are particularly serious.

Another important decision is about the organizational boundaries that can be defined as the boundaries that determine the operations owned or controlled by the focal organization. Here, the key decision is to what extent these activities must be included or excluded from the sustainability plan. For example, companies such as McDonald's, KFC and Subway in food retail, and also Lush Cosmetics or Zara in the fashion industry make use of large networks of franchisees. How should we consider the franchisees with regard to our sustainability plan? Are franchisees within the boundary of the organization or not? Let's provide an example: if our company identifies addressing waste production and increase recycling as a key goal, setting a specific "zero waste" target in, let's say, five years, are we going to include the hundreds of franchises in our plan? Similarly, if we target sustainable supply chains and resource scarcity, should we focus on our Tier 1 suppliers, or should we also consider Tier n suppliers, going upstream in our supply chain, up to the stage where resources are cultivated and extracted (see Chapter 5). These are important decisions that dramatically condition the setting of the sustainability goals and targets, and the formulation of the strategic actions.

Another relevant factor is the definition of the operational boundaries. In this case, we refer to the direct and indirect impact generated by our owned or controlled operations. For example, electronics companies like Sony, Samsung and Apple that acquire and assemble components must decide if the reduction of the environmental impact generated by their suppliers has to be considered in the boundaries of their sustainability plan and if they have to elaborate specific goals, targets, and actions that include their suppliers. In Chapters 5 and 6, we have illustrated the importance of moving upstream in the supply chain in order to address what for many commodities (e.g., crops and minerals) are the real sustainability challenges. Let's take the example of rare earth minerals, often coming from conflicted areas, with the exploitations of workers who are often exposed to very toxic substances for very low pay. All smartphone companies rely on these minerals for making their products. Responding to these social and environmental problems requires extending the operational boundaries in order to create closed loop supply chains (favoring the development of take-back, recovery, and recycling systems) and to participate with coalitions such as the Responsible Minerals Initiative (former Conflict-Free Sourcing Initiative) or the Responsible Business Alliance (former Electronic Industry Citizenship Coalition) that contribute to support companies up to the mining phase. A clear case for deciding operational boundaries can be observed when organizations that are at the stage of elaborating a specific climate plan with the focus on actions to mitigate their impact and adapt to climate risks need to decide if the analysis would be done with direct emissions (Scope 1 emissions) or indirect ones (Scope 2 and 3 emissions) (see Box 6.1. in Chapter 6 for further explanation).

GEOGRAPHIC SPECIFICITIES

A third dimension refers to the geographic extension and coverage of the sustainability plan. Large multinational companies manufacture and distribute their products around the world, but different regions and markets can provide very diverse situations in terms of local infrastructures (e.g., facilities for waste recycling), natural resource availability, policies and regulations, attention from stakeholders including consumer awareness and social movements.

Companies, therefore, must carefully understand specific local conditions in order to elaborate their sustainability plan. Let's consider the example of a company that decides to transform its supply chain into a sustainable one: what regions should be engaged first with the new production? To start this complex process, the company must identify which suppliers have the specific know-how and capabilities to change. It's difficult to think that entire requests for sustainable components or materials (e.g., fibers or crops that are sustainably grown) can be satisfied immediately, or in the short term. Therefore, the company needs to implement a plan considering sourcing from the different regions, collaborating with suppliers to increase their capacity for sustainable productions. Similarly, if we consider the downstream side, let's look at the case of a company that wants to develop a novel green brand: What markets should be targeted first for the distribution of the new product? Consumers in different geographies have different awareness when it comes to sustainability, so what region should be the first to launch the product? Finally, in Box 8.3, we illustrate another short case referring to the question of waste and recycling, and more specifically, the question of plastics.

Box 8.3 Geographic specificities: the case of waste and recycling

According to the most recent studies (Ellen MacArthur Foundation, 2017), about 480 billion plastic drinking bottles are sold around the world every year (this make about 900,000 every minute), and these numbers are expected to sharply grow in the near future. Fewer than half of these bottles are collected for recycling and only a very small percentage (about 7%) is transformed into new bottles. Instead, most of the plastic bottles end their life in the oceans or in landfills. Moreover, in Asian countries such as China, Indonesia, or India, the request for plastic bottles continues to expand as a consequence of the urbanization process, the bad quality of tap water, and new lifestyles.

Major drink brands are responsible for a large amount of this plastic (Coca-Cola alone produces more than 100 billion of plastic bottles a year), and the quantity of PET (polyethylene terephthalate) recycled by these companies is still very limited (Elmore, 2017). Many developing countries do not have waste recovery systems and recycling facilities, and local regulations do not push for the separate collection of packaging waste and sustainable practices. For brands such as Coca-Cola, Pepsi, Nestlé Waters and Danone, a key challenge in their sustainability plans is to reach the ambitious goal of "zero waste." The problem is how can these companies develop credible plans and goals that address packaging waste generated around the world? A solution can be developed through collaborations and partnerships with local organizations, investing in recycling facilities and supporting the development of waste management systems. When planning for sustainability, these companies need to acknowledge that geographical conditions represent obstacles to the possibility of implementing efficient and effective actions.

Source: authors' elaboration on multiple sources.

In these pages, we have analyzed three major dimensions that affect the development of a strategy for sustainability: the environmental and social issues, the organizational and operational boundaries, the geographical conditions. How can a CSO identify priorities when a multitude of issues challenge every firm worldwide, independent of its size, the industry, or the market? A universe of stakeholder campaigns, public debates, standards and regulations, environmental and social crisis populates the day-to-day journey to sustainability. In this very complex and dynamic landscape, it is paramount to identify principles or instruments that help the firm to set priorities, in order to focus the sustainability plan on those issues that really matter for the company and for its stakeholders.

How to identify priorities

In both academic and professional sustainability literature, it is common to use the term materiality to identify and prioritize those issues that matter most across the value chains, from early sourcing to product end of life. In the jargon of sustainability, a material issue can be defined as an issue (often called aspect) that can significantly influence the decisions, actions, and performances of an organization and its stakeholders. Moreover, a material issue has the potential to impact the organization's ability to create value in the short, medium, or long term.

Consultants, private companies, and multi-stakeholders' organizations have developed a variety of methods and tools that can be utilized to assess material issues and set sustainability priorities. A diagnostic tool like the LCA that was introduced in Chapter 6 provides an important support to identify what is really critical with regard to the impacts and effects on social-ecological systems of products and services. For example, undertaking an LCA allows companies to learn at what phase of the value chain carbon emissions are mostly generated. Denim jeans companies like Levi Strauss and Wrangler have understood that a large part of their environmental impact is upstream, at the stage of cotton cultivation, and downstream, when customers wash and dry their clothes. In the case of the Environmental Profit & Loss methodology, Puma has realized that for several products in its portfolio, impacts and effects on social-ecological systems are localized at the stage of Tier 4 suppliers. For companies such as Puma, Levi Strauss and Wrangler, it is, therefore, key to prioritize actions at critical stages of the supply chain.

Among the tools used to identify material issues, the *materiality matrix* has been largely adopted by companies (e.g., it is a common practice in sustainability reports) to accurately chart risks, sense opportunities, set sustainability and refine strategy. Materiality matrices started to diffuse when AccountAbility and the GRI (see later in this chapter) pushed the utilization of this tool for sustainability reporting. Over the years, companies have applied this, and it has become a relevant component of sustainability planning for many of them. While there are multiple forms of materiality matrices, they all share some common features. As illustrated in Figure 8.2, one axis represents the importance of environmental, social, and governance (ESG) issues from the company perspective, while on the other axis, the same issues are assessed in the perspective of stakeholders. Sustainability issues are plotted in the matrix, and those issues that emerge as significant for both the company and the stakeholders are considered material (Eccles et al., 2015). Logically, material issues challenge the company and require more attention in terms of resources and investments, and communication.

Stakeholder engagement is a fundamental part of the process for the construction of the matrix and for the identification of the sustainability priorities. In general, stakeholder theory (Freeman, 1984) has emerged as one of the most important perspectives when approaching corporate sustainability and CSR. A growing number of scholars and practitioners have emphasized the importance of involving stakeholders in the process of integrating sustainability into strategy (Schaltegger et al., 2017). As a first step, stakeholder engagement requires interviews, focus group, surveys, monitoring the news and the social media in order to understand and formalize stakeholder concerns. Companies approach this process in very diverse ways, with different levels of depth and attention ranging from a simple instrument for public relations and reputation management to a structured process for scanning and sensing issues that generate concerns outside the firms and require to be managed carefully. In sum, stakeholder engagement might call for a change in the mind-set of companies: at its full potential, it can become a part of the strategic thinking, supporting the sustainability planning through the selection of the material issues and helping the CSO with the identification of novel opportunities for innovation.

We open this paragraph with a vital question: Where should a company start with its sustainability plan? In order to survive the multitude of risks, pressures, and concerns from the world outside, it seems of paramount importance to initiate the process of planning with a careful identification of its scope and priorities. Sustainability issues are many and permeate the whole value chain across different business units and geographies, and therefore, companies need to establish clear boundaries and focus on material issues. The CSO can utilize environmental assessment tools such as the LCA or the EP&L, or a methodology like the materiality matrix to identify issues that are of high interest to the stakeholders and the company. Stakeholder engagement offers a valuable method to identify the risks

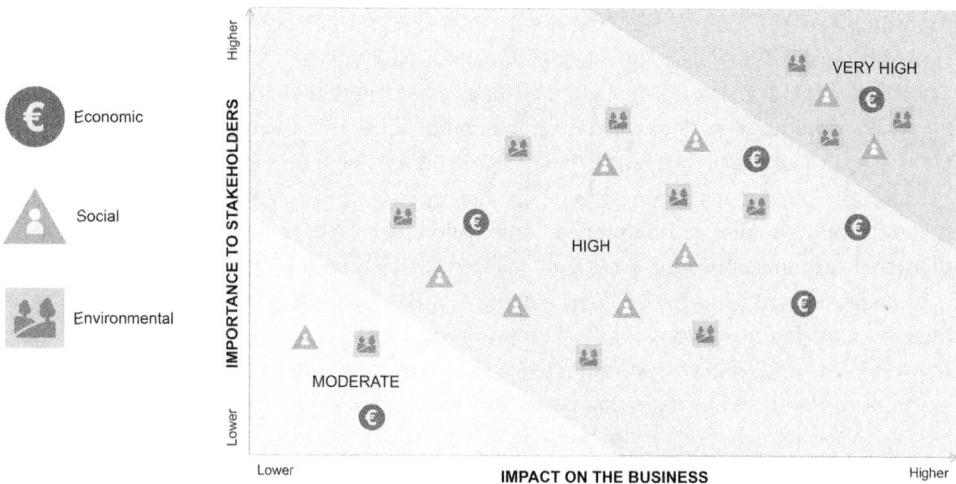

Figure 8.2 The materiality matrix.
Source: authors' elaboration.

and opportunities related to sustainability, helping with the identification of priorities and favoring the alignment of the sustainability strategy with the societal and environmental challenges.

Engaging the organization

In this book, we have often said that companies must experience significant organizational transformation in order to effectively address environmental and social challenges. These changes apply to both the organizational structure and the organizational culture. We have already introduced the CSO, the C-level (top executives) with the formal responsibility to deal with sustainability. It is necessary to note that the CSO has limited possibilities of success-fully implementing the sustainability agenda if:

- she/he is not integrated in the organizational structure, reporting directly to the top management (CEO or Chairman), and
- if the shared values and assumptions of the organizational culture are not aligned with environmental and social goals and with a long-term business orientation.

Surveying and analyzing the organizational structure of sustainability champions, we have seen that most of them have adopted similar decisions. A first response is the establishment of a Sustainability Unit or Department (sometimes called Corporate Social Responsibility, or Corporate Citizen department), under the guidance and coordination of the CSO. Again, there is no "rule" about where the Sustainability Unit is positioned within the organizational chart and at what hierarchical level. There is important evidence that when companies are mature in the sustainability journey, the Sustainability Unit rises in the organization, acquir-ing more resources and competences. For larger firms, the Sustainability Unit covers a broad range of competencies and articulates in areas with specific responsibilities on social and environmental issues: strategy, policy, and goals; stakeholder engagement; monitoring and reporting; health, safety and environment; climate change and energy; etc.

A second major change is the introduction of a Sustainability Committee (or Sustainability Council) for the decision making and coordination of the many functions and business that are involved with the sustainability strategy and actions. This committee can include Board Members, C-level executives, and head of the organizational divisions. The main activity of this group of senior managers is to share and approve the sustainability strategy and to take care of the implementation of the goals and targets with regard to their functional or divisional responsibilities. At a regional level, regional steering committees and operating divisions can be responsible for linking the strategy to the sustainability issues relevant at local level. At the operational level, there are representatives who carry out specific tasks. An example is the person responsible for HS&E issues and management systems who usually operates at the level of production plant. One example of organizational chart that include sustainability management is illustrated in Figure 8.3.

Third, companies with highest level of sustainability integration have often introduced a direct responsibility at the level of the Board of Directors or Executive Board. It is interesting to underline that a growing number of firms have also introduced incentives, both monetary

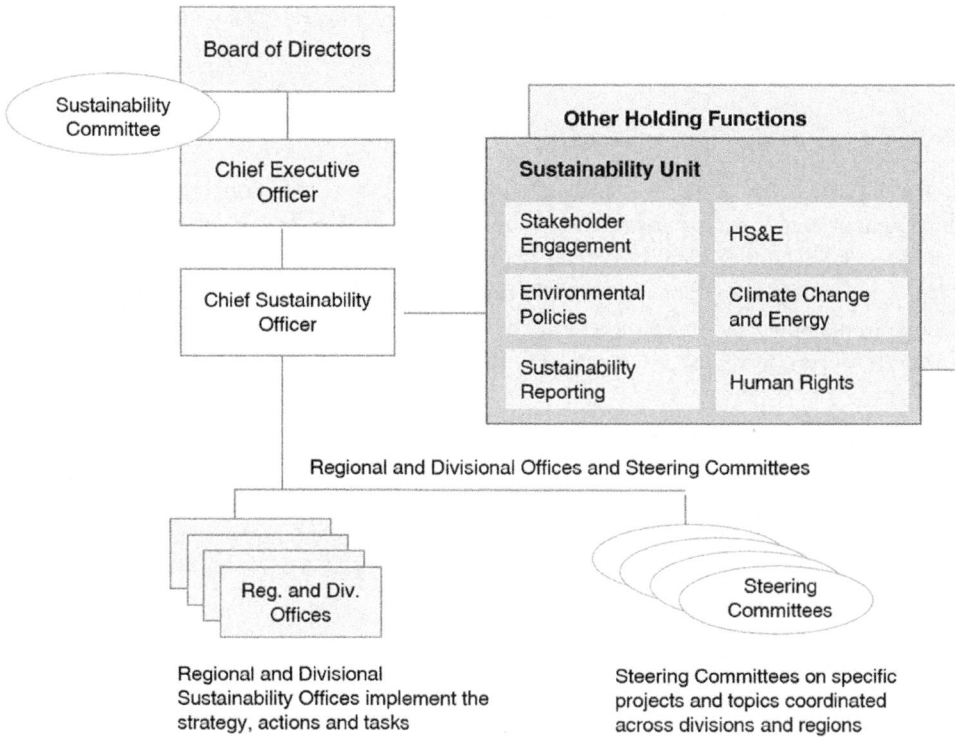

Figure 8.3 Example of organizational charts for sustainability.
Source: authors' elaboration.

and non-monetary, for the management of sustainability and that these incentives can go up to the level of the chairperson, the board members, or the corporate executive team.

The other crucial dimension for successfully integrating sustainability into the organization is about changing people's attitudes to appreciate that environmental and social issues are now key drivers of the company value (Haugh and Talwar, 2010). Often, the CSO gets "trapped" into the formal process of planning and executing, focusing on key actions and initiatives, on leadership structures, on setting goals and targets, and on reporting. While these activities are central, involving the employees is also essential for successfully implementing the strategy. The chief sustainability executive, therefore, has the challenging task to engage managers and employees and make them learn about sustainability. This is about building new knowledge company wide and changing the way employees work across different functions and businesses so that sustainability becomes embedded in the shared values of the organization. Training, educating, inspiring and using gamification are important options to raise awareness, increase advocacy and encourage employees to be more responsible.

To sum up, sustainability asks for relevant organizational changes that include the establishment of specialized departments, steering committee and mechanisms like incentive systems to align the decision making with social and organizational goals. Engaging the company at multiple levels is one of the key responsibilities of CSO, and it is probably an essential component of the sustainability plan.

Defining the goals and the targets

Every strategy needs to set ambitious goals that help with stimulating innovation, favor the engagement of managers and employees, help keep the focus, and ultimately guide the performance. Goals outline what the organization needs to accomplish to move in the direction of its vision. Goals need to be relevant and they need to express a strong commitment within a time frame. Goals can focus on operation and manufacturing activities, can extend downstream to the product/services and upstream to the supply chain, or even consider the entire life cycle.

Targets, on the other hand, are specific and measurable objectives, they are also time-bound and they contribute to reaching the goals. Goals are qualitative or quantitative, while targets must be quantitative. The definition of goals and targets is an essential part of any process of strategic planning; therefore, when companies are preparing the sustainability strategy, it is crucial to set specific goals and stretch them into measurable targets.

In the last decade, a growing number of companies have identified and communicated their sustainability goals as part of their sustainability plans and strategy. Large multinational companies such as Unilever, Ikea, Coca-Cola, Enel, BASF, Henkel, Nike, Nestlé, Toyota and many others have set ambitious long-term goals, usually with a time frame of 10 years or more, which goes to 2020 or even 2030 (Table 8.2). Companies claim promises that are very diverse:

- **specific and complemented with a deadline:** "being carbon free by year 2030," "increase our efficiency three times by 2025," or "use only recycled and sustainably-sourced materials by 2020"
- **specific but without a deadline:** "achieving zero waste in our operations" or "reducing by 50% the environmental impact of our products across the lifecycle", and
- **intentional and vague,** "decarbonize the energy-mix," "reduce water emissions."

It is important to notice that sustainability targets can be provided in relative or absolute terms (see Box 2.2, Chapter 2 for a practical case on the French beauty company L'Òreal):

- **Absolute targets:** these usually indicate more solid commitments by the company since they measure the performance without referring to other variables (e.g., volumes, sales). Absolute targets usually refer to a baseline year against which improvements are quantified and assessed. For example, a company can set a target of reducing 30% of the GHG emissions from its owned and controlled facilities by 2020, having as a baseline the year 2010. This target is established regardless of the fact that the company can increase the volume of production and expand its markets;
- **Relative (or intensity) targets:** these are normalized using variables like sales, employees or volumes (e.g., tons, liters, square meter, etc.). For example, a company that sets

a target of reducing 30% of GHG emissions per unit of product in the next five years. Therefore, relative targets reflect the variations taking place in the business activities and do not guarantee that there will be a reduction in the overall impact generated (e.g., overall GHG emissions may rise even if the intensity goes down).

The definition of the goals and of the targets is not an easy exercise. Goals are established top-down and reflect the long-term vision and purpose of the company. They also incorporate the choices about the priorities (issues) and the boundaries (operational, organizational and geographical) that we have previously illustrated. Targets, instead, result from a more iterative top-down and bottom-up process, where the single departments, functions, and operating units (plants and facilities) are involved providing data and analyzing the possibility of addressing the challenges in the time frame that has been established. Targets are then allocated at various levels in the company and are usually further stretched into sub-targets.

Table 8.2 Example of environmental sustainability goals and targets.

Nike	**Goal.** Minimizing the environmental footprint through the entire life cycle while increasing business by 2020. **Targets.** Six issues are identified: products, materials, carbon and energy, waste, water, and chemistry. • *Products*: Deliver products for maximum performance with minimum impact, with a 10% reduction in the average environmental footprint. • *Materials*: Increase the use of more sustainable materials (materials that reduce the environmental impact). Source 100% of cotton more sustainably (e.g., certified organic, Better Cotton Standard System for recycled cotton). • *Carbon and energy*: 100% renewable energy in owned or operated facilities by the end of 2025. • *Waste*: Eliminate footwear waste to landfill or incineration from manufacturing activities while continuing to reduce overall waste. • *Water*: Reduce water use in the supply chain. Reduce by 20% the freshwater use in textile dyeing and finishing per unit of production. • *Chemistry*: Enable 100% compliance with the company's internal standards on hazardous chemicals manufacturing.
Unilever	**Goal.** Halve the environmental footprint of the making and use of our products as we grow our business by 2020. **Targets.** Four issues are identified and distinguished between products and manufacturing: GHGs, water, waste and packaging, and sustainable sourcing. • *GHG*: Halve the GHG impact of products across the life cycle by 2030. With regard to manufacturing, by 2020, reduce CO_2 emissions from energy in owned facilities to below 2008 levels in spite of the increase in production volumes. • *Water*: Halve the water consumption associated with the use of products by 2020. With regards to manufacturing, by 2020, reduce the consumption from the Unilever's factory network to below 2008 levels in spite of the increase in production volumes. • *Waste and packaging*: By 2020, halve the waste associated with the final disposal of products. By 2020, reduce waste from manufacturing sent to disposal to below the 2008 baseline in spite of the increase in production volumes. • *Sustainable sourcing*: By 2020, source 100% of agricultural raw materials according to sustainability standards (e.g., palm oil, paper and board, soy beans and soy oil, tea, fruits, vegetables, cocoa, sugar, sunflower).

Source: authors' elaboration (www.unilever.com; www.nike.com).

We also need to highlight that environmental goals and targets can be much easier to quantify and measure than social and governance ones. This is the case, for example, of aspects such as carbon emissions, energy, water and raw material consumption, emissions, waste, and chemicals. Differently, social issues can be much more complicated to operationalize, quantify, and measure. A strong support in this direction has been provided by the GRI that have framed guidelines and standards to help with reporting and disclosing information on sustainability goals and targets.

In the last few years, the adoption of goals and targets at the corporate level has been further pushed by the action of policy makers, civil society and the scientific community. With the development of the United Nations SDGs, many companies have started to mobilize and engage, connecting their strategies and sustainability plans with these with the 17 SDGs and the 169 associated targets. To help the companies, GRI, in collaboration with the UN Global Compact, has launched a series of documents focused on bridging the SDG goals and targets with business reporting and performance and helping with incorporating these top global priorities into the sustainability plan. Among the many initiatives that have been growing to provide support to business goal and target setting, today the following example seems particularly relevant:

- The Science-Based Target is an initiative launched by CDP, the WRI, the WWF and the UN Global Compact organizations. Many companies are adopting goals and targets to limit their carbon emission, but to what extent are these strategies and actions aligned with climate science and with the commitments of the Paris Agreement (2015)? This initiative aims at developing tools and methodologies for companies to set goals and targets that align with the scientific consensus about limiting temperature rise well below 2 degrees Celsius in order to fight the climate change. As of January 2018, 341 companies have joined this program. (http://sciencebasedtargets.org/)
- The RE100 initiative was launched in 2014 by The Climate Group in partnership with CDP. Its aim is to support companies that commit to a goal of 100% renewable power, highlighting the existence of a compelling business case for renewables. The RE100 campaign also aims to showcase the business role in promoting the diffusion of renewables along the supply chain. Companies that join the initiative communicate their goal to source 100% of their global electricity consumption from renewable sources by a specified year. They are also invited to disclose information on energy and electricity. As of January 2018, 124 companies have committed to the reach the goal of becoming "100% renewable". (http://there100.org/)
- The WBCSD's Action2020 initiative was guided by the WBCSD in partnership with the Stockholm Resilience Center. It has set a roadmap on how business can positively influence environmental and social trends while strengthening their own organization. It is particularly interesting because it attempts to identify a set of business priorities based on scientific consensus and intensive study on environmental, demographic, and development trends. Action2020 identifies nine priority areas on the basis of the nine planetary boundaries identified by Rockström et al. (2009) and by a scientific review provided by the Stockholm Resilience Centre (http://action2020.org/).

To conclude, setting goals and targets is a fundamental phase of every strategic process. It favors making better decisions and managing things in a better way. Sustainability goals and targets are necessary to integrate the sustainability plan with the corporate strategy and to anchor the sustainability strategy to the execution. This process allows us to focus on the priorities and to establish a bold but realistic roadmap of actions that can be tracked and assessed. Moreover, goals and targets are an essential part of the communication, reporting with the stakeholders. The credibility of the social and environmental strategy is also assessed by looking at the reliability of the goals and targets that the company has established. Therefore, the CSO needs to handle this critical phase of the sustainability plan with great care.

Establish actions, tasks and execution

In order to reach the sustainability goals and implement the sustainability strategy, the planning process should elaborate and display the array of actions (or operational strategies) and tasks that the company must implement. In Chapters 4 to 7, we have introduced and analyzed a framework that allows us to organize the many different actions that companies can adopt to implement a sustainability strategy. We have discussed production and internal processes, supply chain management, product and service innovations, and business model transformations. Each single action requires to be further deployed into tasks and programs that engage different departments of the organizations and different business units. In Box 8.4, we briefly illustrate the case of IKEA's actions with regard to its sustainability strategy "People & Plane Positive" launched in 2012.

Box 8.4   IKEA sustainability actions for "A more sustainable life at home"

In 2012, when IKEA launched "People & Plane Positive," an ambitious and pervasive strategic plan targeting 2020 for a major business transformation toward sustainability, three main goals/commitments were identified (IKEA, 2017: 9):

- a more sustainable life at home,
- resource and energy independence, and
- a better life for people and communities.

How can the company implement each of these commitments? Analyzing the sustainability plan and focusing on the first goal – "a more sustainable life at home" – actions and tasks must not just involve the entire range of products and services offered by IKEA, but must also inspire and enable customers who are a part of the "problem" through their purchasing decisions and their utilization schemes (see Chapter 6). In order to reach this goal, multiple actions were planned:

- innovating products in order to offer a whole range of energy and water efficient solutions (e.g., electric hobs, switch to LED in the lighting range, provide energy efficient home appliances, shower sink accessories, dishwater),
- inspiring and motivating customers with regard to converting waste into resources, providing innovative and easy-to-use solutions for sorting and managing waste at home,
- focusing on sustainable food and encouraging a more sustainable and balanced diet,
- focusing on IKEA customers (IKEA Family members) through multiple initiatives in order to enable them to adopt more sustainable habits, and
- working on the development of new concepts such as "future homes," providing examples of attractive and sustainable forms of living.

All these initiatives further require the deployment of specific plans and programs. Take the case of LED. IKEA decided that LED had to be the only kind of lamp sold in their stores by September 2015. How could everyone make it happen? The company had to work on innovation to improve the efficiency and reduce the costs of the new technology. For example, IKEA reported in its Sustainability Report that in 2014 they succeeded in halving the price of one of the most popular LED bulbs (the 40W equivalent), and this contributed to many more customers purchasing energy efficient light sources. Then, the plan had to set specific targets and establish actions in different geographical markets (e.g., Europe, US, Australia, China), where consumers have different green awareness and attitudes, and work on distribution and communication. It was also necessary to define tasks at the level of single stores, engaging the store managers and employees to push the sales of the new green range of products. Of course, the program of changing to LED had to be associated to a specific budget, resources and responsibilities to be coordinated with the CSO and the Sustainability Units/managers at regional levels, and with other departments.

Source: authors' elaboration on multiple sources.

When defining the sustainability plan, it is also important to remember that social-ecological system transformations might require actions not only to mitigate the impacts and effects of production processes and products/services, but also to adapt to the changes. A common distinction is between two types of operational strategies: adaptation and mitigation. For example, if we consider climate change, ecosystem degradation and resource scarcity in many sectors, from the agro-food to energy and electricity, have implemented plans for mitigation and adaptation. We have broadly discussed mitigation actions in this book, probably less adaptation. Box 8.5 provides the example of the main actions of Nestlé adaptation strategy to climate change.

Box 8.5 Nestlé adaptation to climate change

Nestlé is a world leader in food processing, operating in areas such as wellness, health and nutrition; it is a major player in coffee (Nespresso and Nescafé) and bottled water. Nestlé has developed a broad sustainability strategy that builds on Michael Porter's approach of Creating Shared Value (Porter and Kramer, 2011) focusing on individuals and families, communities and the planet. The attention to resource scarcity (water), ecosystem services and vulnerability to climate change is a major focus of the environmental sustainability plan since the company depends on the quantity and quality of the natural capital. In particular, with regard to tackling climate change adaptation, actions are based on the following areas (Nestlé, 2013):

- engage with governments and farmers to develop vulnerability assessment and action plans with regard to different regions and businesses,
- collaborate with farmers to develop agricultural and cropping techniques and assistance to build climate change resilience (e.g., water stewardship),
- focus on plant breeding programs that can increase climate change resistance (including Genetic Modified Organisms (GMOs)),
- develop information systems to monitor climate change at farm and landscape levels, and
- generate and share knowledge on climate adaptation with stakeholders.

In 2017, for the second year in a row, Nestlé's program to address climate change along the supply chain and manage water and forest risks has won the world leader award given by CDP (CDP, 2017).

Source: authors' elaboration on multiple sources.

Execution is about getting the sustainability plan done, and it is often the key to success. The plan has to identify the resources (money and time allocated) and the responsibilities associated (the persons, either executives or managers) to each action and task. As is the case with other strategic activities, in sustainability, the importance of organizational management systems has been demonstrated to be essential to reach social and environmental goals and targets. Management systems consist of practices and procedures that tie the actions and tasks to specific responsibilities, also introducing a system of indicators to measure the performance of managers and employees, and mechanisms to incentivize the orientation toward sustainability. On this point, Eccles et al. (2012), in a paper published in the *Sloan Management Review*, found that more than 80% of sustainable companies have "management systems" for executing sustainability, while only 20% of traditional companies use the same mechanism.

Monitoring and reporting

In every process of planning, the final step refers to the activity of tracking the results in order to assess if the company is in line with its goals and targets. Companies also derive benefits from activating feedback mechanisms that allow them to review and update the overall strategy. It is, therefore, paramount for the success of the firm to integrate the monitoring and reporting system into multiple phases of the planning process. Turning to corporate sustainability, it, therefore, seems necessary to implement an accounting system for:

- monitoring and measuring the results of the sustainability strategy, and
- disclosing and reporting to stakeholders the information about its environmental and social results.

Over the last 20 years, many different proposals for monitoring, measuring, and reporting have been advanced by scholars and practitioners. Burritt (1997), Bennett and James (1999) and Elkington, (1994, 1997) were among the first to introduce the concept of environmental and social accounting and reporting, proposing novel tools to integrate the traditional financial accounting in order to satisfy the requests of new information requirements to support decision-making processes and communication policies (Perrini and Tencati, 2006).

Today, the landscape for information and accounting systems that support managing sustainability has evolved and matured, and many consulting and IT companies offer specialized products and services. These tools help with checking the implementation of the strategy using quantitative, qualitative, and financial metrics. They gather data and information from multiple internal and external sources (e.g., they can be fully integrated with the Enterprise Resource Planning (ERP) system, the financial accounting system, or with specialized environmental information systems), from different locations (plants, facilities, suppliers) and regions around the world. This data and information is then organized to support decision making and inform the CSO and other sustainability managers who can then assess the level of the implementation of the multiple actions and tasks. Moreover, in order to guarantee that data standardization guidelines and procedures are utilized, the reliability of the information is validated trough internal or external auditing processes. In a nutshell, the company has multiple solutions to integrate this phase into the planning process.

The value of a monitoring system should be based on its capacity to correctly measure the key processes, providing feedback to the organization that can induce the implementation of corrective actions. At the same time, the monitoring system should be able to respond to external reporting needs and pressures. In this book, we have underlined several

times that disclosure of sustainability information has become a relevant issue for companies. Governments, investors, consumers, and NGOs increasingly require transparency and
accountability. A sustainability report is an attempt to provide a response to these demands;
it is a report published by a company or an organization that focuses on the environmental,
social, and economic impacts generated by its activities (processes, products, etc.). Usually
it also discloses information on the strategy and contribution to Sustainable Development,
the mission, vision and values, and the governance system.

Despite studies that show that the number of companies worldwide that are providing
data on a full set of key sustainability indicators is still limited (Cofino, 2014), the number
of firms that are publishing a sustainability report has been increasing dramatically. KPMG
provides an annual survey on corporate responsibility reporting (KPMG, 2017). Based on two
international samples, the N100 (the top 100 companies in the 49 countries researched)
and the G250 (the world's largest 250 companies by revenues according to Fortune 500
rankings), the study shows positive trends (see Figure 8.4) with 93% of the G250 and 75%
of the N100 companies now publishing responsibility reports. These data show that sustainability reporting has become a standard practice in large and mid-sized companies around
the world.

Why has sustainability (non-financial) reporting practices increased? There are several
public and private initiatives, mandatory and voluntary, that have tried to promote the

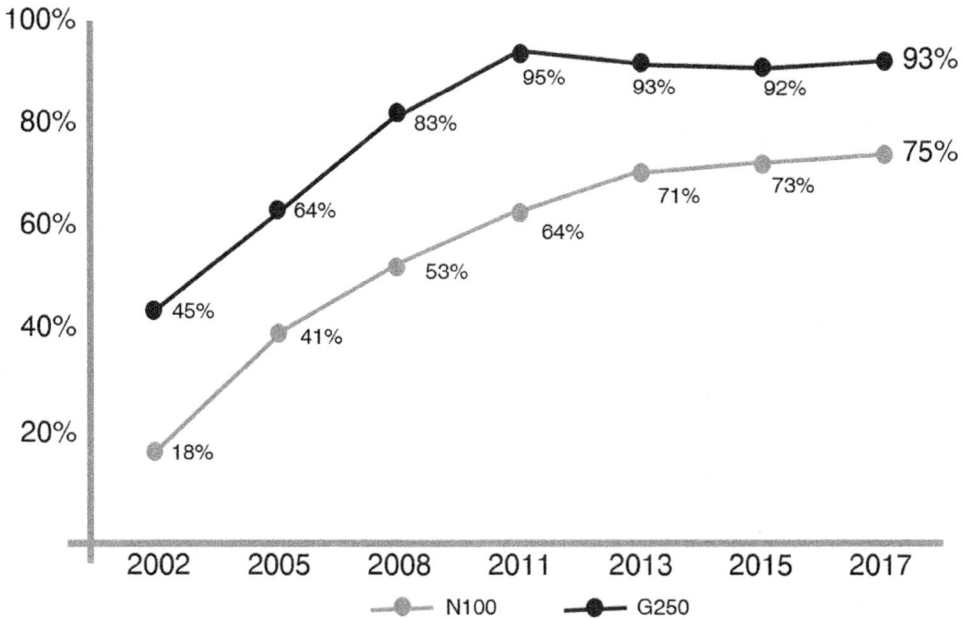

Base: 4900 N100 companies and 250 G250 companies

Figure 8.4 Growth of responsibility reporting.
Source: authors' elaboration.

diffusion of standardized systems for communicating social and environmental information. In the previous chapters, we mentioned the European Directive 2014/95/EU on non-financial and diversity information (EU, 2014) and the Sustainability Reporting Guidelines fostered by the Global Reporting Initiative (GRI 2006, 2015a). Today, these are probably the two most important instruments in terms of relevance and impact, but there are many other tools (standards, guidelines, protocols) that have pushed large firms to disclose and be accountable. For example, several governments (e.g., France, Denmark, South Africa and Japan) have enforced mandatory regulations that require specific categories of companies, for example, listed companies, or companies above a certain size, to publish information on specific environmental and social indicators. In several countries, stock exchanges are also playing an important role, providing requirements and enforcing regulation for mandatory disclosure. In this vein, the Sustainable Stock Exchange Initiative (SSEI) is a peer-to-peer platform that aims at increasing transparency on ESG indicators and Socially Responsible Investing. In 2015, the SSEI introduced its Model Guidance for sustainability reporting. Among the international sustainability frameworks for reporting, two frameworks besides GRI are competing for the market and are, therefore, worth mentioning.

- **The International Integrated Reporting Council:** This global coalition includes regulators, investors, accounting organizations, standard setters, companies, and NGOs and aims at establishing integrated reporting (reporting that integrates financial and non-financial information) as the norm within mainstream public and private sectors.
- **The Sustainability Accounting Standard Board (SASB):** This organization was established in 2011 as an independent standards-setting initiative for sustainability accounting standards. Its goal is to respond to the needs of investors by disclosing high quality and material sustainability information. The idea of standardizing ESG indicators attempts to address the request for effectiveness and comparability of corporate disclosure. SASB currently maintains provisional standards for 79 industries across 11 sectors. The SASB sustainability framework is organized under five sustainability dimensions – environment, social capital, human capital, business model and innovation, business and governance – and 30 sustainability issues.

Currently, the most important framework for reporting is the GRI. The last KPMG survey on this practice (2017) highlights that 63% of N100 and 75% of G250 companies use GRI as a standard (KPMG, 2017: 28). According to the GRI sustainability disclosure database, in 2016, the number organization reports that were officially included was 6,248 (5,551 as of 2013) in over 90 countries.

But what is GRI? How does it work? What is the GRI standard about?

The GRI was founded in 1997 in Boston by a US non-profit organization named the Coalition for Environmentally Responsible Economies (CERES) and the Tellus Institute, an interdisciplinary not-for profit research and policy organization, with the support of the United Nations Environment Programme (UNEP). In two decades, it has become the most important sustainability standard setter, evolving in parallel with the debate on environmental and social challenges and defining the landscape for reporting on non-financial information.

The first version of the GRI guidelines for reporting was launched in 2000, representing the first attempt to provide a global and comprehensive framework for sustainability disclosure. In 2002, the G2 Guidelines were unveiled, on the occasion of the World Summit on Sustainable Development in Johannesburg. The G3, the third generation of Guidelines, were introduced in 2006, with the contribution of over 3,000 experts worldwide and a broad multi-stakeholder perspective (business, NGOs, governments, and academia). This new guidance for reporting contributed to the process of framework standardization becoming the point of reference in the market. In these years, GRI has started to expand its services to include conferences and events, research reports and publications, training programs for certifiers, and spreading globally with new GRI regional Hubs opened around the world (Latin America, Asia, China, North America and Africa). Sectorial Guidelines were first introduced in 2008 (the first sector was the Financial Service), with the goal of capturing the specificities of each industry with regard to sustainability and material issues. An update of the G3 was released in 2011, G3.1, expanding the reporting approach to new relevant issues such as gender, human rights, and community. GRI operated a strong lobbying activity to connect the standard with other disclosure initiatives that were taking off at the UN level (the UN Global Compact) on specific areas such as climate change or water (the CDP guidelines), or with the growing request for standardized performance indicators from the financial community and Socially Responsible Investing (e.g., RobecoSAM responsible for the Dow Jones Sustainability Index). A new generation of guidelines, the G4, was launched in 2013 and provided a new system of requirements for the preparation of sustainability reports by organizations of any size or sector. The last milestone in GRI history was the launch of the GRI Standards in October 2016. The new approach has moved in the direction of providing a whole family of Reporting Standards that incorporate all the principles and concepts of the G4 Guidelines, while increasing the structure flexibility and the clarity requirements, with a set of topic-specific standards (see Box 8.6 for a quick example of the main features of the new GRI Standards – GRI, 2016).

Box 8.6   The GRI Standard: main features

In 2016, GRI introduced the GRI Standards, a set of global standards for sustainability reporting, that consists of a modular and integrated system of documents designed to become the new best practice for reporting publicly on economic, environmental and social impacts. The GRI Standards have been designed to be used together and to support organizations in focusing the reporting activity on material issues. A company that decides to follow the GRI approach has to comply with a list of requirements on the process of preparation of the report and must disclose a series of specific information. The final report can be a standalone document (e.g., paper-based) or can be reference information disclosed in different locations.

The GRI Standards structure consists of the following documents, organized in series.

- **Universal Standards (100 series).** It contains three documents:

  (1) *GRI 101: Foundations* – establishing the reporting "principles" for defining the contents (stakeholder inclusiveness, sustainability context, materiality and completeness) and the quality (accuracy, balance, clarity, comparability, etc.); (2) *GRI 102: General Disclosure* – helping to identify the contextual information that must be included in the report with regard to the "organization's profile, strategy, ethics and integrity, governance, stakeholder engagement practices, and reporting process" (GRI, 2016: 4); and (3) *GRI 103: Management Approach* – a document that focuses on the identification of material topics and how the organization manages these issues.

- **Topic-specific standards.** GRI has developed three series of standards on economic topics (GRI 200), environmental topics (GRI 300) and social topics (GRI 400). Each series provides specific documents that are used to support the organization with reporting on the negative and positive impacts generated by each specific topic. Examples of specific topics are: GRI 201 – Economic performance; GRI 204 – Procurement practices; GRI 303 – Water; GRI 304 – Biodiversity; GRI 305 – Emissions; GRI 4002 – Labor/Management relations; GRI 406 – Non-discrimination; GRI 412 – Human rights assessment; and GRI 413 – Local communities.

For every topic, the standards provide reporting **requirements, recommendations** and **guidance**. For example, in the case of GRI 303 – Water – reporting **requirements** ask the organization to report specific information and explains how to do it (e.g., specific indicators, measures). The **recommendations** indicate actions that are suggested or encouraged, but not required. Finally, **guidance** provides background information, explanations and illustrative examples to support the organization in understanding how to report and disclose.

The GRI website offers a broad set of resources (documents, video, and research reports) to support the diffusion and implementation of the GRI Standards. All the GRI Standards and many tutorials are available free on the GRI website (www.globalreporting.org).

Source: authors' elaboration based on GRI Standards, 2016.

Today, GRI is also collaborating with many governments trying to bridge the SDGs' perspective with the business one. A meaningful disclosure of sustainability should help policy makers and companies in tracking their commitment and progress toward Sustainable Development.

Looking forward, the process of sustainability reporting is a dynamic field, and new challenges have to be addressed. With regard to the contents for reporting, several studies (see Chapter 2) show that new social and environmental topics are acquiring importance for stakeholders. For example, on social issues, the question of inequality of wealth distribution across societies is becoming increasingly important, as is the increasing social conflict and migration. On the environmental side, ecosystem services deterioration and biodiversity loss and the protection of the ocean are acquiring momentum. Similarly, the focus on extended supply chain reporting, and on transparency and traceability, is going to become more important for many industrial sectors in the years to come. Another hot topic is the measurement of the externalities and evaluation of the costs and benefits associated to corporate activities like in the case of the Puma EP&L (see Chapter 5). Another important area of evolution refers to format of reporting. It is important to note that the availability of big data on sustainability (e.g., large data systems to collect relevant information) and the digitalization process (digital technologies for editing and publishing contents) will impact the way companies gather information, produce their reports, and disclose them. For example, companies will have the possibility of updating data and indicators frequently or continuously, and stakeholders will have access to it in every time and from everywhere (GRI, 2015b). Finally, the growing trend for sustainability reporting, as a result of regulation and legal requirements or self-regulations, mandatory or voluntary, looks like a major force to orient business toward sustainability strategies and shape corporate actions.

To conclude, and turning back to the sustainability plan, the introduction of a reporting cycle, integrated with the monitoring and measurement system and with the sustainability planning process, is beneficial to the organization both internally and externally. The systematization of sustainability data and information favors the possibility to benchmark the performance over time, with respect to law, norms, and performance standards and with respect to competitors. Moreover, disclosing standardized and certified information can improve the legitimacy and reputation of the company contributing to build trust and brand loyalty among customers and other stakeholders. Finally, it shows how the company is addressing the risks and opportunities related to social-ecological transformations of this time and how it contributes to the system resilience, responding to the increasing demand by investors and shareholder. Sustainability monitoring and reporting is, therefore, a major component of the process of planning for sustainability. It needs to be carefully designed and implemented by the CSO who must make sure it incorporates the entire value chain of the organization.

Summary

This last chapter provided a major challenge to the authors. In order to put in practice the "Business In Nature" (BInN) view, we have tried to offer a solid framework for sustainability planning with a double objective:

- to help our readers make sense of the complexity of this process that permeates the entire company at any level: from the CEO to the employees working at the shop floor, from Tier n. suppliers to customers, and
- to support managers interested in implementing corporate sustainability strategies, with a guidance that links the development of a sustainability strategy from the vision and the purpose to the pragmatic process of elaborating actions and tasks, up to reporting the results.

In doing so, we have decided to focus on large business organizations, but we promise that the same logic can be applied to any type of organization.

First, we introduced the concept of sustainability plan. We have used the case of Unilever and its SLP that, according to many experts, is one of the most popular and convincing initiatives in the sustainability field (SustainAbility, 2018) to exemplify how companies integrate sustainability into their strategies. Then, we have presented the figure of the CEO who is the C-level manager who has the formal responsibility to take care of sustainability and the importance of sustainable leadership. Further, we have proposed our framework for a sustainability plan. Step-by-step, we have analyzed each phase using several cases and examples (Nike, IKEA, Nestlé, and GRI) to illustrate how companies are managing this process:

- monitoring the external context: understanding the signals,
- defining the scope and the priorities,
- engaging the organization,
- defining the goals and the targets,
- establishing actions, tasks and execution, and
- monitoring and reporting.

In the 21st century, the sustainability challenges require new responses from companies. Preserving or enhancing the resilience of social-ecological systems is paramount for maintaining our well-being. Business organizations must acknowledge the "Business In Nature" (BInN) view. Companies and managers must understand the interdependence with ecosystems and the services and goods that ecosystems provision. Redefining the concept of strategy, and new business models that allow us to incorporate a mix of environmental and social as well as financial value, seems to be major transformations that society and nature are asking of business leaders. In this context, the development of the sustainability plan is a key step for the full and successful integration of environmental and social issues into the organization.

Chapter 8 annexes

Annex 1

A.1.1: Questions for discussion

- What is a sustainability plan and why is it important?
- What are the main phases of the sustainability plan?
- Who is the CSO and what are her/his main responsibilities? Why is the support of the CEO for sustainability plan so critical?
- How do you establish the scope, boundaries and priorities for your sustainability strategy?
- What is the difference between operational and organizational boundaries?
- What is a materiality matrix?
- What is stakeholder engagement? Elaborate on the relation between stakeholder engagement and sustainability strategy.
- What is the difference between sustainability goals and targets? Provide some examples.
- What are absolute and relative targets? Why do we need a baseline to measure our sustainability impacts?
- Imagine that you have been asked to suggest that a company uses absolute or relative targets to measure GHG emissions. What is more credible for an external stakeholder? Why?
- What are science-based targets? How are "planetary boundaries" linked with science-based targets?
- What is sustainability reporting? What is the most important standard for sustainability reporting?
- What are the benefits of disclosing social and environmental information? What are the benefits of a sustainability report? What are the main risks?

A.1.2: Assignments

Assignment 1: Analyzing the sustainability strategy and plan

This assignment is about using a pair-compared design approach to examine corporate sustainability strategies and sustainability plans. Please select a pair of companies from the same industry (e.g., Nike and Adidas, Coca-Cola and Pepsi, Henkel and P&G, Unilever and Nestlé, ExxonMobil and Shell). Have a look at their websites, at their sustainability plans and at the sustainability reports. Use other resource centers like the GRI–Sustainability Disclosure Database, or CDP website, that we have illustrated in this book to collect other material and documents about the companies. If you have access to ASSET4 (Thomson Reuters ESG research data) or Bloomberg ESG data, you can also analyze the information contained in these datasets. Then analyze and compare how the sustainability strategy is organized, investigate the scope and priorities, the sustainability goals and targets, the main policies and actions.

Assignment 2: Analyzing the sustainability strategy and plan

Consider a specific industry - e.g., apparel industry, airlines, energy, utility, electronics, food processing, automotive. Analyze the external megatrends of the industry (understanding the signals) and try to assess how they can affect the sustainability strategy of the companies operating in the industry. This assignment aims at mapping the broad context and the challenges we have introduced in Chapters 1 and 2 in order to identify what matters for an individual company and its sustainability. You can find inspiration looking at the website and sustainability reports of sustainability champions (e.g., Nike, Patagonia and IKEA).

Assignment 3: Elaborate a specific plan to contrast ocean litter

One of the emerging issues in sustainability is the challenge of protecting oceans from plastics and micro-plastics. Imagine you are in the shoes of a CSO of a large multinational company operating worldwide (e.g., an apparel, beverage or leading hospitality firm with departments and operations in industrialized and developing countries) that uses huge amounts of plastic packaging. Elaborate a specific sustainability strategy and a plan to deal with the problem and limit the negative impacts. Please identify the possible actions and think about their feasibility in the short, medium and long term; think about goals and targets; and think about the key partnerships you need to establish. Don't forget that the problem is global, and that in different parts of the world infrastructures and facilities to treat packaging waste are different. Be creative and involve different functions: innovation, marketing and operations. Take inspiration from best practices.

A.1.3: Further readings and additional resources

- Acknowledging and assessing company interdependence with ecosystems and the goods and services ecosystem has become critical for every company. The Natural Capital Protocol offers an interactive database that helps businesses better understand these relations and to find the right instruments to measure and value the natural capital they use. (www.naturalcapitaltoolkit.org/)
- In order to measure and monitor your company's carbon impact, please have a look at the GHG Protocol website. The GHG Protocol has established the standard frameworks to measure and manage GHG emissions from private and public sector operations, value chains and mitigation actions. (www.ghgprotocol.org/)
- CDP has become the most important organization for the disclosure of information on environmental impacts. It works with companies, investors, cities and policy makers. The CDP disclosure system enables organizations to measure and manage their environmental impacts. Today, CDP probably has proprietary of one of the most comprehensive databases on self-reported environmental data in the world. (www.cdp.net/)
- We have introduced the GRI. It offers multiple resources with regard to sustainability reporting, including the GRI Standards, videos and tutorials. (www.globalreporting.org/)

Annex 2

A.2.1: References

Bennett, M. and James, P. (Eds.). (1999). *Sustainable measures. Evaluation and reporting of environmental and social performance*. Sheffield, UK: Greenleaf.

Burritt, R. L. (1997). Corporate environmental performance indicators: cost allocation - boon or bane? *Greener Management International*, 17, 89-100.

CDP. (2017). Global Supply Chain Report 2017. Retrieved from www.cdp.net/en/research/global-reports/global-supply-chain-report-2017.

Cofino, J. (2014, 13 October). 97% of companies fail to provide data on key sustainability indicators. *The Guardian - International edition*. Retrieved from www.theguardian.com.

Earley, K. (2016, 30 September). More than half of all businesses ignore UN's sustainable development goals. *The Guardian - International edition*. Retrieved from www.theguardian.com.

Eccles, R. G., Perkins, K. M. and Serafeim, G. (2012). How to become a sustainable company. *MIT Sloan Management Review*, 53(4), 43-50.

Eccles, R., Krzus, M. and Ribot, S. (2015). *The integrated reporting movement. Meaning, momentum, motives, and materiality*. Hoboken, NJ: John Wiley & Sons Inc.

Elkington, J. (1994). Towards the sustainable corporation: win-win-win business strategies for sustainable development. *California Management Review*, 36(2), 90-100.

Elkington, J. (1997). *Cannibals with forks. The triple bottom line of 21st century business*. Oxford: Capstone.

Ellen MacArthur Foundation. (2017, 13 December). The new plastics economy: Rethinking the future of plastics & catalyzing action. Retrieved from www.ellenmacarthurfoundation.org/publications.

Elmore, B. (2017, 2 May). Plastic bottles are a recycling disaster. Coca-Cola should have known better. *The Guardian - International edition*. Retrieved from www.theguardian.com.

EU (2014, 15 November). Directive 2014/95/EU of the European Parliament and of the Council of 22 October 2014. amending Directive 2013/34/EU as regards disclosure of non-financial and diversity information by certain large undertakings and groups. *Official Journal of the European Union*. Retrieved from http://eur-lex.europa.eu/legal-content/EN/TXT/PDF/?uri=CELEX:32014L0095&from=EN.

Freeman, R. E. (1984). The stakeholder concept. In Pitman Publishing, *Strategic management: A stake-holder approach, 1st ed.* (pp. 24–25). Cambridge: Cambridge University Press.

Global Reporting Initiative, GRI. (2006). G3.1 Sustainability Reporting Guidelines. Retrieved from www. globalreporting.org/resourcelibrary/ 3.1-guidelines-incl-technical-protocol.

Global Reporting Initiative, GRI. (2015a). G4 Sustainability Reporting Guidelines. Reporting principles and standard disclosure. Retrieved from www.globalreporting.org/resourcelibrary/GRIG4-Part1-Re porting-Principles-and-Standard-Disclosures.pdf.

Global Reporting Initiative, GRI. (2015b). Sustainability and reporting trends in 2025. Preparing for the future. Retrieved from www.globalreporting.org/resourcelibrary/.

Global Reporting Initiative, GRI. (2016). GRI Standards. GRI 101: Foundation 2016. Retrieved from www. globalreporting.org/standards/gri-standards-download-center/?g=bc004da6-97e4-430d-acd8-f5b80d9d36c3.

Haugh, H. M. and Talwar, A. (2010). How do corporations embed sustainability across the organiza-tion? *Academy of Management – Learning & Education*, 9(3), 384–396.

IKEA. (2017). People & planet positive IKEA. Group sustainability strategy for 2020. Retrieved from www. ikea.com/ms/de_DE/pdf/reports-downloads/sustainability-strategy-people-and-planet-positive.pdf.

KPMG. (2017). The road ahead. The KPMG Survey of Corporate Responsibility Reporting. Retrieved from https://home.kpmg.com/content.

Laville, S. and Taylor, M. (2017, 28 June). A million bottles a minute: World's plastic binge "as dangerous as climate change". *The Guardian – International edition*. Retrieved from www.theguardian.com.

Miller, K. P. and Serafeim, G. (2014, 20 August). Chief sustainability officers: Who are they and what do they do? (HBS Working Paper 15–011). Retrieved from *Harvard Business School* website www.hbs.edu/ faculty/Publication%20Files/15-011_a2c09edc-e16e-4e86-8f87-5ada6f91d4cb.pdf.

Nestlé. (2013, February). Appendix of The Nestlé Policy on Environmental Sustainability. *Nestlé Com-mitment on Climate Change.* Retrieved from www.nestle.com/asset-library/documents/library/docu ments/corporate_social_responsibility/commitment-on-climate-change-2013.pdf.

Pacheco, P. (2017, 3 August). The long and winding road to sustainable palm oil. Finding a way forward for profits, people and the planet. *Forest News.* Retrieved from https://forestsnews.cifor.org/.

Perrini, F. and Tencati, A. (2006). Sustainability and stakeholder management: The need for new corpo-rate performance evaluation and reporting systems. *Business Strategy and the Environment*, 15(5), 296–308.

Pirker, J., Mosnier, A., Kraxner, F., Havlík, P. and M., Obersteiner, (2016). What are the limits to oil palm expansion? *Global Environmental Change*, 40, 73–81.

Porter, M. E. and Kramer, M. R. (2011). The Big Idea: Creating shared value. How to reinvent capitalism – and unleash a wave of innovation and growth. *Harvard Business Review*, 89(1–2).

Schaltegger, S., Hörisch, J. and Freeman, R. E. (2017). Business cases for sustainability: A Stakeholder Theory Perspective. *Organization & Environment*, 1–22.

Unilever (2013). Sustainable Living Plan. Progress Report 2012. Retrieved from https://ceowaterman date.org/files/endorsing/Unilever_2013.pdf.

WRI & WBCSD. (2004). The Greenhouse Gas Protocol. A corporate accounting and reporting standard. Retrieved from www.ghgprotocol.org/sites/default/files/ghgp/standards/ghg-protocol-revised.pdf.

Annex 3

A.3.1: Case study resources

In order to complement the learning experience, we suggest the following case studies that address the question of developing a sustainability plan, setting goals and targets, imple-menting initiatives, monitoring and reporting.

- Chakroborty, B. and Purkayastha, D. (2016). Championing sustainability and responsible capitalism: Paul Polman at Unilever. Published by IBS Center for Management Research. Reference no. 416-0128-1. Available at The Case Center. Prize Winner.

- Sivakamasundari, S. (2011). Dow's Successful business plan: A clean move towards environmental sustainability. Published by Amity Research Centers. Reference no. 711-039-1. Available at The Case Center.

An interesting and novel case on sustainable banking, sustainability performance measurement and Global Reporting Initiative is:

- Szekely, F., Dossa, Z. and Kaeufer, K. (2015). Triodos Bank: Measuring sustainability performance. Published by IMD. Reference no. IMD-7-1738. Available at The Case Center.

Another case on reporting and materiality matrix is:

- Eccles, R. G., Serafeim, G. and Escoriaza, A. C. (2012), Developing the materiality matrix at Telefonica. Published by Harvard Business Publishing. Reference no. 9-413-088. Available at The Case Center.

Epilogue

Since the turn of the century, there have been increased warnings from the scientific community and the United Nations whose purpose is clear: we need to pay much more attention to the alterations of earth system dynamic processes that are taking place on our planet. During the course of human history, various regional human civilizations have collapsed due to loss of resilience in ecosystem functioning. However, in a social-ecological interconnected and globalized world the potential consequences of crossing our planetary boundaries may threaten the stability of the entire global system. The 13th edition of the Global Risk Report of the World Economic Forum celebrated in Davos (Switzerland) in January 2018, confirmed that environmental risks associated with earth system dynamic alterations are increasing and that they contribute to making other social-related risks even bigger.

> We have been pushing our planet to the brink and the damage is becoming increasingly clear . . . A trend towards nation-state unilateralism may make it more difficult to sustain the long-term, multilateral responses that are required to counter global warming and the degradation of the global environment.
>
> (WEF Global Risk report, 2018: 6)

This statement is also a call for action to the business community to develop innovative and collaborative responses in order to deal with the challenges of our planet.

We need to do more. During the last decades, the practices of corporate sustainability have consistently improved. However, the increase in global activity, the upsurge in human population, and economic improvements in certain world regions that led some countries to exhibit patterns of consumption not previously seen, mean that those achievements are not enough to address the deep environmental emergency we are currently experiencing. We need to move corporate sustainability further, to the point where it becomes part of the solution to this crisis.

This book has been devoted to bridging the boundaries between the warnings of earth system and sustainability science and the managerial practices of corporate sustainability. This book aims to provide an integrated and common form of jargon for business education, simultaneously helping students and practitioners to take a step forward toward what sustainability means for business in the future. Our main objective was to frame a new vision of how companies can incorporate these concepts and practices to favor disruptive and decisive changes in our production, consumption, and living patterns, the "Business In Nature" (BInN) view. In doing so, we hope that we can help companies and managers tackle the problems of the Anthropocene epoch by suggesting novel opportunities of doing business while maintaining or increasing the resilience of our social-ecological systems. The BInN view requires a socio-technological revolution based on green economy and responsible behavior, able to change the logics of production-consumption mechanisms and to facilitate large transformative solutions by the incorporation of new radical ideas and sustainable business models. This transformation should facilitate us to escape from past industrial revolutions and to face the future humanity challenges for this century, helping young people to have the world they deserve.

Index

Printed in Great Britain
by Amazon